GASEOUS NITROGEN EMISSIONS FROM GRASSLANDS

Gaseous Nitrogen Emissions from Grasslands

Edited by

S.C. Jarvis and B.F. Pain

Institute of Grassland and Environmental Research,
North Wyke, Okehampton,
UK

CAB INTERNATIONAL

CAB INTERNATIONAL
Wallingford
Oxon OX10 8DE
UK

Tel: +44 (0)1491 832111
Fax: +44 (0)1491 833508
E-mail: cabi@cabi.org

CAB INTERNATIONAL
198 Madison Avenue
New York, NY 10016-4314
USA

Tel: +1 212 726 6490
Fax: +1 212 686 7993
E-mail: cabi-nao@cabi.org

A catalogue record for this book is available from the British Library, London, UK

ISBN 0 85199 192 0

Library of Congress Cataloging-in-Publication Data

Gaseous nitrogen emissions from grasslands/edited by S.C. Jarvis and
  B.F. Pain.
      p. cm.
    Includes index.
    ISBN 0-85199-192-0 (alk. paper)
    1. Grassland ecology.   2. Nitrogen cycle.   I. Jarvis, S.C.   II. Pain, B.F.
QH541.5.P7G37 1997
577.4'145—dc21   97-13541
                                                                          CIP

Typeset in Garamond by AMA Graphics Ltd
Printed and bound by Biddles Ltd, King's Lynn and Guildford.

Contents

Preface

Grasslands, whether intensively managed, semi-natural or natural, are a dominant feature of many landscapes. In many instances they make substantial
contributions to national economies through their use in the production of
meat, milk and other animal products, but they may also have considerable
effects on atmospheric and aquatic quality. Grassland can act as both source
and sink for many environmentally active agents and, increasingly, the role of
these large tracts of land, often intensively managed, is being questioned in
terms of their impact on the environment.

Over recent decades, nitrogen has been the key to productivity and
flexible farm management; it is also a key controlling factor in sustaining
particular ecological balances in natural/semi-natural habitats. Losses of
nitrogen from all the various components of grassland production systems –
and there are many stages within a production cycle at which this can occur
– can be considerable. Whilst nitrate losses into waters have provided the
impetus to research over the last decade, attention is turning increasingly
towards gaseous losses to the atmosphere, especially of nitrous oxide and
ammonia for which grassland production can provide major sources.

Much research effort has been developing into aspects of these emissions
at a range of levels: understanding process controls, defining losses at a range
of scales from field to global, developing predictive capacity and practical
methods for decreasing emissions and their effects. Emissions can occur at
many stages during a production cycle and, because of the complexity of the
system, the approaches have had to be wide ranging including, for example,
inputs from soil and atmospheric scientists, agronomists, engineers and

modellers, and have embraced those with ecological as well as agricultural interests. This volume reports the proceedings of a conference which attempted to bring these disciplines and interests together. The aim was to review current scientific activities and progress and to exchange information on effects that occur within grassland soils, the fields that are used to produce herbage, the buildings that house the animals, the manures that are produced and applied to land, and to discuss how this information can be scaled to whole systems, and to national and global budgets and inventories.

Over 100 scientists, from 14 different countries, attended the conference to bring together a unique blend of scientific backgrounds and interests. This book contains the information presented at the conference at North Wyke, and provides a state of the art description of our knowledge of the impact that grasslands, and the production systems that are based on them, have on emissions of nitrogen, at a range of geographical scales. The conference and book were designed to cover: (i) controls over soil processes and their interactions; (ii) ammonia fluxes; (iii) nitrous oxide emissions; (iv) specific effects of manures; (v) effects at the large scale (practical and predictive); and (vi) a review of further research requirements. Each of Chapters 1–5 contains a keynote, overview discussion followed by a series of detailed descriptions of findings of recent, specific research and shorter outlines of ongoing, new projects. We believe that this provides a unique description of a highly complex area which will be of interest to agricultural and atmospheric scientists, the agricultural community and policy decision makers.

Acknowledgements

We gratefully acknowledge the invaluable contributions of Mr N.E. Young and Mrs A. Roker in helping to organize the conference on which this book is based. We thank also the many other members of staff at IGER, North Wyke, who contributed to the success of the conference or who provided secretarial services during the editing of this book. The conference was sponsored by IGER, which is supported by the Biotechnology and Biological Sciences Research Council (BBSRC).

S.C. Jarvis and B.F. Pain

Emission Processes and Their Interactions in Grassland Soils

S.C. Jarvis

Institute of Grassland and Environmental Research, North Wyke Research Station, Okehampton, Devon EX20 2SB, UK

Summary. Emissions of nitrogen (N) from grassland soils can, depending on the intensity of management, be substantial with opportunities for major environmental impact on atmospheric and water quality. The N cycle in grassland soils is complex with many unknowns controlling the fate of excess N in mobile pools in the soil. Where there is little opportunity for removal of nitrate (NO_3^-) by plant or microbial biomass, there may be competition for removal by denitrification or leaching. Recent studies have shown that there is considerable opportunity for interaction between these two processes, even at depths below the normal zone of N transformation activity in the soil. Generation or removal of NO_3^- by microbial activities are keys to the controls over the emission of nitric (NO) and nitrous (N_2O) oxide gases to the atmosphere. It is clear that the soil N cycle should be considered as a much more dynamic system than is currently the case. Recent results have demonstrated the difficulties in identifying the specific processes involved in the release of N_2O under a range of conditions which, coupled with the problems of measurement, makes the development of the means of prediction difficult. As the models develop it will therefore become increasingly important not to consider the soil N processes of leaching, nitrification and denitrification as almost mutually exclusive.

Introduction

The way in which nitrogen (N) cycles within and from grassland systems is complex and multicompartmental, with many opportunities for the release of mobile forms into the atmosphere or aquatic systems (Jarvis *et al.*, 1995; Scholefield *et al.*, 1993). This discussion centres on the processes in soil which control the forms of N and their fate, particularly with respect to the emission of nitrous oxide (N_2O). This gas is of key importance in relation to greenhouse and global warming effects and also with respect to ozone chemistry: grasslands have been shown to play a major role, on both national and international scales, in being responsible for major proportions of the total N_2O emission to the atmosphere (see Chapters 28 and 31). Grassland soils differ from other managed systems in the nature and extent of the internal, recycled N from mineralization and excreta (which are in addition to the external inputs) added to the pool of available and potentially available materials held within the black box shown in Fig. 1.1. The N transformation processes which operate in soils are well known and are not peculiar to grasslands but the balance of the flows along the various pathways, and the impact of animal production management on the competition between the processes involved, have a major effect on the extent and form of emission (Fig. 1.2). In practical terms, this distribution of effects is important in deter-mining whether, for example, fertilizer recommendation policies result in or can be directed towards ammonia (NH_3) volatilization, nitrate (NO_3^-) leached or N_2O emitted (see Chapter 6).

At an early stage in the sequence of events (Fig. 1.2), NH_3 volatilization is an important emission process. Grassland-based agriculture and ruminant production is responsible for the bulk of the increase in atmospheric concen-trations and the resultant deposition of N that occurs (Buijsman *et al.*, 1987). As far as direct losses from soils are concerned, the biophysicochemical controls over the two stages of the process (hydrolysis, followed by volatili-zation) are well known and have been used extensively in models to describe fluxes from land surfaces (Hutchings *et al.*, 1996; see also Chapter 7). Imme-diate sources of NH_3 are applied urea fertilizer (other ammoniacal forms of fertilizer release only relatively small amounts of NH_3 – see for example van

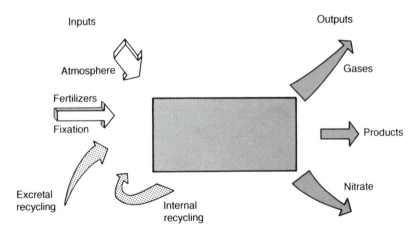

Fig. 1.1.   Nitrogen flows in grassland systems.

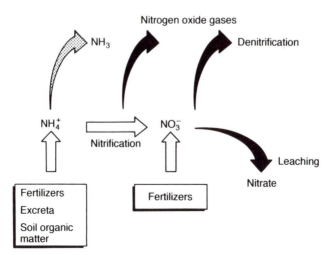

Fig. 1.2.   Nitrogen emissions from grassland soils.

der Weerden and Jarvis, 1997) and excreta (predominantly urine) either deposited directly at grazing (Jarvis *et al.*, 1995) or through application of stored manures (see Chapters 2 and 3). In the latter case, the loss rates can be substantial. Volatilization removes the possibility of this form entering the soil mineral N pool as NH_4^+ and therefore has 'knock-on' effects in reducing transfer into NO_3^- pools with the various associated opportunities for N to escape as gas or leachate. In any consideration of the fate of N added to soils, it is therefore important that the complete chain of events is taken into account, especially when attempts are being made to manipulate management to reduce emissions: the consequences of one such action, i.e. changing fertilizer form (see Chapter 6), have already been noted. Further discussion of NH_3 fluxes are given elsewhere and the remainder of this discussion addresses the interactions between the remaining components of the sequence and their impact on the fluxes of N to the environment as indicated by the solid arrows shown in Fig. 1.2.

Central to the control of emissions is the generation and removal of NO_3^- (Fig. 1.2). Nitrate can be supplied directly through fertilizer addition or through nitrification of ammonium (NH_4^+) derived from fertilizer, dung, urine or mineralization of soil organic matter. Nitrification is of itself important, not only as a supplier of NO_3^- but also through the release of N_2O and its transmission into the atmosphere (Bremner and Blackmer, 1977). Not only is it likely that nitrification and denitrification can occur concurrently in adjacent microsites in soils under certain conditions (Robertson and Tiedje, 1987; see Chapter 5), there is also the strong possibility that they are closely linked or coupled in some way. Many of the intermediates in each process are common to both, so that, for example, the nitric oxide (NO) released during nitrification (see Chapter 4) could be consumed during the denitrification process (Remde and Conrad, 1991), perhaps in preference to NO_3^- (Davidson, 1992). Whilst NO is of particular interest because it occupies a position central to the redox gradients of both processes, similar opportunities with other intermediates may also exist. Knowledge, therefore, of either process alone (and there are extensive literatures associated with both) may not be sufficient to provide the predictive capacity that is currently required for the controls over, and the extent of, emissions of N_2O.

For both NH_4^+ and NO_3^-, removal by uptake by the grass sward or immobilization into the soil microbial biomass is a powerful mechanism for reducing the potential for emissions. The grass plant is highly efficient at removing NO_3^- and NH_4^+ even at very low concentrations in the soil solution and is therefore, at many times of the year, an important component of the system in limiting the availability of substrates for the removal processes. Because the majority of grasslands support perennial crops, there are longer opportunities for this to occur than with tillage crops. However, because of the timing and amounts of many of the excretal returns in relation to weather conditions and growing period, excesses will remain and, potentially, can be lost.

Excess NO_3^- is, therefore, available to be lost by leaching or through denitrification when uptake by plants or microbes cannot reduce this pool quickly enough. As a process, denitrification provides opportunity for much N within an animal production system to be lost to the atmosphere (Jarvis et al., 1996a). The general conditions promoting maximum activity of denitrification are well known, (see Chapters 2, 3 and 5), as are the considerable problems in being able to measure the net outcome of the process as dinitrogen (N_2) or as N_2O. However, there is a growing body of evidence that is demonstrating the important role that denitrification has in reducing NO_3^- and generating N_2O (and NO) throughout many phases of grassland managements (Jarvis and Pain, 1994). An improved understanding and knowledge of the total outputs from denitrification activities is a key prerequisite for improving N use efficiency. Increased understanding of the process itself (and, as indicated earlier, of its interaction with others) is an essential component of this in order to make major progress. For example, the importance of antecedent conditions on denitrification is well demonstrated by Dendooven et al. (see Chapter 2), and knowledge of this may offer opportunities to provide a sounder basis for future manipulation. It will also be important to improve the predictive capacity for the ratio of the products of denitrification, i.e. $N_2 : N_2O$. The approach taken by Scholefield et al. (see Chapter 3), whereby both N_2 and N_2O generation can be determined under carefully controlled laboratory conditions, will allow better empirical models to be developed in order to make more confident estimates of losses. Of course, the more fundamental information provided by the studies of Dendooven et al. (see Chapter 2) allow a more mechanistic interpretation of how the process is controlled.

Much of the peak denitrification activity in soils can be associated with a coincidence of large inputs of substrate (i.e. fertilizer) and a subsequent rainfall event or when large returns of N are made in excreta as dung or urine (Allen *et al.*, 1996; see Chapter 19) or slurry (see Chapter 22). Addition of excreta not only provides sources of N, but also mobile available carbon (C) sources and substantial changes in soil moisture conditions. Another particularly vulnerable period in temperate grasslands occurs in the autumn, when there may be an accumulation of unutilized N (e.g. from the excreta of grazing animals) at the time when sward growth is slow, but soil moisture has increased to create anaerobic conditions. At this time there is also, depending upon soil type and drainage, the potential for leaching to occur, and in intensively managed systems this can be substantial (Scholefield *et al.*, 1993). Removal of NO_3^- into excess rainfall follows a typical trend of relatively high concentrations being transported initially, followed by ever decreasing amounts throughout a typical pattern of a drainage season (Scholefield *et al.*, 1993). This is likely to have resulted from a changing pattern of NO_3^- removed from different components of the range of pores in the soil, the differential generation of 'new' NO_3^- through mineralization and subsequent nitrification of N originating from soil organic matter, and some interaction with denitrification. The latter will be difficult to define but again will be dependent upon the location of the NO_3^- in relation to the architecture of the soil and the location of the denitrifying bacteria. Once removed from the 'active' zone of the soil there may still be further interactions which change the fate of the transferred N. Subsoil denitrification has been indicted in a number of studies and has been considered to be of environmental benefit in reducing NO_3^- that would have otherwise been transferred into waters. Results, however, have been varied and whilst Parkin and Meisinger (1989) indicated that subsoil impact was insignificant below 1.6 m, others (McCarty and Bremner, 1992) have found denitrification activity and associated microbial populations to 1.5–2 m depth below agricultural soils. Nitrous oxide emissions have been observed in soil samples from 2 to 20 m (Lind and Eiland, 1989; Yeomans *et al.*, 1992).

Recently, Jarvis and Hatch (1994) (Table 1.1) found substantial potential for denitrification down to 7 m below long-term grass and arable systems: this potential was increased substantially by an increased C supply. Other studies (Obenhuber and Lawrance, 1991) have shown that even deeper material (6–12 m and into the aquifer) also had denitrification capacity. The extent to which this occurs and the implications that it might have for N_2O emission as compared with NO_3^- leached are not known. It would seem likely that, if denitrification processes did reduce NO_3^- moving to depth, the conditions would be such that the major product would be N_2 and not N_2O and thus be a benign outcome. However, there is little or no information upon which to substantiate this conclusion.

Table 1.1. Rates of potential denitrification to depth below grazed grass/clover swards. (Calculated from Jarvis and Hatch, 1994.)

Depth (m)	Potential denitrification (kg N ha^{-1} day^{-1})	
	+N	+N+C
0–0.5	15.3	20.9
0.5–1	5.4	8.9
1–2	15.4	30.6
2–4	4.7	69.7
4–6	3.8	35.0
	44.3	164.6

A series of unknowns therefore exists which clearly influence the fate of soil N in grassland soil and which has important implications for environmental impact and any potential remedial actions which could be imposed through changes in management. Further information is also required, so that current predictive models can be modified to take full account of these effects. This chapter discusses some preliminary findings from a number of recent studies which particularly address:

1. The extent of removal of NO_3^- in soil columns to 1 m by denitrification and effects on N_2O release.
2. The role of denitrification and nitrification processes in determining N_2O emissions in both laboratory and field systems.

Experimental Methods

Full details of all methods and materials are given elsewhere: the following is a brief summary description of each of the studies described.

Experiment a: leaching/denitrification interactions (from Jarvis, Dixon and Hatch, 1996, unpublished results)

Intact cores (12), 19 mm diameter × 1.2 m deep, were collected from a freely draining loam soil (Crediton Series: dystric/eutric cambisol) and housed indoors in a naturally ventilated building. Deionized water was applied to the surface at a steady rate until field capacity was reached and drainage from the base of the column started. At this point, NO_3^- (as KNO_3) was applied at either the surface, or at 50 cm or 80 cm from the surface at a rate equivalent to 120 kg N ha^{-1}. Bromide as KBr was also applied at the same rate: there were three replicates for each application depth and three replicate control cores with no addition. Application of water to the surface was continued to maintain a realistic drainage rate, and was distributed over the period of the experiment to be equivalent to the average drainage through the profile in its field location. NO_3^-, Br and dissolved N_2O in the leachate were measured over an annual period. The head of the columns was also capped periodically and N_2O emission determined from the accumulation of N_2O in the head space.

Experiment b: fertilizer and grazing effects on N_2O/NO release
(Williams, Jarvis and Dixon, 1996, unpublished results)

Three treatments on a poorly undrained silty clay soil at North Wyke (Halstow series: gleyic cambisol) were investigated: i.e. + fertilizer N (120 kg N as NH_4 NO_3), cut sward; + fertilizer and grazed sward, (densely stocked sheep); control – no fertilizer and cut sward. The experiment was specifically timed for autumn when denitrification potential was high (i.e. high water content and warm soils). For gas measurements, six replicate circular enclosure chambers (0.126 m^2) were randomly positioned each day and samples taken after a 30-min enclosure for N_2O (gas chromatography) and NO (chemiluminescence) on two to four occasions per day over a period of 24 days. Soil samples (0–10 cm) were collected at the same time for the determination of extractable NO_3^- and NH_4^+.

Experiment c: laboratory studies of the effects of urine on N_2O/NO release
(from Lovell and Jarvis, 1996)

Turves ($15 \times 15 \times 15$ cm deep) were cut in spring from the same poorly drained soil as used in experiment b when the soil moisture content was 39.2% and fitted into storage boxes with sealable lids. There were eight replicates of experiment with a urine treatment and eight with water only (control): four of each treatment were used for gas analysis and four for soil sampling. Boxes were sealed for 1–2 h, and 5 ml head space were removed by syringe and N_2O measured by gas chromatography. Fluxes of NO were measured by chemiluminescence using a flow-over system for the boxes. Gas measurements were made over a 39-day period and soil samples were taken periodically for NO_3^-, NH_4^+ and moisture contents.

Experiment d: N_2O flux after addition of N in excreta or fertilizer to soil
in winter (Williamson and Jarvis, 1996, unpublished results)

Eight plots, each 6 m^3, were established on a grass sward on the previously described poorly drained soil and randomly allocated to one of eight treatments: cow dung $\pm$ DCD (dicyandiamide used as a nitrification inhibitor), cow urine $\pm$ DCD, dairy waste water (dirty water) $\pm$ DCD and $KNO_3 \pm$ DCD. Two enclosure chambers (design as before) were used for the *in situ* measurement of N_2O fluxes. Duplicate samples of head-space gas were withdrawn from each chamber and the accumulation of N_2O over an hour determined periodically over 37 days.

Experiment e: impact of NH_4^+ and NO_3^- additions on NO fluxes
(from Bisson, 1994)

Four perspex chambers ($70 \times 35 \times 30$ cm) were used to monitor NO fluxes from the poorly drained soil during early spring when the soil moisture was close to saturation and soil temperature was $< 10°C$. Two chambers were used to determine fluxes from unamended soil (controls) and two to examine the impact of adding 100 kg N ha^{-1} as KNO_3 on day 7 after measurements started and that of a further 100 kg N ha^{-1} as $(NH_4)_2SO_4$ on day 14. NO in each chamber was monitored regularly throughout each day during the measurement period.

Results

NO_3^- leaching and denitrification interactions (experiment a)

Preliminary data for NO_3^-, and emitted and dissolved N_2O are shown in Fig. 1.3. Under the moisture regime that was maintained for the columns, distinctive elution patterns for NO_3^- (Fig. 1.3a) were demonstrated (there were even clearer ones for Br – data not shown) and indicated that the flow through the cores was by piston flow. Peak concentrations from NO_3^- added at depth appeared and decayed more quickly than those added higher in the profile. Mineralization presumably was responsible for the relatively high concentration of NO_3^- shown in the control treatment. The amounts of dissolved N_2O in the leachate (Fig. 1.3b) were small and always less than 30 µg N per day. There were very small quantities in the surface-applied treatment which did not differ substantially from the control. The greatest amounts of N_2O leached were from the 50-cm treatment, the high concentrations in solution appearing some long period after the NO_3^- fronts in the leachate (Fig. 1.3a). Patterns of N_2O in the head space (Fig. 1.3c) were very erratic and showed little relationship with the treatments. Total amounts of N_2O in all treatments emitted from the soil surface were much greater than those contained in the leachate.

Emission of N_2O and NO from soil

Experiment b: fertilizer and grazing effects

Emission of N_2O (Fig. 1.4a) was increased by fertilizer addition and reached a peak after 2 days. Grazing further increased the N_2O emissions by over twofold on each occasion. Again the maximum rates were observed on day 2 and declined significantly by day 4. Nitric oxide emissions varied from 0.8 to 70 µg N m^{-2} h^{-1} (Fig. 1.4). Emissions were significantly increased by fertilizer and, additionally, by grazing. The patterns of NO emission, however, were

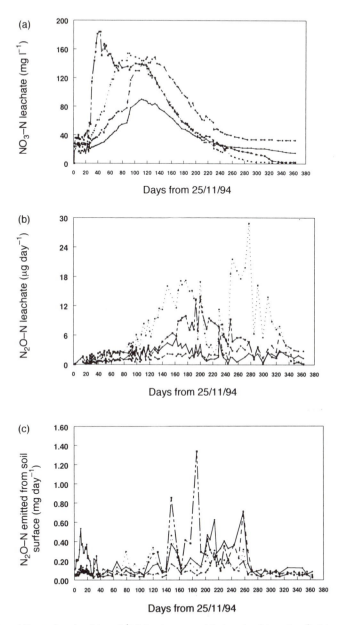

Fig. 1.3. Nitrate leached (mg l^{-1}) (a), nitrous oxide leached (µg day^{-1}) (b) or nitrous oxide emitted from soil surface (mg day^{-1}) (c) from 1.2 m intact soil columns after addition of 120 kg ha^{-1} NO$_3$-N to the soil surface (···○···), at 50 cm (···■···) or 80 cm (···□···) depths. (From Jarvis *et al.*, 1996, unpublished results.)

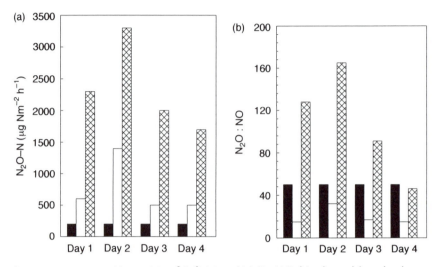

Fig. 1.4. Nitrous oxide (μg N m^{-2} h^{-1}) (a) and $N_2O : NO$ (b) released from background soils (■), fertilized sward (□) or fertilized and grazed sward (▧). (From Williams *et al.*, 1996, unpublished results.)

different from those for N_2O. Thus the ratio $N_2O : NO$ for the background soil remained remarkably constant over the 4-day period, that for the fertilizer alone treatment was decreased and that for the grazed plus fertilizer treatment was substantially increased over that for the control.

Experiment c: urine effects on N_2O/NO release

Nitrous oxide emission rose steadily from day 2 (Fig. 1.5) to reach a peak of 64 μg N m^{-2} min^{-1} at day 13 and then decreased so that rates were returned to background levels by day 38 after adding urine to the turves in control environment conditions. There was a similar trend in NO fluxes: in this case the maximum was reached on day 9 and, on average, the rates were *c.* 100× lower than those for N_2O. Soil mineral N was measured on days 1, 6, 14 and 42 after adding urine and on all occasions urine-treated soil had a total content that was *c.* 10× that of the control. Nitrate contents increased with time with added urine and NH_4^+ decreased, indicating that a steady nitrification conversion was occurring (the ratio NH_4^+-N : NO_3^--N declined from 123 on day 2 to 0.1 on day 38).

Experiment d: N_2O from field-applied excreta

All forms of excreta led to increased N_2O fluxes (Fig. 1.6): added KNO_3 produced the biggest flux, followed by urine, waste water and dung. Rainfall influenced the temporal changes in the rates of flux. The addition of DCD significantly reduced N_2O fluxes from added KNO_3 and urine: those from

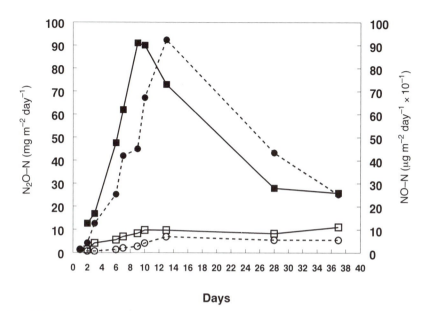

Fig. 1.5. Nitrous oxide (○,●) and NO (□,■) released from control soils (open symbols) or after addition of urine (closed symbols). (From Lovell and Jarvis, 1996.)

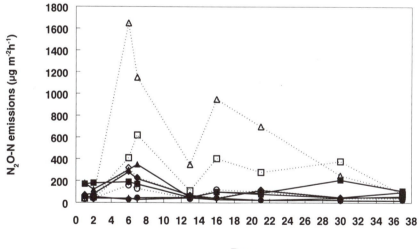

Fig. 1.6. Nitrous oxide released after addition of dung (○,●), urine (□,■), KNO₃ (△,▲) or dirty water (◇,♦) (or no addition (+) and with (solid symbols) or without (open symbols) DCD added. (From Williamson and Jarvis, 1996, unpublished results.)

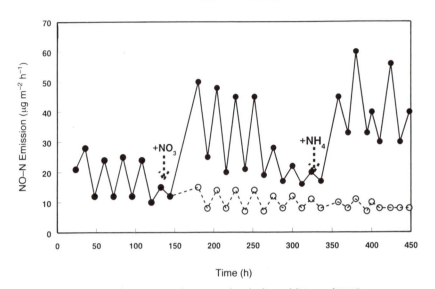

Fig. 1.7. Nitrous oxide emission from grassland after addition of KNO_3 or $(NH_4)_2SO_4$. (From Bisson, 1994.)

dung were also decreased but not significantly. There was no effect of DCD with waste water.

Experiment e: impact of NH_4^+ and NO_3^- additions on NO fluxes

The soil at the time of this experiment had high moisture contents ($c.$ 46%) and low temperature (< 10°C). The addition of KNO_3 to the soil in the field in April caused a rapid elevation in NO flux of up to four times that seen in the control chambers and then declined but still remained above the control levels (Fig. 1.7). In both treatments, there was a very marked diurnal pattern: this was also displayed when $(NH_4^+)SO_4$ was added on day 14, which resulted in similar elevations in NO to those seen with KNO_3. The addition of NO_3^- and NH_4^+ resulted in substantial changes in the soil N pools which remained at high levels throughout the measurement period and masked any subtle changes in pool size occurring as the result of N transformations (denitrification or nitrification).

Discussion

Emissions of N to the environment occur in all systems whether natural or agricultural. Ultimately, the fate of all N contained within biological systems is to be denitrified and returned to the atmosphere to maintain the balances within the global N cycle. Transmission of excessive amounts from agricultural management to the atmosphere and into waters is a result of excesses in mobile pools within the soil reacting to particular combinations of biological activities and environmental conditions. The route by which the N is transferred and the quantities involved is the net result of those interacting factors and the competition between all the processes which can remove or transform NH_4^+ or NO_3^-. Nitrate is a key component in this because of: (i) its mobility and (ii) its accessibility and ability to provide, under many conditions, an oxygen source for facultative anaerobes.

Much of the excess NO_3^- in freely draining soils is removed by leaching although this, as noted earlier, may subsequently be removed by subsoil denitrification. The possible consequences of this are removal of the environmental impact of NO_3^- but with the possibility of further release of N_2O. The conditions of the intact soil core experiment were such that little N_2O was evolved from the surface, even from a surface application of NO_3^-. This could have been because either: (i) leaching rate was controlling the residence time of NO_3^- in potential denitrification zones; (ii) there was little potential for denitrification to occur through an inappropriate microbial population or a lack of C substrate, or (iii) that the conditions were such that the $N_2 : N_2O$ ratio of the products was high. Other, unpublished information on the same soils indicates that adding C in at depth can stimulate a great deal of denitrification and N_2O generation (Clough *et al.*, 1996, unpublished results) which is not expressed as a markedly increased flux at the soil surface. As it is a highly soluble gas, there was also the possibility that N_2O may be removed in solution, but the amounts involved in this case were small. However, the fact that there were increases in the amounts removed in this way for NO_3^- added in at 50 or 80 cm depth, indicates that denitrification took place at some depth below the surface.

Whether antecedent conditions influence denitrification activity at depth in the same way as they do within the surface soil (see Chapter 2) is not known. The impact of these and of the other factors which influence the $N_2 : N_2O$ ratio (NO_3^- content; O_2 moisture tension, etc., see Chapter 3), will decrease as N_2O diffuses up through the profile and meets other opportunities for further reduction to N_2. Closer to the surface, the N_2O fluxes to the atmosphere will be regulated to a much greater degree by the interactions between moisture/temperature regimes and spatial variability (Velthof *et al.*, 1996), especially in the case of denitrification.

How denitrification interacts with nitrification is as yet unquantified. In many soils both processes can occur at the same time (De Klein and Logtestijn, 1994; see Chapter 5) and this makes prediction of rates of emission and mechanistic explanation of the controls difficult. In our laboratory study (Lovell and Jarvis, 1996) where soil water was below field capacity and the system was generally aerobic, nitrification would ordinarily be assumed to be the most important. However, there was high CO_2 generation which may have generated anaerobic conditions within microsites and stimulated denitrification activity. The high levels of NH_4^+ in the treated soil at the start of the incubation fell considerably (i.e. nitrification) between days 6 and 14: this was coincident with the peak NO emission rate which has been considered to be derived mainly from nitrification and linearly dependent on available NH_4^+ (Skiba et al., 1993). Nitrous oxide emission reached a maximum at a later stage, perhaps because denitrification was contributing N_2O at this time. This in turn would have been reduced as the soil became more aerobic with a reducing CO_2 concentration.

In the field, any clarity between the roles of these two processes becomes even less distinct. Thus when excreta are applied, a number of studies indicate that substantial amounts of N_2O are released (Allen et al., 1996; see Chapter 19) and that ambient conditions suggest that both nitrification and denitrification could be contributing. The preliminary data from Williamson and Jarvis (unpublished results) demonstrate that even under waterlogged, cold soil conditions, the addition of a nitrification inhibitor reduced N_2O fluxes from urine and dung. The likely mechanisms are by: (i) reducing production by direct inhibition of nitrification and (ii) a secondary effect by reduced substrate (NO_3^-) availability for denitrification. The effects of DCD with added NO_3^- are surprising, as the assumption was that there would be little impact of DCD on denitrification. Skiba et al. (1993) have also reported a decrease in N_2O emission from KNO_3 applied with DCD. The possibility that the N_2O fluxes in this case were derived from denitrification of added NO_3^- plus nitrification and subsequent denitrification of NH_4^+ displaced by added $K+$ does not make the task of assessing relative contributions of nitrification and denitrification to N_2O fluxes any easier.

Similar complex interactions must have also occurred in the sheep grazing experiment (experiment b). Added fertilizer (as $NH_4 NO_3$) increased N_2O substantially above that of the control soil and the soil conditions were those which should have promoted denitrification. Grazing further increased N_2O emission, presumably because of the concentrations of N in dung and especially urine patches. Other field studies have shown that both excretal types can provide hot spots of substantial N_2O release (e.g. de Klein and Logtestijn, 1994; Allen et al., 1996; Müller et al., 1997; see Chapter 19). Again there is opportunity for nitrification to be an important contributor to this, either directly from the process itself or indirectly through the provision of NO_3^- for subsequent denitrification.

Changes in NO fluxes were also substantial. Meixner et al. (see Chapter 4) have described the environmental importance of this gas. Fluxes in the extensive subtropical grassland systems and fertilized cultivated soils were of the same order (up to 18 and 140 μg N m^{-2} h^{-1}, respectively) as those recorded in intensively managed grazing systems (up to 70 μg N m^{-2} h^{-1}). The lower emission rates in the grazed than in the fertilized treatments is again surprising, as the expectation was that if nitrification was enhanced because of the additional NH_4^+ derived from excreta there would also have been a concurrent increase in NO emission (Skiba *et al.*, 1993). It is, however, clear from other studies (e.g. Bisson, 1994) that the patterns and ratios of N_2O and NO emissions cannot easily be predicted from a knowledge of the soil conditions alone. As well as soil factors, the plant sward may have also influenced fluxes to different extents in the two treatments through differential effects of cutting/grazing on uptake of NH_4^+ or NO_3^- and physical effects of the herbage on the diffusion of the gases to the top of the canopy.

Bisson (1994) showed that under controlled laboratory conditions with soil under defined moisture status and with or without a nitrification inhibitor present, N_2O appeared to come from a combination of nitrification and denitrification when soil was not saturated and that thereafter it was released predominantly through denitrification. Prediction of NO release was more difficult. Bisson's results indicated that although NO appeared to be released primarily from nitrification, denitrification could become a source when a period of very low activity was followed by the sudden imposition of conditions favourable to NO_3^- reduction. Although in pure culture (Anderson and Levine, 1985) the ratio of NO : N_2O was > 1 for nitrifiers and > 1 for denitrifiers, any attempts to use ratios of this nature as 'fingerprints' for the process involved will be fraught with difficulty. As well as the displacement in time between periods of maximum emission (Fig. 1.5), the marked diurnal patterns in NO emission (Fig. 1.7) further confound the problem.

Conclusion

Much information is currently being gathered on emission of N compounds to the environment. A basic need for national and international policy agencies is the provision of the current status with regard to emissions and the managerial steps required to reduce those emissions when they are seen to be too high. Because of the interactive nature of the problems involved (i.e. NH_3 to the atmosphere, N_2O and NO to the atmosphere and NO_3^- and NO_2^- to waters), it is important that any changes that are imposed are based on a mechanistic understanding of the processes involved. Whilst it is becoming more easy, for example, to measure fluxes of gases to the atmosphere as techniques and instrumentation improve (see Chapter 36), there is still much opportunity to improve our conceptual thinking of the way that the processes involved are operating. It is clear that the soil N cycle should be considered as a much more dynamic system than is currently the case. As demonstrated by the data in this paper, it is usually difficult to distinguish between the activities of different processes and as our understanding increases and the use of new techniques (such as the use of natural abundance stable isotopes) provides new information, it will become more and more important not to consider soil N processes such as leaching, nitrification and denitrification as almost mutually exclusive. Further consideration of the sources and role of NO in this may be an important way forward. This paper is limited to discussion of leaching denitrification and nitrification. In many grasslands and for much of the time, the balance of mineral N in the soil is strongly influenced by mineralization/immobilization (Jarvis *et al.,* 1996a). Supplies of NO_3^- and NH_4^+ through this route (which can be considerable – see Gill *et al.,* 1995), and their linkages with the loss processes is another important area which merits attention if we are to manage N in an effective efficient way to reduce environmental impact.

Acknowledgements

I would like to acknowledge the substantial input that all my various co-workers made in the development and conduct of the various experiments described here. The work was, in part, funded by MAFF, London; the Institute of Grassland and Environmental Research is supported by the Biotechnology and Biological Sciences Research Council.

References

Allen, A.G., Jarvis, S.C. and Headon, D.M. (1996) Nitrous oxide emission from soils due to inputs of nitrogen from excreta return by livestock ungrazed grassland in the UK. *Soil Biology and Biochemistry* 28, 597–607.

Anderson, I.C. and Levine, J.S. (1985) Relative rates of nitric oxide and nitrous oxide production by nitrifiers, denitrifiers and nitrate respirers. *Applied Environmental Microbiology* 51, 938–945.

Bisson, G.D. (1994) Sources of nitric and nitrous oxide in grassland soil. PhD Thesis, Reading University.

Bremner, J.M. and Blackmer, A.M. (1977) N_2O emission from soils during nitrification of fertilizer N. *Science* 199, 295–296.

Buijsman, E., Maas, H.F.M. and Asman, W.A.H. (1987) Anthropogenic NH_3 emissions in Europe. *Atmospheric Environment* 21, 1009–1022.

Davidson, E.A. (1992) Sources of nitric oxide following wetting of dry soil. *Soil Science Society of America Journal* 56, 95–102.

DeKlein, C.A.M. and Van Logtestijn, R.S.P. (1994) Denitrification and N_2O emission from urine-affected grassland soil. *Plant and Soil* 163, 235–242.

Gill, K., Jarvis, S.C. and Hatch, D.J. (1995) Mineralization of nitrogen in long-term pasture soils: effects of management. *Plant and Soil* 163, 235–242.

Hutchings, N.J., Sommer, S.C. and Jarvis, S.C. (1996) A model of ammonia volatilization from a grazing livestock farm. *Atmospheric Environment* 30, 589–599.

Jarvis, S.C. and Hatch, B.F. (1994) Potential for denitrification at depth below long term grassland. *Soil Biology and Biochemistry* 26, 1629–1636.

Jarvis, S.C. and Pain, B.F. (1994) Greenhouse gas emissions from intensive livestock systems, their estimation and technologies for their reduction. *Climatic Change* 27, 27–38.

Jarvis, S.C., Scholefield, D. and Pain, B.F. (1995) Nitrogen cycling in grazing systems. In: Bacon P.E. (ed.) *Nitrogen Fertilization in the Environment*. Marcel Dekker, New York, pp. 381–420.

Jarvis, S.C., Stockdale, E.A., Shepherd, M.A. and Poulson, D.S. (1996a) Nitrogen mineralization in temperate agricultural soils: processes and measurement. *Advances in Agronomy* 57, 187–235.

Jarvis, S.C., Wilkins, R.J. and Pain, B.F. (1996b) Opportunities for reducing the environmental impact of daily farming managements: a systems approach. *Grass and Forage Science* 51, 21–31.

Lind, A.M. and Eiland, F. (1989) Microbial characterization and nitrate reduction in subsurface soils. *Biology and Fertility of Soils* 8, 197–203.

Lovell, R.D. and Jarvis, S.C. (1997) Effects of urine on soil microbial biomass, methanogenesis, nitrification and denitrification in grassland soils. *Plant and Soil* 186, 265–273.

McCarty, G.W. and Bremner, J.M. (1992) Availability of organic carbon for denitrification of nitrate in subsoils. *Biology and Fertility of Soils* 14, 219–222.

Müller, C., Sherlock, R.R. and Williams, P.H. (1997) Factors influencing N_2O emission from dung pasture soil affected by animal urine or ploughing. *European Journal of Soils Science* (in press).

Obenhuber, D.C. and Lawrance, R. (1991) Reduction of nitrate in aquifer microcosms by carbon additions. *Journal of Environmental Quality* 20, 255–258.

Parkin, T.B. and Meisinger, J.J. (1989) Denitrification below the crop rooting zone. *Journal of Environmental Quality* 18, 12–16.

Remde, A. and Conrad, R. (1991) Metabolism of nitric oxide in soil and denitrifying bacteria. *FEMS Microbiological Ecology* 85, 81–94.

Robertson, G.P. and Tiedje, J.M. (1987) N_2O sources in aerobic soils: nitrification, denitrification and other biological processes. *Soil Biology and Biochemistry* 19, 187–193.

Scholefield, D., Tyson, K.C., Garwood, E.A., Armstrong, A.C., Hawkins, J. and Stone, A. (1993) Nitrate leaching from grazed grassland lysimeters: effects of fertilizer input, field drainage, age of sward and patterns of weather. *Journal of Soil Science* 44, 601–613.

Skiba, U., Smith, K.A. and Fowler, D. (1993) Nitrification and denitrification as sources of nitric oxide and nitrous oxide in a sandy loam soil. *Soil Biology and Biochemistry* 25, 1527–1536.

Velthof, G.L., Jarvis, S.C., Stein, A., Allen, A.G. and Oenema, O. (1996) Spatial variability of nitrous oxide fluxes in mown and grazed grasslands in a poorly drained clay soil. *Soil Biology and Biochemistry* 28, 1215–1225.

van der Weerden, T.J. and Jarvis, S.C. (1997) Ammonia emission factors for N fertilizers applied to two contrasting grassland soils. *Environmental Pollution* 95, 205–211.

Yeomans, J.C., Bremner, J.M. and McCarty, G.W. (1992) Denitrification capacity and denitrification potential of subsurface soils. *Communications in Soil Science and Plant Analysis* 23, 919–927.

Controls over Denitrification and its Gaseous Products in a Permanent Pasture Soil

L. Dendooven[1], P. Splatt[2], E. Pemberton[2], S. Ellis[3] and J.M. Anderson[2]

[1]IACR-Rothamsted, Department of Soil Science, Rothamsted Experimental Station, BBSRC Institute of Arable Crop Research, Harpenden, Hertfordshire, AL5 2JQ, UK; [2]Department of Biological Sciences, University of Exeter, Prince of Wales Road, Exeter, Devon EX4 4PS, UK; [3]ADAS Boxworth, Cambridge CB3 8NN, UK

Summary. A research project was started to investigate controls over the denitrification process and its gaseous products in a soil from permanent pasture. The experimental procedure was such that experimental data would be used to develop a mechanistic denitrification model. The experiments were carried out with soil slurries to minimize variations in nitrate (NO_3^-) concentrations under strict anaerobic conditions, so that the soil sample represented an anaerobic micro-site with no differences in NO_3^- concentrations, carbon (C) substrate or soil microbial biomass. The ratio between carbon dioxide (CO_2) produced and NO_3^- reduced and thus the nitrous oxide (N_2O) produced was affected by the period of aerobiosis-anaerobiosis, the composition of the C substrate and the proportion of denitrifiers in the soil microbial biomass. The NO_3^- concentration largely affected the $N_2O : N_2$ ratio, as no N_2 was formed within 5 days when 2000 mg NO_3^--N kg^{-1} was added but when 20 mg NO_3^--N kg^{-1} was added more N_2 was formed than N_2O within a day. The *de novo* synthesis, the persistence of the reduction enzymes and their affinity for NO_3^-, NO_2^- and N_2O was an important control of the $N_2O : N_2$ ratio. The percentage of N_2 in the ($N_2 + N_2O$) produced ranged from 66 to 0%, depending on the enzyme status of the soil microbial biomass. The antecedent water regime of the soil affected the $N_2O : N_2$ ratio independently from the water content or the NO_3^- concentration. The availability of easily decomposable organic C, i.e. glucose, reduced the $N_2O : N_2$ ratio. The exposure of a soil microbial population to frequent anaerobiosis will select denitrifiers in relation to other soil bacteria and thus the kinetics of NO_3^-, NO_2^-, N_2O and N_2 and the dynamics of the reduction enzymes.

Introduction

Denitrification is the biological reduction of nitrate (NO_3^-) under anaerobic conditions by facultative anaerobic organisms or denitrifiers. Nitrate is reduced to nitrite (NO_2^-) by NO_3^- reductase, NO_2^- to nitrous oxide (N_2O) by NO_2^- reductase and N_2O to dinitrogen (N_2) by N_2O reductase (Knowles, 1982). Nitric oxide (NO) is also formed but it was not considered in this study as it is a very reactive gas under strict anaerobic conditions.

Proportions of the gaseous products of the denitrification process are of considerable environmental importance, as N_2O indirectly (through the formation of NO and NO_2) depletes the stratospheric ozone layer (Cicerone, 1987), enhances the formation of acid rain (HNO_3) (Tanner, 1990) and reduces radiative heat loss from the soil surface, thereby contributing to the greenhouse effect (Lashof and Ahuja, 1990). Ratios of $N_2O : N_2$ from denitrification are considered to be largely controlled by oxygen (O_2) and NO_3^- concentrations, C substrate availability, enzyme status and pH (Firestone, 1982). Increases in the concentrations of O_2 and NO_3^- and decreases in pH increase the proportion of N_2O produced. This research investigated controls over the denitrification process and its gaseous products in permanent pasture soil to account for short-term changes.

Materials and Methods

Experimental site

The experiments were carried out with soil from a permanent pasture at IGER, North Wyke, Devon, UK. The soil, a clayey pelostagnogley, is currently classified as belonging to the Hallsworth Series (Clayden and Hollis, 1984). The inorganic fraction contained 36.6% clay, 47.7% silt, 13.9% fine sand and 1.8% coarse sand (Armstrong and Garwood, 1991). The organic C content of the upper 10 cm was 5.3%, the organic N content 0.62% and the pH (H_2O) 6.0.

Conditioning of the soil

Soil samples were collected between February 1992 and December 1993 from the 0–10 cm layer of an experimental plot that had been grazed by cattle but had received no N fertilizer for at least 10 years. The moist soil was sieved (5 mm) and conditioned for a maximum of 5 days at 25°C.

Assays for enzyme activity

A detailed description of the assays for enzyme activity is given by Dendooven and Anderson (1995a). Experiments were done with soil slurry to eliminate microspatial factors which act as variable controls under field conditions and under strict anaerobic conditions to eliminate possible effects of altering O_2 concentrations. Soil was incubated with and without acetylene (Balderstone *et al.*, 1976) to assess the N_2 production and with or without chloramphenicol (assumed to inhibit the *de novo* synthesis of the reduction enzymes involved in the denitrification process) to study the dynamics of the reduction enzymes involved in the denitrification process.

Statistical analysis

Regression coefficients were calculated and covariance analysis was carried out with the SAS statistical package (SAS Institute, 1988), using the general linear model procedure. The link between the carbon dioxide (CO_2) production and the denitrification process and the characteristics of the reduction enzymes involved in the denitrification process were investigated using the DETRAN model (Dendooven *et al.*, 1994; Dendooven and Anderson, 1995b).

Results and Discussion

Link between CO_2 produced and NO_3^- reduced

The link between the CO_2 produced, a measure of soil microbial activity, and the amount of NO_3^- reduced ($CO_2 : NO_3^-$) or N_2O produced ($CO_2 : N_2O$) was not constant. For the pasture soil, $CO_2 : N_2O$ in the assays for enzyme activity ranged from 0.96 to 2.27. The precedent period of aerobiosis or anaerobiosis largely affected $CO_2 : N_2O$ (Dendooven and Anderson, 1995a). The ratio increased from 0.99 for soil anaerobically conditioned for 3 days to 1.25 for the same soil subsequently incubated aerobically for 3 days. After 70 days of aerobic conditioning, the $CO_2 : N_2O$ ratio had increased to 1.43. Anaerobic conditioning of this soil for 3 days decreased that ratio again to 1.10. The composition of organic material affects the $CO_2 : N_2O$ ratio. The application of glucose to the pasture soil decreased the ratio from 2.14 to 1.94 (Dendooven *et al.*, 1996c). It is difficult to ascertain how the C substrate affected the $N_2O : CO_2$ ratio but is presumably due to a more efficient use of C. Composition of the soil microbial biomass affects the $CO_2 : N_2O$ ratio. In a well-drained forest with a soil microbial biomass less well adapted to anaerobiosis, the ratio was 12.5 (Dendooven *et al.*, 1996b). Saturating the soil with distilled water for 6 months decreased the ratio to 3.33, a value still higher than values observed in the pasture soil (Pemberton *et al.,* unpublished results).

Concentration of NO_3^-

The concentration of available NO_3^- affects the $N_2O : N_2$ ratio (Dendooven *et al.,* 1994). In an anaerobic incubation experiment no N_2 was measured within 5 days when 2000 mg NO_3^--N was added. No N_2 was detected after 1 day when 200 NO_3^--N was applied but more N_2 than N_2O was produced after 5 days. When only 20 mg NO_3^--N was added, more N_2 than N_2O was produced after 1 day.

The dynamics of NO_3^- under anaerobiosis often followed a specific pattern (Dendooven *et al.,* 1996c, Ellis *et al.,* 1996). The NO_3^- concentration decreased sharply within the first 5 h of anaerobiosis and increased again sharply at 12 h and then gradually decreased. The amount of N that could not be accounted for after 5 h often exceeded more than 50 mg NO_3^--N in the pasture soil. In an experiment with an arable soil stripped of most of its organic material and inoculated with *Pseudomonas denitrificans* L., as much as 100 mg NO_3^--N was unaccounted for (Ellis *et al.,* 1996). The same phenomenon was observed in a forest soil where the amount of NO_3^--N that could not be accounted for was lower (2 mg NO_3^--N) but lasted for 38 h. This phenomenon depended on the period of aerobiosis–anaerobiosis, the C substrate and the soil microbial biomass. As a consequence, using the concentration of NO_3^--N in a soil to predict denitrification could be biased.

Characteristics of the reduction enzymes

The percentage of N_2 in the total gas production ($N_2O + N_2$) as affected by the status of the reduction enzymes ranged from 56 and 0% in the assay for enzyme activity (Dendooven and Anderson, 1995a). The characteristics of the reduction enzymes which will affect the amount of N_2 produced are: (i) the affinity for the oxides of N; (ii) the *de novo* synthesis of the enzymes when anaerobiosis is induced; and (iii) the persistence of the reduction enzymes under aerobic conditions.

The affinity for the oxides of N is defined as a weighting factor for competition among electron acceptors (Cho and Mills, 1979). Using the DETRAN model (Dendooven *et al.,* 1994) to estimate those affinities using a 'least-square' curve fitting technique (Dendooven and Anderson, 1995b), and assuming that the affinity for NO_3^- was 1, gave an affinity for NO_2^- of approximately 1000 and of 0.75 for N_2O. In a forest soil the affinity for NO_2^- was lower and between 120 and 160 (Dendooven *et al.*, 1996b).

When anaerobiosis is induced reductase enzymes are formed. The *de novo* synthesis of NO_3^-, NO_2^- and N_2O does not occur at the same time but differs substantially. In the pasture soil, NO_3^- reductase was in excess while the *de novo* synthesis of NO_2^- reductase started after 5 h and that of N_2O reductase after 16 h (Dendooven and Anderson, 1994). In a well-drained forest soil, NO_3^- reductase was not in excess and the *de novo* synthesis of the enzymes started after approximately 20 h.

Under aerobic conditions, the concentrations of the reduction enzymes decrease. The persistence of NO_3^- reductase is high but low for N_2O reductase (Dendooven and Anderson, 1995a). As a result, the percentage of N_2 in the $(N_2O + N_2)$ produced decreased from 49 to 28% for a soil anaerobically conditioned for 3 days and then aerobically incubated for 3 days.

Oxygen

Nitrous oxide reductase is more sensitive to O_2 than the other reduction enzymes (Betlach and Tiedje, 1981).

Antecedent water regime of the soil

The antecedent water regime of a soil has a large effect on the denitrification process (Dendooven *et al.*, 1996a). Intact cores aerobically conditioned for 100 days were submerged in water for 6 h, drained for 2 h resulting in a gravimetric water content of 40.4%, and then amended with 100 mg NO_3^--N kg^{-1}. The ratio $N_2O : N_2$ as measured after 1 day was 4.15. For cores submerged in water for 96 h, drained for 6 h with a gravimetric water content of 43.6% and then amended with 100 mg NO_3^--N kg^{-1}, $N_2O : N_2$ was 1.18.

The aerobic conditioning of the cores decreased the concentration of the reduction enzymes. Upon waterlogging anaerobiosis was induced. A short-term period of submerging was not sufficient to start *de novo* synthesis of N_2O reductase but NO_3^- and NO_2^- were fully de-repressed. During 1-day incubation, *de novo* synthesis of N_2O reductase started but it was only towards the end that it was fully de-repressed. A long-term period of anaerobiosis (96 h) gave a fully de-repressed N_2O reductase at the start of the 1 day incubation, so more N_2O was reduced to N_2 than in the short-term incubated soil.

Composition of the C substrate

The effects of easily decomposable organic material on $CO_2 : N_2O$ have been mentioned earlier but glucose also affected $N_2O : N_2$ (Dendooven *et al.*, 1996c). The percentage of N_2 in the $(N_2O + N_2)$ produced decreased from

32 to 24% when glucose was applied. The addition of glucose increased the reduction potential of the soil and that was instantly matched by a de-repression of NO_3^- and NO_2^- but not of N_2O reductase.

pH

A decrease in soil pH generally decreases the proportion of N_2 produced. (For more details see Chapter 17.)

Composition of soil microbial population

Groffman and Tiedje (1989) suggested that denitrifiers are selected in relation to other soil bacteria in soils with water regimes inducing frequent anaerobiosis. Pasture soil which is often waterlogged for long periods has a soil microbial biomass adapted for anaerobic conditions, while a well-drained forest soil has a less well adapted microbial population. Saturating the forest soil for 6 months changed the characteristics of the denitrification process (Pemberton *et al.*, 1996, unpublished results). The $CO_2 : N_2O$ ratio decreased from 12.5 to 3.3 in the assay for enzyme activity, $N_2O : N_2$ decreased from 3.1 to 1.1 and the dynamics of the reduction enzymes changed. The $CO_2 : N_2O$ in the forest soil was still higher than in the pasture soil and the question remained as to whether that difference was due to the proportion of the denitrifiers that was still higher in the pasture as compared to the forest soil and/or due to a difference in composition of the soil organic matter.

Acknowledgements

The research was funded by BBSRC and EU-COPERNICUS Project CIPA-CT93-0250 (L.D.), by EU-CORE project, phase III, Grant 9027 (E.P.) and by BBSRC (Global Adaptation to Environmental Change) (S.E.). IACR-Rothamsted receives grant-aided support from BBSRC, UK.

References

Armstrong, A.C. and Garwood, E.A. (1991) Hydrological consequences of artificial drainage of grassland. *Hydrological Processes* 5, 157–194.

Balderstone, W.L., Scherr, B. and Payne, W.J. (1976) Blockage by acetylene of nitrous oxide reduction in *Pseudomonas perfectomarinus*. *Applied and Environmental Microbiology* 31, 504–508.

Betlach, M.R. and Tiedje, J.M. (1981) Kinetic explanation for accumulation of nitrite, nitric oxide, and nitrous oxide during bacterial denitrification. *Applied and Environmental Microbiology* 42, 1074–1084.

Cho, C.M. and Mills, J.G. (1979) Kinetic formulation of the denitrification process in soil. *Canadian Journal of Soil Science* 59, 249–257.

Cicerone, R.J. (1987) Changes in stratospheric ozone. *Science* 237, 35–42.

Clayden, B. and Hollis, J.M. (1984) Criteria for differentiating soil series. *Soil Survey of England and Wales.* Technical Monograph No. 17, Harpenden.

Dendooven, L. and Anderson, J.M. (1994) Dynamics of reduction enzymes involved in the denitrification process in pasture. *Soil Biology and Biochemistry* 26, 1501–1506

Dendooven, L. and Anderson, J.M. (1995a) Maintenance of denitrification potential in pasture soil following anaerobic events. *Soil Biology and Biochemistry* 27, 1251–1260.

Dendooven, L. and Anderson, J.M. (1995b) Use of a 'least-squares' optimization procedure to estimate enzyme characteristics and substrate affinities in the denitrification reactions in soil. *Soil Biology and Biochemistry* 27, 1261–1270.

Dendooven, L., Splatt, P., Anderson, J.M. and Scholefield, D. (1994) Kinetics of the denitrification process in a soil under permanent pasture. *Soil Biology and Biochemistry* 26, 361–370.

Dendooven, L., Duchateau, L. and Anderson, J.M. (1996a) Denitrification as affected by the previous water-regime of the soil. *Soil Biology and Biochemistry* 28, 239–245.

Dendooven, L., Pemberton, E. and Anderson, J.M. (1996b) Denitrification potential and the dynamics of the reduction enzymes in a Norway spruce plantation. *Soil Biology and Biochemistry* 8, 151–157.

Dendooven, L., Splatt, P. and Anderson, J.M. (1996c) Denitrification as affected by C substrate availability. *Soil Biology and Biochemistry* 28, 141–149.

Ellis, S., Dendooven, L. and Goulding, K.W.T. (1996) Quantitative assessment of soil nitrate uptake and N_2O evolution during denitrification. *Soil Biology and Biochemistry* 28, 589–595.

Firestone, M.K. (1982) Biological denitrification. In: *Nitrogen in Agricultural Soils.* Stevenson, F.J. (ed) *Agronomy Monograph,* 22. American Society of Agronomy, Madison, Wisconsin. John Wiley & Sons, New York, pp. 289–326.

Groffman, P.M. and Tiedje, J.M. (1989) Denitrification in north temperate forest soils: spatial and temporal patterns at the landscape and seasonal scales. *Soil Biology and Biochemistry* 21, 613–620.

Knowles, R. (1982) Denitrification. *Microbiological Review* 46, 43–70.

Lashof, D.A. and Ahuja, D.R. (1990) Relative contributions of greenhouse gas emissions in global warming. *Nature* 344, 529–531.

SAS Institute (1988) *Statistic Guide for Personal Computers.* Version 6.03, Edn. SAS Institute, Cary.

Tanner, R.L. (1990) Sources of acids, bases, and their precursors in the atmosphere. In: Adriano, D. and Johnson, A. (eds) *Acidic Precipitation, Volume 3: Sources, Deposition, and Canopy Interactions.* Springer-Verlag, New York, pp. 1–19.

Determination of Controls over Denitrification Using a Flowing Helium Atmosphere System

3

D. Scholefield and J.M.B. Hawkins

Institute of Grassland and Environmental Research, North Wyke, Devon EX20 2SB, UK

Summary. During the last 20 years considerable progress in the study of denitrification has been made using the ^{15}N and acetylene inhibition (AI) techniques with field chambers and in laboratory-based soil core incubations. Most progress in determination of the controlling variables has resulted from laboratory incubations of soil slurries and re-packed, sieved soils. Despite the widespread use of the closed incubation AI method, attempts to account for the variability in rate of denitrification measured in intact soil cores from grassland have met with only limited success. There is a need to improve our capability to predict rates of denitrification in the field and, in particular, to provide information to enable the management of nitrogen (N) for minimal nitrous oxide (N_2O) emission from farms. The latter will require a better understanding of the effects of the major controls on the $N_2O : N_2$ ratio. A technique is described in which intact soil cores are incubated in a flowing atmosphere of helium (He) and oxygen (O_2), after first purging the soil and incubation vessel free from dinitrogen (N_2). This technique allows the simultaneous and independent assessment of N_2O and N_2 fluxes without recourse to AI. Square-section cores are extracted from random locations and assembled, without air gaps, to make composite turves in the incubation vessel. Dinitrogen-free amendments may be added to each vessel via a metered irrigation assembly. The technique was used to measure the effects of nitrate supply, soil water content, temperature and pH in a clay-loam under grassland management, receiving simulated additions of fertilizer nitrate. Typically, the peak efflux of N_2O and N_2 occurred 2 and $3\frac{1}{2}$ days after fertilizer addition, respectively. The $N_2O : N_2$ ratio increased with decreasing temperature, pH and water content, and with increasing amount of nitrate added. Additionally, the effects of antecedent soil conditions and gas flow rate on the $N_2O : N_2$ ratio are reported, both of which may be important in the field. At present, application of the technique is limited to soils receiving medium to high rates of fertilizer N, but improvements to gas seals on vessels and valves would extend the application to soils under extensive management and in semi-natural environments.

Introduction

Since the mid-1970s there has been considerable progress in the study of denitrification using both ^{15}N and acetylene inhibition (AI) techniques (Tiedje et al., 1989). Most progress in quantifying the effects of the controlling variables has been achieved using laboratory incubations of soil slurries and re-packed, sieved soils (e.g. Knowles, 1981). However, despite the wide-spread use of the closed incubation AI method developed by Ryden et al. (1979), comparatively little progress has been made in predicting the out-come of these controls (temperature, nitrate (NO_3^-) concentration, soil water content, pH and carbon (C) supply) in intact cores under field conditions (e.g. Jarvis et al., 1991). There are several possible reasons for this, including slow diffusion of acetylene to active sites in wet, clay soils, poor representation of spatial and temporal variability in sampling strategies, and the inability to simulate natural aerobicity and oxygen gradients with core incubations. In view of the shortcomings of existing methodology, a novel soil core incuba-tion technique has been developed, based on that used by Wickramasinghe et al. (1978) and Galsworthy and Burford (1978). The technique involves replacing the soil atmosphere in intact cores by a mixture of oxygen (O_2) and helium (He) to enable direct and simultaneous measurement of both nitrous oxide (N_2O) and dinitrogen (N_2) as products of denitrification. This paper reports briefly on the design of the incubation system, the operation of the technique and its use to determine the effects of the major controls of denitrification on the flux of N_2O and N_2.

Methods

Soil was sampled from an unfertilized experimental plot under grazed pasture at the Institute of Grassland and Environmental Research, North Wyke, Devon, UK. The soil is a stagno-Dystric Gleysol with a pH of 5.7 and contains very small concentrations of NO_3^--N. Purpose-built square-section corers (45 mm × 45 mm) were used to obtain the intact, 100-mm-deep soil samples required with the technique. Each extracted core was reduced in length to 90 mm by carefully paring its base with a scalpel. The grass sward was clipped to 25 mm height prior to incubation but was otherwise left intact. One hundred and fifty cores (c. 30 kg soil) were incubated per experimental run, with 25 cores in each of 6 vessels.

The incubation vessels were made from 12-mm-thick ABS plastic sheet with internal dimensions 225 mm × 225 mm × 120 mm, resulting in an enclosed volume of 6.07 l of which 4.56 l was occupied by soil. The cores were assembled in a 5×5 array, without air gaps, and supported clear of the bottom of the vessel on a steel mesh to allow gas to be forced through the cores from beneath. At the start of each experiment, the original soil atmosphere was replaced by a mixture of O_2 in He, with the O_2 content adjusted to that measured in the field using gas diffusion probes inserted to 90 mm depth. Thus the original N_2 was replaced with He to enable any N_2 produced by denitrification to be detectable. This purging procedure normally took about 17 h with a gas flow rate of 180 ml min^{-1}. Nitrate, glucose, water and any other amendments were applied to the surface of each core via the He-purged irrigation assembly. The gas flow was then switched from 'flow through' to 'flow over' mode and the flow rate reduced to 20 ml min^{-1}. Effluent gases from each of the six vessels in the system were piped through a 6-way valve, a gas meter and a water trap to waste (Fig. 3.1).

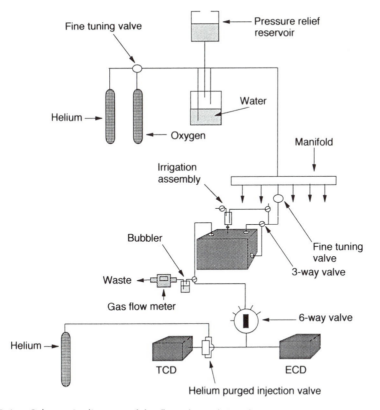

Fig. 3.1. Schematic diagram of the flow-through incubation system.

The valve was used to switch flow from each vessel sequentially to two gas chromatographs.

One chromatograph (Pye 104; 5Å molecular sieve) was fitted with a He-purged injection valve and a katharometer detector for measurement of N_2, while the other (Pye 4500; Porapak Q) had an electron-capture detector for measurement of N_2O. The minimum time to measure both N_2 and N_2O (and O_2, carbon dioxide (CO_2) and nitric oxide (NO) if required) in effluent from all six vessels was 47 min but readings were normally repeated every 2–3 h. The system was operated in a room with temperature controlled to $\pm 1°C$. An experimental run was continued until concentrations of N_2O and N_2 had returned to near background, which normally took 6–8 days. The limits of detection for N_2 and N_2O were 16 and $0.05\ \mu l\ l^{-1}$, respectively, which is equivalent to 50 and $0.17\ g\ N\ ha^{-1}\ d^{-1}$, respectively, at a flow rate of 20 ml min^{-1}.

Each of the major controls of denitrification was varied systematically over a naturally occurring range while attempting to keep the other controls non-limiting. Thus a set of standard incubation conditions was defined as follows: the soil was collected at or near field capacity and its water content adjusted to 42% by weight; N was applied as KNO_3 at a rate equivalent to 100 kg N ha^{-1} (or 50 kg N ha^{-1}, occasionally); the temperature was set at 20°C; C was applied as glucose at a rate equivalent to 394 kg ha^{-1}; and the pH was that of the field soil (5.7).

The main reason for sampling from the field at or near field capacity was to ensure that the soil used for each run had experienced a similar history of aerobicity. However, the effects of antecedent soil aerobicity were investigated by conducting standard incubation runs (50 kg N ha^{-1}) on cores pre-incubated either aerobically or anaerobically for 7 days. The anaerobic pretreatment consisted of continuous purging with He alone in flow-through mode at 150 ml min^{-1}. For the series of runs investigating the effects of soil water content, samples were taken periodically as the soil dried naturally during spring and early summer. The effects of quality of C substrate were not investigated specifically: 'standard' runs with and without added glucose were undertaken to assess whether denitrification was C limited under these conditions. The effects of NO_3^- were investigated by a series of runs in which KNO_3 was added at rates equivalent to 0, 25, 50, 100, 150 and 200 kg N ha^{-1}. The effects of temperature were investigated by conducting standard runs with the room temperature controlled at 5, 10, 15, 20, 25 and 30°C. The effects of pH were investigated by making adjustments to measured soil pH with additions of KOH. The amounts of KOH required to obtain a range of soil pH .tween 5.1 and 9.4 were determined in a separate experiment. The KOH amendments were made to the soil surface as dilute solutions one day before the start of the incubation to allow time for equilibration.

Results and Discussion

The typical pattern of denitrification produced by the system is shown in Fig. 3.2. During the first day after addition of the amendments, the rate of efflux of N_2O increased almost linearly with time, to peak at *c.* 10 kg N ha^{-1} day^{-1} after 30 h. The N_2 peak was smaller but broader and occurred almost a day later than the N_2O peak. Efflux of N_2O declined to a very small rate after 4 days, whereas efflux of N_2 continued at rates greater than 1.0 kg N ha^{-1} day^{-1} to the end of the run. Of the 100 kg N ha^{-1} added, 22.4 and 29.3 kg N ha^{-1} could be accounted for as N_2O and N_2, respectively. This pattern of gas efflux is similar to those obtained by others working with re-packed, air-dried, sieved soils (e.g. Nommik, 1956; Wickramasinghe *et al.*, 1978). In these earlier studies, however, efflux of both gases tended to peak later. This difference could have been due to reduced activity of reductase enzymes in the dried soil at the start of the incubation.

Several studies have pointed to the importance of antecedent soil conditions in controlling $N_2O : N_2$ ratios during denitrification. Firestone and Tiedje (1979) demonstrated a 2–3 day lag in the de-repression of N_2O reductase compared to NO_3^- reductase after the onset of anaerobiosis. In more recent

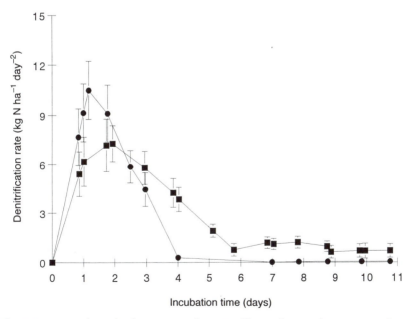

Fig. 3.2. Typical graph of N_2O (●) and N_2 (■) efflux with time during an incubation of an intact clay loam at 40% water content and 20°C and amended with NO_3^- and glucose.

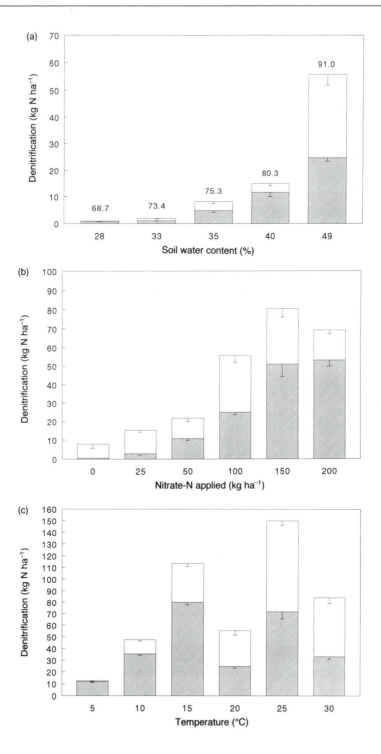

studies with the present soil, Dendooven *et al.* (1996) confirmed the importance of soil preconditioning as a potent cause of variability in N_2O production in the field. In this study, the 7-day aerobic preconditioning resulted in a
$N_2O : N_2$ ratio of 1.7, whereas after 7 days of anaerobism the $N_2O : N_2$ ratio
was 0.15; there was 77% more N_2O after the preconditioning. Systematic
studies of the effects of soil water content are thus difficult to perform. In this
study we tried to rationalize the preconditioning by performing successive
runs, with soil drying out naturally in the field during spring from its wettest
state of about 49% (w/w) reached during winter. With increasing water
content from 28 to 49% (w/w) there was more than a 50-fold increase in
denitrification (Fig. 3.3(a)). This is consistent with data reported from a study
of 18 repacked soils (Aulakh *et al.*, 1992). The $N_2O : N_2$ ratios in Fig. 3.3(a)
generally increased with decreasing water content, which is assumed to be a
result of a progressively disproportionate increase in the inhibition of N_2O
reductase by O_2 (Knowles, 1981).

Figure 3.3(b) shows that denitrification in this soil responds to increasing
supply of NO_3^- within the range 0–150 kg N ha^{-1}, with the $N_2O : N_2$ ratio
increasing progressively from close to zero at 0 kg N ha^{-1} input of NO_3^- to 3.3
at an input of 200 kg N ha^{-1}. At this highest rate of NO_3^- input, however, N_2
production was much reduced so that the overall amount of N lost through
denitrification was smaller than that lost at 150 kg N ha^{-1}. Strong concentrations of NO_3^- are thought to increase the $N_2O : N_2$ ratio through inhibition of
N_2O reductase (Nommik, 1956; Blackmer and Bremner, 1978), but Dendooven *et al.* (1994) concluded that specific inhibitory effects of NO_3^- need
not be invoked: the dynamics of the different N compounds involved in
denitrification could be explained satisfactorily by a competitive Michaelis–
Menton type model in which each compound is ascribed an affinity or
weighting factor as a measure of relative power as an electron acceptor. Such
a model, however, would not predict the much reduced N_2O production at
low applications of NO_3^- (Fig. 3.3(b)). The $N_2O : N_2$ ratios at low NO_3^- applications are also much smaller than those obtained by Wickramasinghe *et al.*
(1978).

The effects of temperature on denitrification have been well studied in
laboratory incubations of dried and sieved soils (Bremner and Shaw, 1958;
Stanford *et al.*, 1975). The denitrification rate was found to increase continuously through the range 3–65°C, with a Q_{10} of about 2 between 15 and 35°C.
The $N_2O : N_2$ ratio was found by Nommik (1956) to be large at low temperature and to decrease progressively over the temperature range. The results of

Fig. 3.3. (opposite) Effects of water content (a), NO_3^- (b) and temperature (c) on
the efflux of N_2O (shaded area) and N_2 (clear area) from an intact clay loam
amended with glucose.

this study are consistent with this early work (Fig. 3.3(c)), with a Q_{10} of 2 for the initial rate of denitrification over the range studied, and a progressive reduction of the $N_2O : N_2$ ratio from 16.7 at 5°C to 0.67 at 30°C.

Increasing pH from 5.1 to 9.4 was associated with both decreasing denitrification and $N_2O : N_2$ ratio. The latter effect is consistent with the results of previous studies (Bremner and Shaw, 1958; van Cleemput and Patrick, 1974), whereas the former is not. There are several possible explanations for decreasing denitrification with increasing pH. One is that the denitrifying population had become selectively adapted to low pH, a possibility which is supported by results of Parkin et al. (1985). Another involves the dynamics of NO_2^- which is short-lived at low pH, but may accumulate at higher pH to inhibit microbial activity (Monoghan and Barraclough, 1992).

The addition of glucose had only a small effect on denitrification in the present soil, with 105 and 94 kg N ha^{-1} lost from the +glucose and −glucose treatments, respectively. Although the $N_2O : N_2$ ratios were not very different (1.4 and 1.0 for the +glucose and −glucose treatments, respectively), there was a pronounced effect on the timing of the N_2 peak, which was eluted after 3 days with glucose and 5 days without glucose addition. While many studies have shown that available C normally limits denitrification potential, these results are consistent with those obtained by others working with the C-rich pasture soils of Devon, UK (Bijay-Singh et al., 1988; Elliott et al., 1991; Jarvis et al., 1991).

Conclusions

This simulation of the effects of fertilizer inputs to an intact agricultural soil under a range of defined conditions represents the most appropriate application of the novel flow-over technique at its present stage of development. The study has shown that there is considerable potential for decreasing N_2O emissions through the 'management' of $N_2O : N_2$ ratios. Thus, application of high doses of fertilizer to a low pH soil in early spring should be avoided, particularly if the application is preceded by a period of dry weather. The study has shown also that more detailed research on the effects of antecedent soil conditions is required if accurate management models of denitrification are to be forthcoming.

References

Aulakh, M.S., Doran, J.W. and Mosier, A.R. (1992) Soil denitrification – significance, measurement, and effects of management. In: Stewart, B.A. (ed.) *Advances in Soil Science*. Springer-Verlag. New York, pp. 1–57.

Bijay-Singh, Ryden, J.C. and Whitehead, D.C. (1988) Some relationships between denitrification potential and fractions of organic carbon in air-dried and field-moist soils. *Soil Biology and Biochemistry* 20, 737–741.

Blackmer, A.M. and Bremner, J.M. (1978) Inhibitory effect of nitrate on reduction of nitrous oxide to molecular nitrogen by soil microorganisms. *Soil Biology and Biochemistry* 10, 187–191.

Bremner, J.M. and Shaw, K. (1958) Denitrification in soil. II. Factors affecting denitrification. *Journal of Agricultural Science* 51, 40–52.

van Cleemput, O. and Patrick, W.H. (1974) Nitrate and nitrite reduction in flooded gamma-irradiated soil under controlled pH and redox potential conditions. *Soil Biology and Biochemistry* 6, 85–88.

Dendooven, L., Splatt, P., Anderson, J.M. and Scholefield, D. (1994) Kinetics of the denitrification process in a soil under permanent pasture. *Soil Biology and Biochemistry* 26, 361–370.

Dendooven, L., Duchateau, L. and Anderson, J.M. (1996) Gaseous products of the denitrification process as affected by the antecedent water regime of the soil. *Soil Biology and Biochemistry* 28, 239–245.

Elliott, P.W., Knight, D. and Anderson, J.M. (1991) Variables controlling denitrification from earthworm casts and soil in permanent pastures. *Biology and Fertility of Soils* 11, 24–29.

Firestone, M.K. and Tiedje, J.M. (1979) Temporal change in nitrous oxide and dinitrogen following onset of anaerobiosis. *Applied and Environmental Microbiology* 38, 673–679.

Galsworthy, A.M. and Burford, J.R. (1978) A system for measuring the rates of evolution of nitrous oxide and nitrogen from incubated soils during denitrification. *Journal of Soil Science* 29, 537–550.

Jarvis, S.C., Barraclough, D., Williams, J. and Rook, A.J. (1991) Patterns of denitrification loss from grazed grassland: effects of N fertilizer inputs at different sites. *Plant and Soil* 131, 77–88.

Knowles, R. (1981) Denitrification. In: Clark, F.E. and Rosswall, T.D. (eds) *Terrestrial Nitrogen Cycles.* Ecological Bulletin 33, Stockholm, pp. 315–329.

Monaghan, R.M. and Barraclough, D. (1992) Some chemical and physical factors affecting the rate and dynamics of nitrification in urine-affected soil. *Plant and Soil* 143, 11–18.

Nommik, H. (1956) Investigations on denitrification in soil. *Acta Agriculturae Scandinavica* 6, 195–228.

Parkin, T.B., Sextone, A.J. and Tiedje, J.M. (1985) Adaption of denitrifying populations to low soil pH. *Applied Environmental Microbiology* 49, 1053–1056.

Ryden, J.C., Lund, L.J. and Focht, D.D. (1979) Direct measurement of denitrification loss from soils: 1. Laboratory evaluation of acetylene inhibition of nitrous oxide reduction. *Soil Science Society of America Journal* 43, 104–110.

Stanford G., Dzienia S. and Vander Pol R. A. (1975) Effect of temperature on denitrification rate in soils. *Soil Science Society of America Journal* 39, 867–870.

Tiedje, J.M., Simkins, S. and Groffman, P.M. (1989) Perspectives on measurement of denitrification in the field including recommended protocols for acetylene based methods. *Plant and Soil* 115, 261–284.

Wickramasinghe, K.N., Talibudeen, O. and Witty, J.F. (1978) A gas flow-through system for studying denitrification in soils. *Journal of Soil Science* 29, 527–536.

Biogenic Emission of Nitric Oxide from a Subtropical Grassland: Miombo-type Savanna (Marondera, Zimbabwe)

4

F.X. Meixner[1], Th. Fickinger[1], L. Marufu[1,2], L. Mukurumbira[2], E. Makina[2], F.J. Nathaus[1] and D. Serça[3]

[1]Max-Planck-Institut für Chemie, Abteilung Biogeochemie, Postfach 3060, D-55020 Mainz, Germany; [2]Grasslands Research Station, Soil Productivity Research Laboratory, Marondera, Zimbabwe; [3]Laboratoire d'Aerologie-O.M.P., Université Paul Sabatier Toulouse III, 14 Avenue Edouard Berlin, F-31400 Toulouse, France

Summary. While pyrogenic emissions of nitric oxide (NO) from African savanna fires have a proven, but seasonally limited, impact on the chemistry of the regional (and presumably the global) atmosphere, only little is known about the importance of the more persistent biogenic fluxes of NO from natural and agriculturally managed African savanna ecosystems. During August to December 1994, a comprehensive field study was performed to quantify the diurnal as well as the seasonal fluxes of NO and carbon dioxide (CO_2), nitrogen dioxide (NO_2) and ozone (O_3) from and to a Miombo savanna type ecosystem. A set of dynamic chambers was applied to grassland and some other biomes at the site. Fluxes were controlled by soil variables (e.g. temperature, water and nutrient content). Therefore, marked effects on the spatial and temporal pattern of the fluxes were observed when considering: (i) natural versus agriculturally managed savanna soils; (ii) grass cover versus tree savanna; and, above all (iii) dry versus wet season. These and the results of a companion paper (see Poster 2) emphasize the importance of natural and agriculturally managed African savanna ecosystems as biogenic sources of NO in southern Africa.

Introduction

Tropospheric carbon monoxide (CO), methane (CH_4) and hydrocarbons (C_mH_n) are oxidized by hydroxyl and other radicals through different catalytic cycles (Crutzen, 1987). Depending on the ambient mixing ratio of the key catalyst NO_x (NO_x = nitric oxide (NO) + nitrogen dioxide (NO_2)), tropospheric ozone (O_3) may be generated or destroyed within these cycles (Chameides *et al.*, 1992). Whereas the present chemical environments of polluted areas will always generate O_3, the situation in the more remote parts of the globe is quite different. There, a so-called 'supercritical' NO_x mixing ratio of only ≈30 pptv switches the system to destruct O_3 (NO_x < 30 pptv) or to generate ozone (NO_x > 30 pptv). Consequently, the present extent and future development of NO_x sources in the non-industrialized regions will trigger the potential increase of global tropospheric O_3 and, therefore, deserves scientific attention. Pyrogenic emissions of NO from African savanna fires have a proven, but seasonally limited, impact on the chemistry of the regional and presumably on the global atmosphere (Andreae *et al.*, 1996). Only little is known about the importance of the more persistent biogenic fluxes of NO from African savanna ecosystems. First (short-term) measurements from west and southern African savannas (e.g. LeRoux *et al.*, 1995; D. Serça *et al.*, unpublished results) report a vast range of NO emission rates (0.05–100 ng N m^{-2} s^{-1}). The large variability is mainly due to very different fire, and soil water and nitrogen (N) conditions/status, which were encountered during the individual measurements. However, model-based constructions of global soil biogenic NO sources (Yienger and Levy II, 1995; Potter *et al.*, 1996) identified tropical and subtropical savannas among the most important source areas, due to their enhanced soil temperatures and their large geographical extent. As a further step in this direction, a comprehensive *in-situ* study was performed over a subtropical Miombo savanna type ecosystem. During August to December 1994, diurnal as well as seasonal NO fluxes were measured from and to grassland, forest soil and some agriculturally managed plots at the site.

Materials and Methods

Site

Field experiments were performed at the Grasslands Research Station, Marondera, Zimbabwe (18°11′ S, 31°28′ E). Soils are strongly leached acidic alfisols developed from granites. The local climate is characterized by wet and dry seasons, where more than 80% of the mean annual rainfall (846 mm) occurs between November and March. Three representative plots were selected within a fenced subsection (360 × 250 m) of the Grasslands Research

Station: (i) a natural Miombo woodland plot (undisturbed for > 50 years); (ii) a grassland plot, converted from Miombo forest by clearing of trees and shrubs (1976); and (iii) some agricultural managed plots (groundnut, maize, sorghum, fallow). Groundnut, maize and sorghum plots were subdivided into non-fertilized and fertilized plots. The latter received a single application of 200 kg ha^{-1} NPK fertilizer (6% P, 5.8% K, 2% S), as is normal for commercial farms in this area. Since the majority of small farmers grow maize and apply domestic organic fertilizer, another maize plot fertilized with 'manure' was laid out.

Flux measurements

To address spatial heterogeneity of the soil, net vertical NO fluxes were determined by simultaneous operation of four dynamic chambers at each plot. Applying dynamic chamber techniques, observed NO fluxes must be corrected for gas-phase reactions of the NO-NO$_2$-O$_3$ triad which occur in the air sample while passing through the chamber and corresponding tubing systems (Remde *et al.*, 1993). Therefore NO, NO$_2$ and O$_3$ mixing ratios were simultaneously analysed in any air sample. Furthermore, a fifth (dynamic) chamber, sealed at the bottom by a PTFE-foil against the soil surface, served as a 'blank' chamber to check the corrections already noted and to quantify corresponding losses to chamber walls. The chambers had a cylindrical design and each enclosed 0.086 m^2 of soil had a total volume of 25.7 l, and consisted of a stainless steel frame (0.32 m diam.), polycarbonate walls and a movable lid. To avoid disturbances of soil structure and texture, steel frames were inserted on all plots 30 cm deep into the soil, 4 weeks before any flux measurements started. Thorough turbulent mixing inside the chamber was achieved by continuous operation of two (PTFE-coated) fans inside each chamber. To guarantee NO flux measurements independent of the flow rate through the chamber (Ludwig, 1994), the chamber air volume was exchanged every 26 s by flushing each chamber with ambient air at a flow rate of 60 l min^{-1}. A valve system switched the main sampling stream alternatively to the ambient air intake and to the outlet of each chamber (successively) every 2 and 3 min. Every flux measurement cycle (five dynamic chambers) was completed in 25 min. Successive NO flux measurements were performed on the Miombo and grassland plots during 1–2 days, followed by those on the agricultural plots during 4–7 days. Commercial analysers were used for the measurement of NO and NO$_2$ mixing ratios (Ecophysics; CLD 780TR and PCL 760), as well as of O$_3$ mixing ratio (Thermo Instruments, model 49). In addition, air and soil temperatures, relative humidity, volumetric soil water content (Time Domain Reflectrometry) and NO$_2$ photolysis rate were continuously recorded inside and outside the chambers. Soil samples (0–3 cm) for the determination of ammonium (NH_4^+) and nitrate (NO_3^-) contents were taken from the different plots.

Flux calculations

The net vertical NO flux (F_{NO}) from the soil enclosed by a dynamic chamber can be estimated by considering the mass balance equation for the dynamic chamber (Remde et al., 1993). The following calculation scheme was applied to determine F_{NO}:

$$F_{NO} = \frac{M_N}{V_m} \cdot (m_{out} - m_{in}) \cdot \frac{Q}{A} - \sum F_K \tag{1}$$

where Q is the flow rate through the dynamic chamber (1 min^{-1}), A is the soil surface area covered by the chamber (m^2), M_N is the atomic weight of nitrogen (14.0067 kg kmol^{-1}), V_m is the molar gas volume (24.465 m^3 kmol^{-1}; 1013 hPa, 298 K), and m_{out} and m_{in} the observed NO mixing ratios of the sample air at the outlet of each chamber (m_{out}) and the ambient air (m_{in}), respectively. For the actual flux calculation, the average of two successive individual measurements of m_{out} was used, as well as the temporal average of two individual measurements of m_{in} (before and after the two measurements of m_{out}, respectively). To account for non-stationary conditions, all flux measurements were rejected when the difference in m_{in} before and after the two measurements of m_{out} exeeded 0.5 ppbv. In equation (1) the term $\sum F_K$ represents fluxes due to sorption effects of trace gases onto the walls of the chamber and intake lines and fluxes caused by chemical reactions of the triad $NO-NO_2-O_3$ in the air sample on its passage through the chamber and intake lines. A comprehensive discussion of $\sum F_K$ is given in Remde et al. (1993) and, in more detail, by Ludwig (1994). During the Marondera experiment, the average correction due to gas-phase reactions of the triad $NO-NO_2-O_3$ was in the order of 30–70% of the flux which is given by the difference $m_{out} - m_{in}$ alone, while the correction for the losses due to wall effects was < 1% of F_{NO}.

Results and Discussion

Grassland

During the dry season, the daily variation in soil temperature (measured in chamber no.4, d = 1 cm) was between 15°C (night) and up to 51°C (early afternoon), while volumetric soil water content was < 1.5%. Individual net NO fluxes ranged from −1.3 to 1.2 ng N m^{-2} s^{-1}. The overall dry season average was 0.0 ± 0.4 ng N m^{-2} s^{-1} and is below the detection limit (0.4 ng N m^{-2} s^{-1}) of our NO flux measurement technique. A completely different picture could be observed during the wet season (Fig. 4.1). After the first major rainfall (13 Oct.) daily mean NO fluxes increased by a factor of 10 after 10 days, declining slightly within the next month. The rainfall of 25 November, however, raised daily mean NO fluxes by a factor of up to 30 (chamber no.3). During this specific period, there was a pronounced difference between fluxes observed by chamber nos 1 and 2 (enclosing soil covered with dry grass) and chamber nos 3 and 4, which enclosed bare soil only. While no diurnal variation in NO flux could be observed during the dry season, strong diurnal variations were found during the wet season (not shown here). As soon as the volumetric soil water content (1–5 cm) was in the order of 10% (but lower than 25%), the NO flux correlated strongly with the soil temperature close to the surface (d = 1 cm). A similar result was reported by Ludwig (1994). Both findings support the hypothesis of Galbally and Johansson (1989), as well as that of Remde *et al.* (1993), that those microbial processes which finally control the net NO release from soils are located within the surface layer of a few centimetres thickness.

Miombo forest

Figure 4.2 shows the seasonal variation of daily mean NO fluxes at the Miombo plot and a similar behaviour was found for the grassland plot.

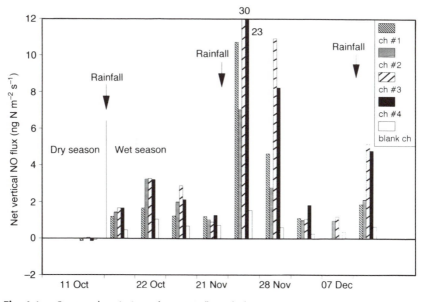

Fig. 4.1. Seasonal variation of net NO flux (daily mean) at the 'grassland plot', Grassland Research Station, Marondera (Zimbabwe).

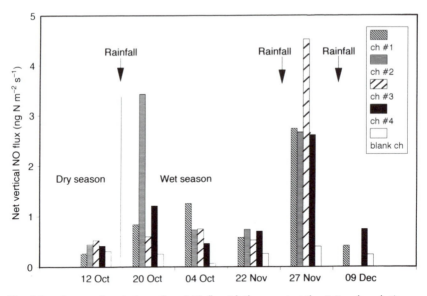

Fig. 4.2. Seasonal variation of net NO flux (daily mean) at the 'Miombo plot', Grassland Research Station, Marondera (Zimbabwe).

Individual dry season NO fluxes ranged between −0.4 and +1.7 ng N m^{-2} s^{-1} and averaged 0.4 ± 0.4 ng N m^{-2} s^{-1}. During the wet season, NO fluxes were also increased after major rainfalls, but only by a factor of 5–10. After prolonged dry periods, NO emissions declined to those of the dry season (22 November). During the wet season the lowest NO fluxes of all investigated plots were found at the Miombo plot (0.5–4.5 ng N m^{-2} s^{-1}). The volumetric soil water content there was very close to that of the grassland plot. The lower NO fluxes at the Miombo plot could be due partially to the lower soil temperature caused by shadowing of the Miombo trees. However, more recent laboratory studies on soil samples of both the grassland and the Miombo plot have shown a higher NO consumption and a lower NO production rate for the Miombo soil (W.X. Yang, 1996, personal communication).

Agricultural plots

Maize, sorghum and groundnuts were planted on the corresponding agricultural sub-plots between 11 and 13 October, while fertilizer was applied on 13 October. With the beginning of the wet season (13 October) all agricultural plots were frequently irrigated. Hence soil moisture was generally high

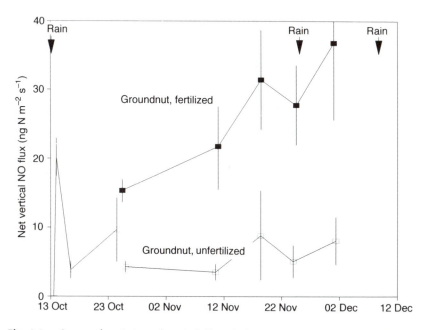

Fig. 4.3.   Seasonal variation of net NO flux (daily mean) at the 'groundnut plot'.

(7–17%) during the entire wet season. The soil temperature varied within the range 13–30°C. As shown in Fig. 4.3, NO fluxes increased from 15 ng N m^{-2} s^{-1} (end of October) to 37 ng N m^{-2} s^{-1} (beginning of December) at the fertilized groundnut plot. Thus they exceeded those of the unfertilized plot by a factor of 2–4.5. As far as the sorghum plot was concerned, NO fluxes ranged from 4.5 to 11 ng N m^{-2} s^{-1} (unfertilized plot), and from 21 to 33 ng N m^{-2} s^{-1} (fertilized plot), resulting in a two- to eightfold enhancement of NO fluxes (Fig. 4.4). In Fig. 4.3, the first data point of the fertilized groundnut plot (13 October) corresponds to 0.25 ng N m^{-2} s^{-1} and was measured just 3 h before the onset of that rain event which terminated the 1994 dry season. Within 12 h, NO flux increased by a factor of 80. This increase is far beyond those NO flux enhancements observed for sudden soil wetting alone and must also be attributed to the very recent fertilizer application.

Conclusion

During the dry season, NO fluxes for all the selected plots were comparable and, on average, well below 1 ng N m^{-2} s^{-1}. The lack of a diurnal variation of

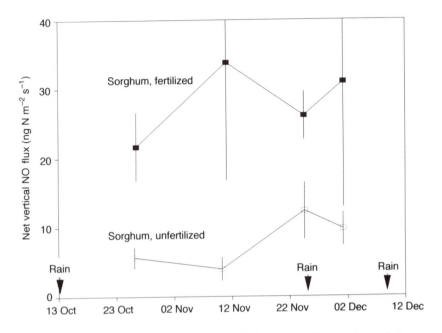

Fig. 4.4. Seasonal variation of net NO flux (daily mean) at the 'sorghum plot'.

the NO flux with soil temperature in connection with the extremely low soil water content indicates only low microbial activity in the soil. Cárdenas *et al.* (1993) also observed low NO fluxes (around 1 ng N m^{-2} s^{-1}) for similar low soil moisture conditions (< 2%) at a managed 'natural' grassland savanna site in Venezuela. For semi-arid savanna sites in South Africa, comparable NO emissions of ≤ 1 ng N m^{-2} s^{-1} are reported by Serça *et al.* (1996), Levine *et al.* (1996) and Parsons *et al.* (1996) during the dry season. This may indicate that the NO input to the atmosphere from (south) African savannas may be generally very small during this period. The transition from dry to wet season was characterized by strong 'pulsing' effects of the NO flux which might be strongly linked to a drastic change in soil moisture. Enhanced soil moisture is considered to be responsible for the generally higher NO fluxes during the wet season at all plots. Similar results have been observed in other studies (e.g. Johansson *et al.*, 1988; Johansson and Sanhueza, 1988; Davidson *et al.*, 1991). During the wet season, NO fluxes at all selected plots strongly fluctuate within a range of 0.5 ng N m^{-2} s^{-1} up to 62 ng N m^{-2} s^{-1}. Soil water content seems to be the predominant factor which controls the NO emission flux (see also Poster 2). Besides the soil water content, the fertilization of agriculturally managed soils also led to a strong increase in the NO emissions (provided soil moisture was not limiting).

Acknowledgements

We are grateful to the Zimbabwean National Research Council and to the Department of Research and Specialist Services for the permission to conduct our field experiments in Zimbabwe. We also thank the Grasslands Research Station for the personal and logistic support and all colleagues of the Soil Productivity Research Laboratory for their cooperation. Thanks are also due to the Max Planck Society, Germany, which financed the field experiment.

References

Andreae, M.O., Atlas, E., Cachier, H., Cofer III, W.R., Harris, G.W., Helas, G., Koppmann, R., Lacaux, J.P. and Ward, D.E. (1996) Trace gas and aerosol emissions from savanna fires. In: Levine, J.S. (ed.) *Biomass Burning and Global Change*, vol. 1. MIT Press, Cambridge, MA, pp. 278–295.

Cárdenas, L., Rondón, A., Johansson, C. and Sanhueza, E. (1993) Effects of soil moisture, temperature, and inorganic nitrogen on nitric oxide emissions from acidic tropical savannah soils. *Journal of Geophysical Research* 98, 14, 783–14,790.

Chameides, W.L., Fehsenfeld, F., Rodgers, M.O., Cardelino, C., Martinez, J., Parrish, D., Lonneman, W., Lawson, D.R., Rasmussen, R.A., Zimmerman, P., Greenberg, J., Middleton, P. and Wang, T. (1992) Ozone precursor relationships in the ambient atmosphere. *Journal of Geophysical Research* 92 (D5), 6037–6055.

Crutzen, P.J. (1987) Role of the tropics in atmospheric chemistry. In: Dickinson, R.E. (ed.) *The Geophysiology of Amazonia*. John Wiley, New York, pp. 107–132.

Davidson, E.A. (1991) Fluxes of nitrous oxide and nitric oxide from terrestrial ecosystems. In: Rogers, J.E. and Whitman, W.B. (eds) *Microbial Production and Consumption of Greenhouse Gases: methane, nitrogen oxides, and halomethanes*. American Society for Microbiology, Washington DC, pp. 219–235.

Galbally, I.E. and Johansson, C. (1989) A model relating laboratory measurements of rates of nitric oxide production and field measurements of nitric oxide emissions from soils. *Journal of Geophysical Research* 94, 6473–6480.

Johansson, C. and Sanhueza, E. (1988) Emission of NO from savanna soils during rainy season. *Journal of Geophysical Research* 93, 14,193–14,198.

Johansson, C., Rodhe, H. and Sanhueza, E. (1988) Emission of NO in a tropical savanna and a cloud forest during the dry season. *Journal of Geophysical Research* 93, 7180–7192.

LeRoux, X., Abbadie, L., Lensi, R. and Serça, D. (1995) Emission of nitrogen monoxide from African tropical ecosystems: control of emission by soil characteristics in humid and dry savannas of West Africa. *Journal of Geophysical Research* 100, 23,133–23,142.

Levine, J.S., Cofer III, W.R., Cahoon Jr, D.R., Winstead, E.L., Sebacher, D.I., Sebacher, S., Scholes, M., Parsons, D.A.B. and Scholes, R.J. (1996) The impact of wetting and burning on biogenic soil emissions of nitric oxide (NO) and nitrous oxide

(N$_2$O) in savanna grasslands in south Africa. *Journal of Geophysical Research* 101, 23,689–23,697.

Ludwig, J. (1994) Untersuchungen zum Austausch von Stickoxiden zwischen Biosphäre und Atmosphäre. PhD Thesis, University of Bayreuth, Bayreuth, Germany.

Parsons, D.A.B., Scholes, M., Scholes, R.J. and Levine, J.S. (1996) Biogenic NO emission from savanna soils as a function of soil nitrogen and water status. *Journal of Geophysical Research* 101, 23,683–23,688.

Potter, C.S., Matson, P.A., Vitousek, P.M. and Davidson, E.A. (1996) Process modeling of controls on nitrogen trace gas emissions from soils worldwide. *Journal of Geophysical Research* 101, 1361–1377.

Remde, A., Judwig, J., Meixner, F.X. and Conrad, R. (1993) A study to explain the emission of nitric oxide from a marsh soil. *Journal of Atmospheric Chemistry* 17, 249–275.

Serça, D., Delmas, R., LeRoux, X. and Labroue, L. (1996) Emission of nitrogen monoxide from African tropical systems: spatial variability at local and regional scale. *Journal of Geophysical Research* (in press).

Yienger, J.J. and Levy II, H. (1995) Empirical model of global soil-biogenic NO$_x$ emissions. *Journal of Geophysical Research* 100, 11,447–11,464.

Concurrent Nitrification and Denitrification in Compacted Grassland Soil

5

M.K. Abbasi[1], Z. Shah[2] and W.A. Adams[1]

[1]Soil Science Unit, Institute of Biological Sciences, University of Wales, Aberystwyth, UK; [2]Department of Soil Science, NWFP Agricultural University, Peshawar, Pakistan

Summary. In mid Wales, previous field and laboratory studies on pasture soils compacted by grazing animals have suggested that nitrification and denitrification may occur concurrently. Laboratory experiments were conducted using intact soil cores equilibrated at −5 kPa matric potential to provide further evidence on this possibility. Ammonium-nitrogen (NH_4^+-N) applied to the soil surface disappeared progressively from the mineral N pool. Nitrate (NO_3^-)-N concentration increased to some extent and the occurrence of denitrification under these conditions was confirmed in separate experiments. Potential denitrification in soil exceeded nitrification by around fivefold, so substantial accumulations of NO_3^- would be unlikely to occur. It was concluded that concurrent nitrification and denitrification can occur at shallow depths in pasture soils in western areas of the UK and together may constitute a significant N loss pathway in pastoral agriculture.

Introduction

The soils of mid Wales receive high rainfall, typically in excess of 1000 mm per year and a substantial proportion of the area is grazed intensively by sheep and cattle. High rainfall increases soil susceptibility to compaction which may be further enhanced by intensive stocking with cattle (Davies *et al.*, 1989). In compacted soils, the aerobic and anaerobic zones may occur in close proximity in the upper few centimetres of the soil, so both nitrification and denitrification could occur concurrently and adjacently. Direct evidence of concurrent nitrification-denitrification in soil was provided by Starr *et al.* (1974) in an ammonium (NH_4^+) enrichment soil column. Interaction between the two processes in Histosols has been studied recently by Zanner and Bloom (1995). They reported that the high water-holding capacity of these soils increases the potential for concurrent nitrification and denitrification and can result in the removal of native and applied NH_4^+ from the system. It was thought that the lower uptake of N and lower concentration of nitrate (NO_3^-)-N in the herbage on Denbigh series soils compacted by dairy cattle may have resulted from denitrification (Yulun, 1987; Davies *et al.*, 1989). Adams and Akhtar (1994) concluded that both nitrification and denitrification had occurred simultaneously when NH_4^+-N was applied to waterlogged cores taken from compacted pasture soil.

Several observations have pointed to the possibility that concurrent nitrification and denitrification may act as a significant N loss pathway operating at shallow depths in pasture soils. Denitrification alone was the key factor in the experimental work of Davies *et al.* (1989) where NH_4NO_3 was used as the N fertilizer source; however, Yulun (1987) in a similar experimental system used urea and found that both denitrification and nitrification could be important. The initial aim of the research reported here was to test the hypothesis that nitrification and denitrification can occur concurrently and adjacently.

Materials and Methods

Site description

The soil used in these studies was collected from grazed pasture land about 5 km south east of Aberystwyth, Wales. The soil was a cambic stagnogley and classified as Cegin series (Grid Ref. SN 626772). The land had been in pasture for 17 years and was grazed all year round by sheep, and by cattle for part of the growing season. The site had an altitude of 180 m. The soil had developed from glacial till and the surface soil (0–8 cm) had silty clay loam particle size class and was stone free. The soil had a pH (H_2O) of 4.8. The main pasture species were *Agrostis capillaris* L., *Lolium perenne* L., *Agrostis stolonifera* L. and *Trifolium repens* L.

Experiment 1: fate of added NH_4^+-N in soil cores over 56 days of incubation

Intact soil cores (5.3 cm in diameter) were removed at random from an area of visual uniformity. Grass was cut to soil level and the cores cut to 6 cm length. The sides of the cores were smeared with a slurry prepared from the soil removed below 6 cm deep to fill cracks created during sampling. Soil water content was standardized by saturation, followed by equilibration at −5 kPa over 2 days. The soil cores were divided into two groups. One group received a surface application of 5 ml of $(NH_4)_2HPO_4$ solution supplying 200 mg N kg^{-1} soil (on a moist soil basis), while 5 ml of distilled water was applied as a control treatment to the remaining cores. Cores were incubated at 20°C and duplicate cores were analysed at 0, 14, 28 and 56 days. Soil moisture conditions were maintained throughout the incubation by adding distilled water carefully to the core surface at intervals of about twice a week. After each incubation interval, duplicate cores were sectioned into 0–2, 2–4 and 4–6 cm depths. Each part was broken down separately and mixed thoroughly by hand. Mineral N was extracted with 1 M KCl and determined by the method of Keeny and Nelson (1982).

Experiment 2: fate of added NO₃-N in intact soil cores

In this experiment a 5 ml solution of $KNO_3 + KH_2PO_4$ was applied to the surface of the cores at a rate of 200 mg of N and 220 mg of phosphorus (P) kg^{-1} soil, respectively. Control cores were treated with 5 ml of distilled water. Acid-washed sand was placed in the incubation pots underneath the cores, to collect any leachate leaving the cores during the incubation. Cores were incubated at 20°C and triplicate cores analysed at 0, 7, 14, 28 and 56 days. The incubation and analytical methods were the same as in experiment 1. The acid-washed sand was extracted with distilled water and analysed for NO₃ by ion chromatography.

Experiment 3: nitrification potential of the soil

The nitrification potential of the same grassland soil was examined on sieved samples from 0–2, 2–4 and 4–6 cm depths. Two millilitres of a solution of $(NH_4)_2HPO_4$ supplying 200 mg N kg^{-1} was mixed thoroughly with 20 g of moist soil. Pots containing soil were incubated at 20°C, during which soil moisture and aerobic conditions were maintained. Duplicate samples were removed from the incubator at 7, 14, 21 and 28 days and analysed for total mineral N, NH₄-N and NO₃-N by the methods described earlier.

Experiment 4: denitrification potential of the soil

Sieved samples of 30 g moist soil from 0–2, 2–4 and 4–6 cm sections were saturated with 50 ml boiled distilled water. A 10 ml solution of $KNO_3 + KH_2PO_4$ supplying 200 mg N and 220 mg P kg^{-1} soil, respectively, was applied to each bottle. Once all the additions had been made to the bottles, they were flushed with oxygen-free N_2. Sealed bottles were incubated at 20°C. Contents of duplicate bottles were analysed at zero time for total mineral N, NH_4^+-N and NO_3^--N and after 2, 4 and 6 days of incubation.

Experiment 5: soil core incubation for N₂O measurement using acetylene inhibition

Soil cores 6 cm deep and 5.3 cm wide with grass removed were equilibrated at −5 kPa. Five millilitres of KNO_3 + KH_2PO_4 solution at the rate of 200 mg of N and 220 mg of P kg^{-1} moist soil was carefully pipetted onto each core. A total of three treatments, in duplicates, were arranged as follows:

T1 = soil cores without sampling rings at −5 kPa water potential, T2 = soil cores with sampling rings at −5 kPa water potential and T3 = soil cores with sampling rings and totally saturated. The cores were placed in 1-litre Kilner jars fitted with gas-tight lids and a gas sampling port. Fifty millilitres of air was removed from the head space of each jar and acetylene was injected to produce a concentration in the head space of approximately 5% (v/v). Each jar was then incubated at 20°C for up to 72 h. Gas samples were removed using a 1-ml disposable syringe and nitrous oxide (N_2O) concentrations determined by gas chromatography using an electron-capture detector.

Results

Fate of added NH₄⁺-N

The NH_4^+-N concentration in the control soil increased to some extent at 14 days, but at 28 and 56 days its concentration was less than at zero time in all three depths (Table 5.1). In the NH_4^+-N treated cores, NH_4^+-N concentrations decreased progressively over all depths and after 56 days the concentration in the 0–2-cm depth had fallen from 791 to 46 mg kg^{-1}. The concentration of NO_3^--N in NH_4^+-N treated soil increased substantially over the first 14 days of incubation, especially in the 0–2-cm-depth range. The increase was less than 20% of the decrease in NH_4^+-N. After 14 days the concentration of NO_3^--N remained reasonably constant over all three depth ranges despite the continuing fall in NH_4^+-N. The fate of added NH_4^+-N is summarized in Table 5.2. Of the 277 mg kg^{-1} of NH_4^+-N added to the cores, only 43 mg kg^{-1} was recovered after 56 days. This amounted to 85% of the applied N. The rate of N loss was greatest over the first 14 days, equivalent to around 1 kg ha^{-1} day^{-1}.

Table 5.1. Changes in NH_4^+-N and NO_3^--N in soil cores at different depths over a period of 56 days with or without added ammonium nitrogen.

Treatment	Depths (cm)	Incubation periods (days)			
		0	14	28	56
		mg NH_4^+-N kg^{-1} soil			
+NH_4^+-N/control	0–2	791/37	388/43	178/18	46/22
	2–4	80/31	74/28	56/11	19/20
	4–6	53/24	36/48	14/9	12/9
		mg NO_3^--N kg^{-1} soil			
+NH_4^+-N/control	0–2	9/9	79/15	65/5	70/5
	2–4	2/4	28/8	28/4	28/1
	4–6	4/3	14/14	13/5	14/1

Table 5.2. Fate of added NH_4^+-N in 0–6-cm soil cores over 56 days of incubation.

Time (days)	Total mineral N (mg N kg^{-1})		Amount of added N unaccounted for (%)	Rate of N lost (mg N kg^{-1} day^{-1})
	Recovered	Lost		
0	277	0	0	—
14	155	122	44	9
28	101	176	64	6
56	43	234*	85	4

*234 mg N kg^{-1} = 23 kg N ha^{-1}.

Fate of added NO_3^--N

The initial concentration of NO_3^--N in amended soil was 845 mg in the 0–2 cm depth and this comprised about 94% of the NO_3^--N in the 0–6-cm core (Table 5.3). By 14 days the total NO_3^--N in the 0–6 cm layer had fallen by about 10% but the proportion in the top 2 cm had decreased by 44%. The pattern of change in NO_3^--N concentration with depth over the total incubation period suggests that the main zone of denitrification was below 2 cm. A small amount (19 mg kg^{-1}) of NO_3^--N was leached from the cores and, after taking this into account, more than 60% of the added NO_3^- was still unaccounted for at the end of the incubation.

Nitrification and denitrification potential

Nitrification and denitrification potentials were determined on sieved soil samples from 2 to 6 cm (Table 5.4). Nitrification rate increased over the

Table 5.3. Changes in NO_3^--N concentration in soil cores at different depths over a period of 56 days with or without added NO_3^--N.

Treatments	Depth (cm)	Incubation periods (days)				
		0	7	14	28	56
		mg NO_3^--N kg^{-1} soil				
Control	0–2	5 ± 1	11 ± 1	9 ± 3	7 ± 2	17 ± 12
	2–4	9 ± 2	6 ± 2	3 ± 1	17 ± 11	2 ± 1
	4–6	6 ± 2	2 ± 1	2 ± 2	5 ± 5	6 ± 3
+NO_3^--N	0–2	854 ± 55	513 ± 43	479 ± 39	301 ± 120	167 ± 70
	2–4	47 ± 24	178 ± 30	233 ± 10	136 ± 58	75 ± 41
	4–6	10 ± 6	73 ± 17	120 ± 6	119 ± 20	72 ± 43

$\pm$ Values are the standard error of the means.

Table 5.4. Nitrification and denitrification potential of sieved soil from different depths with added NO_3^--N.

	Nitrification with added NH_4^+-N				Denitrification with added NO_3^--N		
	Depths (cm)			Time	Depths (cm)		
Time (days)	0–2	2–4	4–6	(days)	0–2	2–4	4–6
	mg NO_3^--N kg^{-1} soil				mg NO_3^--N kg^{-1} soil		
0	6 ± 1	2 ± 1	4 ± 2	0	248 ± 4	232 ± 6	214 ± 6
7	20 ± 3	9 ± 1	11 ± 4	2	146 ± 7	187 ± 3	183 ± 0
14	81 ± 2	29 ± 1	24 ± 1	4	59 ± 3	142 ± 7	200 ± 3
21	129 ± 18	92 ± 2	54 ± 4	6	2 ± 1	89 ± 7	117 ± 5
28	231 ± 10	149 ± 16	92 ± 1	—	—	—	—
0–28	*8.0	5.3	3.1	0–6	*41	24	16

*Mean rates per day over stated period.
$\pm$ Values are the standard error of the means.

Table 5.5. Nitrous oxide production from 0–6-cm soil cores with NO_3^--N added in the presence of acetylene.

Treatments*	mg N_2O kg^{-1} soil at different incubation periods			Rate of N_2O production (mg kg^{-1} day^{-1})	Loss of N (mg N kg^{-1} day^{-1})
	4 h	24 h	72 h		
T_1	0.11	12.3	54.7	18.3	11.6
T_2	0.35	9.0	64.2	21.4	13.6
T_3	0.0	6.8	60.0	20.0	12.7
Mean	0.15	9.4	59.6	19.9	12.6

*T_1 = soil cores without sampling rings; T_2 = soil cores with sampling rings; T_3 = soil cores with ring totally saturated with water.

incubation period but decreased with depth. Nitrification potentials ranged from 8.0 mg N kg^{-1} for the 0–2 cm to 3.1 mg N kg^{-1} for the 4–6 cm depth. Denitrification potentials summarized over the 6-day incubation ranged from 41 mg N kg^{-1} for the 0–2 cm to 16 mg N kg^{-1} for the 4–6 cm depth. Thus denitrification potential was about five times greater than nitrification potential. Furthermore, the decrease in the potential rate of the two processes with depth was similar.

Nitrous oxide measurement in the presence of acetylene

The N_2O concentrations were very low after the first 4 h, but after 72 h a considerable and similar amount of N_2O was produced in all the three treatments (Table 5.5). The cores at −5 kPa matric potential produced a similar amount of N_2O to the fully saturated sample. The mean rate of N_2O-N loss was 12.6 mg kg^{-1} day^{-1}, which is about half the mean rate of denitrification in the sieved soil in experiment 4.

Discussion

When intact soil cores, equilibrated at −5 kPa water potential, were incubated following the application of NH_4^+-N to their surface, a substantial proportion of the applied N was lost from the mineral pool over 56 days. Nitrate-N increased to a maximum of around 80 mg kg^{-1} in the 0–2-cm layer over the first 14 days: the increases in the 2–4-cm and 4–6-cm layers were much less than this. The decrease in NH_4^+-N greatly exceeded the NO_3^--N accumulation. The soil used was moderately acidic, so loss of NH_3-N through volatilization would have been minimal. Furthermore, there was no evidence of significant net immobilization of N from the control soil. Thus it was concluded that concurrent nitrification and denitrification occurred in the cores during incubation (Starr *et al.*, 1974). Subsequent experiments showed that N was lost when NO_3^- was applied to cores in a similar soil moisture condition to the first experiment. Furthermore, using the acetylene block technique it was shown by the evolution of N_2O that denitrification was the process which accounted for the loss of N.

Much more work is needed both to quantify the significance of this N-wasting system in field conditions and also to establish the proximity of the two processes and their detailed microbiology. It is evident from the work reported here that if the two processes are carried out by autrotrophic nitrifiers in aerobic sites and facultative anaerobes in anoxic sites, respectively, then nitrification is potentially the rate-limiting process. This being so, NO_3^- would only be expected to accumulate to modest concentrations, as was observed in the first experiment. However, the rate of translocation of

NO_3^--N between sites could be limiting. With the NO_3^--N gradients of around 50 mg kg^{-1} generated in the incubated cores, then diffusive transport of NO_3^--N over distances of 5 mm could make about 1 kg N ha^{-1} day^{-1} available for reduction. The relative location of the two processes may be complex and the work reported by Knowles (1978), Kuenen and Robertson (1994) and Zanner and Bloom (1995) open the possibility that significant gaseous loses of N from NH_4^+ sources deposited on the pasture may occur through intimately associated nitrification-denitrification processes.

References

Adams, W.A. and Akhtar, N. (1994) The possible consequences for herbage growth of waterlogging compacted pasture soils. *Plant and Soil* 162, 1–17.

Davies, A., Adams, W.A. and Wilman, D. (1989) Soil compaction in permanent pasture and its amelioration by slitting. *Journal of Agricultural Science, Cambridge* 113, 189–197.

Keeney, D.R. and Nelson, D.W. (1982) Nitrogen-inorganic forms. In: Page, A.L., Miller, R.H. and Keeney, D.R. (eds) *Methods of Soil Analysis, Part 2. Chemical and Microbiological Properties.* American Society of Agronomy, Madison, WI, pp. 643–698.

Knowles, R. (1978) Common intermediates of nitrification and denitrification, and the metobolism of nitrous oxide. In: Schlessinger, D. (ed.) *Microbiology.* American Society for Microbiology, Washington DC, pp. 367–371.

Kuenen, J.G. and Robertson, L.A. (1994) Combined nitrification-denitrification processes. *FEMS Microbiology Reviews* 15, 109–117.

Starr, J.L., Broadbent, F.E. and Nielsen, D.R. (1974) Nitrogen transformations during continuous leaching. *Soil Science Society of America Proceedings* 38, 283–289.

Yulun, Z. (1987) Some aspect of nitrogen transformation in soils. MSc thesis, University College of Wales, Aberystwyth, UK.

Zanner, C.W. and Bloom, P.R. (1995). Mineralization, nitrification, and denitrification in Histosols of Northern Minnesota. *Soil Science Society of America Journal* 59, 1505–1511.

Rainfall and Temperature Effects on Nitrogen Losses from Fertilizer Types on Grassland in the Netherlands and the UK

6

D.W. Bussink[1] and O. Oenema[2]

[1]Nutrient Management Institute (NMI), Research Station for Cattle, Sheep and Horse Husbandry PR, Runderweg 6, NL-8219 PK Lelystad, The Netherlands; [2]Department of Soil Science and Plant Nutrition, PO Box 8005, NL-6700 EC Wageningen Agricultural University, The Netherlands

Summary. Differences exist between the Netherlands (NL) and the United Kingdom (UK) in the use of urea (U) and (calcium) ammonium nitrate ((C)AN) fertilizers on grassland. Rainfall and temperature patterns affect the amount of gaseous and leaching nitrogen (N) losses from U and (C)AN and their agronomic efficiency. This study aimed to: (i) examine how these losses relate to the observed U/(C)AN efficiency in NL and UK; and (ii) provide a simple model to choose the most appropriate N fertilizer, based on straightforward and uniform statistical analysis of existing data from numerous field trials. Rainfall and temperature for the first 3 days after fertilizer application defined the differences in U efficiency between countries. By aggregating NL and UK data, a simple regression equation was derived, which was used to develop a decision support model. This model showed that under prevailing weather conditions in NL, only once every 5 and 7 years will it be profitable for the farmer to apply U instead of (C)AN, for the first and second cut, respectively.

Introduction

Grasslands receive large amounts of nitrogen (N) fertilizer in western Europe. In the Netherlands (NL) most of it is applied as (calcium) ammonium nitrate ((C)AN). In the UK and especially in Eire, urea (U) is used as N fertilizer as well as (C)AN. Weather conditions affect N losses of N fertilizers and thereby their agronomic efficiency. Rainfall and low temperatures after fertilizer

©CAB INTERNATIONAL 1997. *Gaseous Nitrogen Emissions from Grasslands*
(eds S.C. Jarvis and B.F. Pain)

application, suppress ammonia (NH$_3$) volatilization from U (Freney *et al.*, 1983), and improve U efficiency. On grassland, extremes in efficiency of 6.5 and 60% of applied U have been reported (Velthof *et al.*, 1990) for situations of, respectively, 25 mm and 0 mm of rainfall within 3 days after fertilizer application and average daily temperatures of 17 and 14°C. In addition, rainfall increases nitrate (NO$_3^-$) leaching, denitrification (Jordan, 1989) and nitrous oxide (N$_2$O) losses from (C)AN, thereby decreasing (C)AN efficiency. Velthof *et al.* (1996, unpublished results) showed that N$_2$O losses on grassland from applied (C)AN increased considerably when total rainfall and mean soil temperature increased, while N$_2$O losses from applied U increased only slightly.

Studies where the sum of gaseous and leaching N losses after U and (C)AN application were measured simultaneously are limited. These losses are, however, reflected in grass dry matter (DM) and N yield (NY), because they determine N availability for uptake. In numerous field trials, U and (C)AN efficiency has been measured. In a review study, Watson *et al.* (1990) concluded that on average U was as effective as (C)AN when applied to grassland in spring, but was less effective in summer. Field trials in the NL showed that over the whole season U was, on average, 15% less effective than (C)AN (Van Burg *et al.*, 1982). Differences in weather conditions between the UK and the NL, causing different N losses from both fertilizers, may result in the contrasting effects.

Losses of N from fertilizer have to be minimized. This study defines criteria to improve fertilizer N use efficiency and thereby decrease N losses from grasslands in the UK, the NL and Eire. Data from a large number of fertilizer trials on grassland sites in these countries were analysed uniformly to test whether differences in rainfall and temperature patterns define the differences in U efficiency. The results were used to construct a decision support model to enable farmers to use the most profitable fertilizer.

Materials and Methods

Available data from representative grassland areas in western Europe were analysed, i.e. datasets from the NL, Northern Ireland (NI) (Stevens *et al.*, 1989) and southern England (En) (Peake and Levington, 1993, personal communication). The details of the datasets and rainfall and temperature data are described elsewhere (Bussink and Oenema, 1996). To predict the chance of a certain amount of rainfall after U application, a probability curve was derived using rainfall statistics for the NL in spring (over the period 1930–60). The probability of an amount of 'X' mm within 'Z' days was calculated. The efficiency of U and (C)AN in field trials with respect to DM yield is expressed in terms of AURY (apparent-urea relative yield), i.e.

$$AURY = \frac{(DM\ yield\ with\ U) - (DM\ yield\ of\ 0\ N\ plot)}{(DM\ yield\ with\ (C)AN) - (DM\ yield\ of\ 0\ N\ plot)} \times 100\% \quad (1)$$

Replacing DMY by NY in equation (1) gives AURNY (apparent-urea relative NY).

Our dataset with N fertilizer trials contains information on treatment effects in similar experiments conducted at different times and places. To analyse treatment effects and distinguish between sources of variability, the method of residual (or restricted) maximum likelihood (Anon., 1994) was used. To test whether differences in rainfall and temperature patterns adequately explain the observed differences in agronomic U efficiency and U use between NL, NI and En, three similar linear mixed models were used.

Results

Aggregation of NL, NI and En data

Analysis of NL, NI and En data for the first and later cuts showed that the best relationships with short-term weather conditions were obtained for the amount of rainfall (R3) and the average temperature (T3) within 3 days after fertilizer application (Bussink and Oenema, 1996). Agronomic U efficiency increased with R3 and decreased with T3. If rainfall and temperature patterns in NL, NI and En differed, then the mean AUR(N)Y were also different. Aggregation of recorded R3 and T3 data in NL, NI and En clearly showed systematic mean differences (Table 6.1). In the NL, low AUR(N)Y were found together with relatively low R3 and T3 in the first cut, and relatively large R3 and T3 in later cuts. Apparently, the increased R3 (because of less NH_3 volatilization) in later cuts was counteracted by the increase in T3 (because of increased NH_3 volatilization). In NI, a relatively large AURY was combined with a relatively large R3 and a modest T3 in the first cut. In the second cut,

Table 6.1. The average AURY and AURNY and the average R3 and T3 in NL, NI and En.

Country	Cut	AURY (%)	AURNY (%)	R3 (mm)	T3 (°C)	3-day periods without rain (%)
NL	First	92.3	86.4	3.63	4.67	15.6
NL	Later	92.3	86.4	8.90	14.21	24.4
NI	First	100.2	99.5	7.38	6.4	23.3
NI	Second	97.3	95.2	7.42	9.1	16.7
En	First	100.9	—	8.94	—	10.0
En	Second	91.1	—	5.70	—	21.7

—No data available.

the mean T3 was higher, which resulted in a decrease in AUR(N)Y. High temperatures and low rainfall promote NH_3 volatilization from U and reduce N losses through denitrification and NO_3^- leaching from (C)AN. Both counteracting processes lead to low AUR(N)Y.

Combining all trials and all cuts led to the following relationship between AURY and R3 and T3 being developed:

$$AURY = 89.48\ (\pm0.781) + 2.188\ (\pm0.148) \times R3 - 1.091\ (\pm0.070) \times T3 \qquad (2)$$
$$(R^2_{adj} = 98.9\%)$$

This relationship clearly indicates that rainfall and temperature define the agronomic efficiency of U in a similar way in NL, NI and En.

Economic evaluation of the use of (C)AN and U

Maximizing the difference between the harvesting value (product quality in combination with yield) of herbage and the costs of fertilizer input determines the choice of fertilizer type by the farmer. To facilitate decision analysis, an empirical relationship was developed based on equation (2) to predict the profitability of U use.

$$U_p = CAN_p + DM_p \times (DMY_{fert} - DMY_{0N}) \times (0.8948 + 0.02188 \times R3 \qquad (3)$$
$$- 0.01091 \times T3)/Nappl - (2.94 - 1/(\text{fraction of N contained in CAN})) \times lime_p$$

where: U_p, CAN_p and DM_p are the prices of U, (C)AN and herbage DM (Dfl. kg^{-1} N), respectively; $DMY_{fert} - DMY_{0N}$ is the DMY difference between the fertilized and the 0N plot (kg DM ha^{-1}); and $lime_p$ is the price of the limestone (Dfl. kg^{-1}). The effect on soil acidification of U use, in relation to (C)AN use, is taken into account by valuing the limestone contained in (C)AN.

Examples of equation (3) are shown in Fig. 6.1 for DMY_{fert} and DMY_{0N} for 3000 and 1500 kg ha^{-1} cut^{-1} in the first (T3 = 5°C) and second cut (T3 = 12°C), respectively. In the NL, herbage DM has a value of Dfl. 0.20 kg^{-1} and U is about Dfl. 0.15 kg^{-1} N cheaper than (C)AN (27%) (Anon., 1996). Figure 6.1 indicates that R3 must exceed 6 mm and 9.5 mm for the first and second cut, respectively, to obtain equal profits for U and (C)AN. In the NL the probability of R3 > 6 mm and R3 > 9.5 mm is 20% and 15%, respectively (Fig. 6.2), indicating that U usage for the first and second cut is only profitable once every 5 and 7 years, respectively. Assuming that the R3 values in Table 6.1 for NI and En are representative for the whole of NI and En, then usage of U is profitable for the first cut, because R3 is > 6 mm (Fig. 6.1). For later cuts it is generally not profitable to use U. Improving the rainfall probability curve by using data from local weather stations will greatly improve the accuracy of the predictions with equation (3).

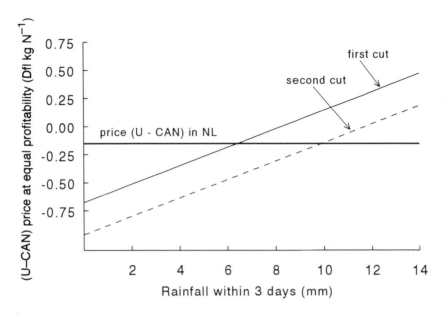

Fig. 6.1. The effect of R3 on the price difference needed between U and (C)AN to obtain equal profits for different temperatures, based on data from the NL and NI.

Discussion

Analysis of NL, NI and En data showed that rainfall and temperature have a marked and uniform effect on the agronomic U efficiency. These results confirm other findings (e.g. Van Burg *et al.*, 1982; Black *et al.*, 1987; Herlihy and O' Keeffe, 1988; Watson *et al.*, 1990; Lloyd, 1992). Thus far, however, no link has been made between rainfall and temperature patterns and differences in observed efficiencies between countries. This study clearly shows that rainfall and temperature not only define AUR(N)Y in field trials within a country but also the observed differences in the use of fertilizer types between countries. Therefore, the conclusion of Watson *et al.* (1990) that U is generally as good as (C)AN early in the growing season, but less effective in summer for conditions in northwest Europe is too inflexible. It is concluded that AUR(N)Y values are similar in the NL and UK where there are similar rainfall and temperature patterns. Variations in AURY between countries, either within or between seasons, can be mainly explained by these two variables (R3 and T3).

These results are based on agronomic trials, which reflect the net effect of gaseous and leaching N losses. However, to quantify the N losses precisely, all N losses from the individual processes (NH_3 volatilization, denitrification, N_2O emission and NO_3^- leaching) should be quantified, preferably

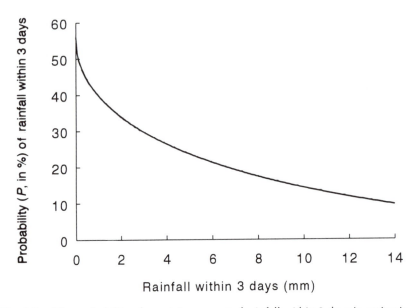

Fig. 6.2. The probability of a certain amount of rainfall within 3 days in spring in the NL.

simultaneously. This may result in an improved decision analysis model, in which the judgement of which N fertilizer to use is based not only on economical but also on environmental constraints.

Acknowledgements

The authors would like to thank R.J. Stevens and H. Peake who have kindly provided their datasets which have made this comparative study possible and G. André for his assistance with the statistical analysis. We thank Hydro Agri for funding this study.

References

Anon. (1994) *GENSTAT 5, Release 3. Reference Manual.* Clarendon Press, Oxford, 796 pp.
Anon. (1996) *Monthly Journal of Price Statistics.* LEI-DLO, Den Haag, 34–3, 31 pp, (In Dutch).
Black, A.S., Sherlock, R.R. and Smith, N.P. (1987) Effect of timing of simulated rainfall on ammonia volatilization from urea, applied to soil of varying moisture content. *Journal of Soil Science* 38, 679–687.

Bussink, D.W. and Oenema, O. (1996) Differences in rainfall and temperature define the use of different types of nitrogen fertilizer on managed grassland in UK, NL and Eire. *Netherlands Journal of Agricultural Science* 44, 317–338.

Freney, J.R., Simpson, J.R. and Denmead, O.T. (1983) Volatilization of ammonia. In: Freney, J.R. and Simpson, J.R. (eds) *Gaseous Loss of Nitrogen from Plant-Soil Systems. Developments in Plant and Soil Science*, Vol. 9. Martinus Nijhoff, The Hague, pp. 1–32.

Herlihy, M. and O'Keeffe, W.F. (1988) Evaluation and model of temperature and rainfall effects on response to N sources applied to grassland in spring. *Fertilizer Research* 13, 255–267.

Jordan, C. (1989) The effect of fertiliser type and application rates on denitrification losses from cut grassland in Northern Ireland. *Fertilizer Research* 19, 45–55.

Lloyd, A. (1992) Urea as nitrogen fertilizer for grass cut for silage. *Journal of Agricultural Science, Cambridge* 119, 373–381.

Stevens, R.J., Gracey, H.I., Kilpatrick, D.J., Camlin, M.S., O'Neill, D.G. and McLaughlan, R.J. (1989) Effect of date of application and form of nitrogen on herbage production in spring. *Journal of Agricultural Science, Cambridge* 112, 329–337.

Van Burg, P.F.J., Dilz, K. and Prins, W.H. (1982) Agricultural value of various nitrogen fertilizers; results of research in the Netherlands and elsewhere in Europe. In: *Netherlands Nitrogen Technical Bulletin 13.* Agricultural Bureau Netherlands Nitrogen Fertilizer Industry, 51 pp.

Velthof, G.L., Oenema, O., Postmus, J. and Prins, W.H. (1990) In-situ field measurements of ammonia volatilization from urea and calcium ammonium nitrate applied to grassland. *Meststoffen* 1/2, 41–46.

Watson, C.J., Stevens, R.J., Garrett, M.K. and McMurray, C.H. (1990) Efficiency and future potential of urea for temperate grassland. *Fertilizer Research* 26, 341–357.

Posters 1–5

I

Measurements of Soil Atmosphere Gases with a Gas-flow Soil Core Method: Some Applications

K. Vlassak, M. Swerts and R. Merckx

K.U. Leuven, Faculty of Agricultural and Applied Biological Sciences, Kardinaal Mercierlaan 92, B-3001 Heverlee, Belgium

Both dinitrogen (N_2) and nitrous oxide (N_2O) result from denitrification in varying amounts. Measuring N_2 production against the ambient concentrations in the atmosphere is a major problem hampering the study of denitrification. A breakthrough was the introduction of the acetylene (C_2H_2) inhibition technique. This technique has, however, several drawbacks. We have developed an automated technique that overcomes the problems involved with the C_2H_2 inhibition technique (Swerts *et al.*, 1995). The technique can measure changing production rates frequently and easily. Some applications of this technique are given below.

The major factors influencing the total denitrification rate are oxygen (O_2) concentration, carbon (C) availability and NO_3^- concentration (Swerts *et al.*, 1996a). The dynamics of the denitrification process were studied using the newly developed method. The influence of changes in available organic C on carbon dioxide (CO_2), N_2O, nitric oxide (NO) and N_2 production was studied by changing the soil C content through aerobic pre-incubations of different length up to 21 days. The aerobic pre-incubation caused an increase in nitrate (NO_3^-) concentration and a decrease in available C content. As an example, Fig. P1.1 shows the time course of the N_2, N_2O, NO and N total

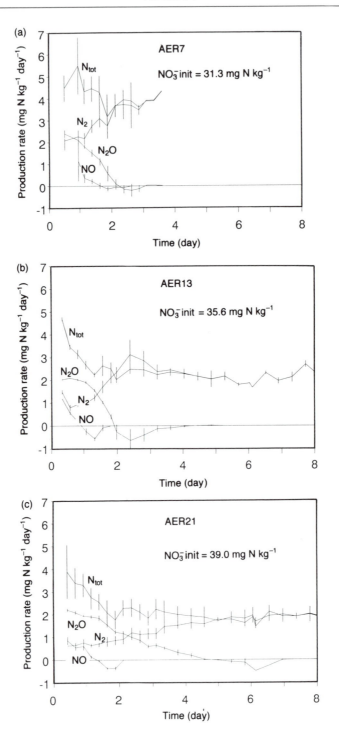

(N_2 + N_2O + NO) production rates for the treatments with 7 (AER7), 13 (AER13) and 21 (AER21) days of aerobic pre-incubation.

It can be concluded that available C content dominated both CO_2 and total N gas (N_2 + N_2O + NO) production during anaerobiosis. Both CO_2 and total N production rates decreased with increasing length of aerobic pre-incubations. This was in spite of the higher initial NO_3^- concentration.

Biological N_2 fixation was measured immediately following denitrification under anaerobic conditions in soils amended with glucose (Swerts *et al.*, 1996b,c). Failure to take account of N_2 fixation could lead to a serious underestimation of denitrification.

As soon as anaerobicity in the soil was established, denitrification started with N_2O as the predominant gas produced (Fig. P1.2(a)). After 40 h, N_2O production started to decline rapidly and the N_2O present was further reduced to N_2. Initial NO_3^- and NH_4^+ concentrations were 88.5 and 0 mg N kg^{-1}. After 50 h, both NO_3^- and NH_4^+ concentrations were zero. Net N_2 fixation was assumed to begin at the point at which the N_2 production became negative.

The changes in microbial activity were reflected in the CO_2 production rates (Fig. P1.2(b)). The first CO_2 production rate peak reached its maximum at the time of maximum denitrification rates, i.e. after 45 h, slightly after the maximum in N_2O production rate at 40 h, but just before the maximum in N_2 production and N_2O consumption rates, 49 h from the start. The minimum value in CO_2 production rate, between the two maxima, corresponds to the time of zero denitrification at 54 h. Therefore the second CO_2 production rate peak probably corresponds with the activity of N_2 fixing organisms. Both the maximum in CO_2 production rate and the maximum in N_2 fixation rate were reached after 84 h. After N_2 fixation faded out, due to N_2 shortage, volatile fatty acid (VFA) production increased (Fig. P1.2(c)).

Measuring N_2, N_2O and CO_2, as well as the NO_3^-, NH_4^+ and VFA concentrations, helps to clarify the picture of the processes involved. During the initial stage of the experiment, denitrification was the major biochemical process. Due to the high amount of easily available C (82.9 mg kg^{-1} of soil) the microbial activity increased, as can be seen from the CO_2 production rates, and N_2O production rates and total denitrification rates increased accordingly. The high N_2O : N_2 ratio is surprising under the given circumstances. A more profound discussion of the latter results can be found in Swerts *et al.* (1996b,c). At the end of denitrification an increasing activity by a second microbial population, accompanied by N_2 fixation was monitored. The experiments illustrate clearly the interactions of C availability, microbial

Fig. P1.1. (opposite) Time course of N_2, N_2O, NO, and N_{tot} production rates (average ± SD in mg N kg^{-1} day^{-1}) for AER7(a), AER13(b), and AER21(c) during the anaerobic incubation, under an He atmosphere, at 25°C (clayey silt loam soil).

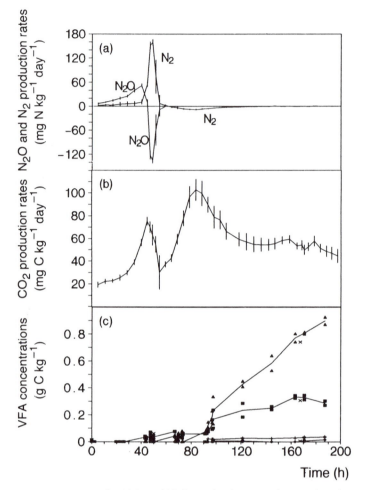

Fig. P1.2. Time course of: (a) N_2 and N_2O production rates (averge ± SD in mg N kg^{-1} day^{-1}); (b) CO_2 production rates (average ± SD in mg C kg^{-1} day^{-1}); (c) VFA concentrations (g C kg^{-1}) of butyrate (▲), acetate (■), propionate (+) and isobutyrate (◆) during the anaerobic incubation, under a He atmosphere at 25°C, of a clayey silt loam soil, after addition of 50 mg KNO_3-N and 4200 mg glucose-C kg^{-1}.

population and NO_3^- availability as influencing factors on denitrification and its $N_2O : N_2$ ratio and on fermentation.

Acknowledgements
This research was conducted in the framework of the Impulse programme 'Global Change', supported by the Belgian Government – Prime Minister's Service – Federal Office for Scientific, Technical and Cultural Affairs.

References

Swerts, M., Uytterhoeven, G., Merckx, R. and Vlassak, K. (1995) Semicontinuous measurement of soil atmosphere gases with gas-flow soil core method. *Soil Science Society of America Journal* 59, 1336–1342.

Swerts, M., Merckx, R. and Vlassak, K. (1996a) Influence of carbon availability on the production of NO, N_2O, N_2 and CO_2 by soil cores during anaerobic conditions. *Plant and Soil* 181, 145–151.

Swerts, M., Merckx, R. and Vlassak, K. (1996b) Denitrification followed by N_2-fixation during anaerobic incubation. *Soil Biology and Biochemistry* 28, 127–129.

Swerts, M., Merckx, R. and Vlassak, K. (1996c) Denitrification, N_2 fixation and fermentation during anaerobic incubation of soils amended with glucose and nitrate. *Biology and Fertility of Soils* 23, 229–235.

Laboratory Studies on the Release of Nitric Oxide from Subtropical Grassland Soils: The Effect of Soil Temperature and Moisture

W.X. Yang[1,2] and F.X. Meixner[1]

[1]*Max-Planck-Institut für Chemie, Abteilung Biogeochemie, Postfach 3060, D-55020 Mainz, Germany;* [2]*Laboratory of Material Cycling in Pedosphere, Institute of Soil Science, Academia Sinica, PO Box 821, Nanjing 210008, PR China*

Biogenic nitric oxide (NO) from soil contributes a large component of the atmospheric NO_x budget on a global scale (IPCC, 1994). However, little is known about the importance of NO emissions from subtropical soils, especially from African savanna grassland (Yienger and Levy II, 1995). In this context, an extended study of the biogenic emission of NO from a Miombo type savanna system (Zimbabwe) was performed in 1994, which is described elsewhere (Fickinger *et al.*, 1996). At the end of that field experiment, soil samples were taken from both the actual experimental plots ('site 1', 0–2 cm) and an adjacent site ('site 2', 0–2, 2–5, 5–10, 10–20 cm).

In the laboratory, soil samples were placed into a chamber for incubation under controlled temperature and moisture conditions. To investigate the influence of soil moisture, soil samples were wetted by spraying prescribed amounts of water with a syringe. NO and water (H_2O) mixing ratios were measured before air entered and after passing through the chamber. The flux of NO and water vapour out of chamber was calculated as:

$$F = \frac{Q}{A} \cdot (m_o - m_i) \cdot \frac{M}{V}$$

As shown in Fig. P2.1(a), the NO emission rate from soil is exponentially related to soil temperature under constant soil water content. Following the suggestion of Williams *et al.* (1992) and Yienger and Levy II (1995), this relationship can be expressed as $F_{NO} = A \times \exp(B \times T_s)$, where A and B are fitting parameters. For different soil depths, Fig. P2.1(a) shows that the emission rate of NO from the top surface layer (0–2 cm) of the soil is much higher

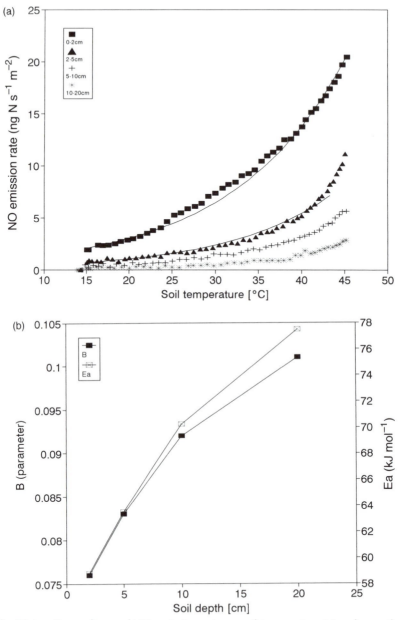

Fig. P2.1. Dependence of NO emission rate on soil temperature (a) and on soil water content (c) and the activation energy (E_a) and fitting parameter (*B*) for NO emission (b) in different soil depths. (d) The comparison of NO flux between model estimation and *in situ* field measurement in a grassland soil ('site 1').

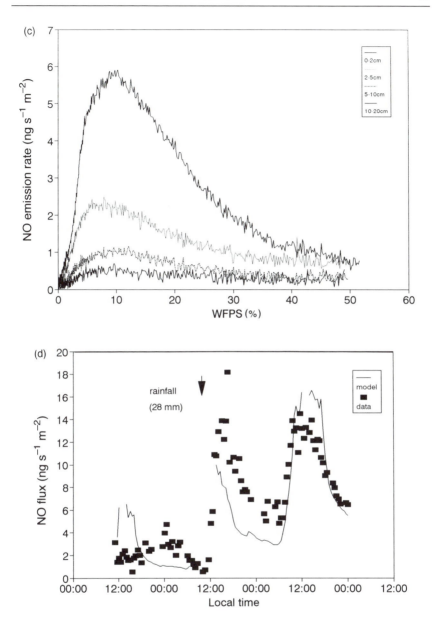

than that from the other three layers at the same temperature. This suggests that the top soil layer contributes most to the NO emission into atmosphere. According to the Arrhenius equation, the activation energy (E_a) for NO emission and the parameter B for each layer of soil depth can be determined. The results indicate that E_a and B decrease with increasing soil depth (Fig. P2.1(b)).

Under identical soil temperature conditions ($T_s = 25°C$), the effect of soil water content on the emission rate of NO can be seen in Fig. P2.1(c). The maximum NO emission occurs at about 10% of water-filled pore space (WFPS) for all four layers of soil depth. A dramatic increase in F_{NO} was observed, as long as soil water content was lower than approximately 10% (WFPS), while for WFPS > 10% F_{NO} is gradually decreasing. Interestingly, the WFPS for maximum NO emission of this study was much lower than 60% as reported by Davidson (1993), but it is more or less consistent with the observations of Parsons *et al.* (1996).

Having parameterized the effect of soil temperature and moisture on NO emission by the procedures mentioned above (see also Grundmann *et al.*, 1996), the model of Galbally and Johansson is extended (1989) to:

$$F_{NO} = \sqrt{D_e \cdot k \cdot \rho_s} \cdot (m_c - m_a) \cdot \frac{M_N}{V} \cdot \exp\left(B \cdot (T_s - 25)\right) \cdot \exp\left(\frac{(a\theta_2 - \theta_m)(\theta - \theta_m)}{(\theta_m - \theta_1)(\theta - a\theta_2)}\right) \cdot \frac{(\theta - \theta_1)}{(\theta_m - \theta_1)}$$

where a is a fitting parameter. The quantities k and m_c, which were used as input parameters, were determined from linear regression analysis of NO net release rate versus the observed NO mixing ratio at the outlet of the chamber. This procedure is described in detail by Remde *et al.* (1993). Soil gas diffusivity (D_e) can be estimated according to an empirical formula presented by Millington (1959). Results show (Fig. P2.1(d)) that under identical soil temperature and water content as observed in the field, the laboratory NO flux reproduces the field NO flux to within 10–30%. Soil water content (followed by soil temperature) was identified to be the most important factor determining the soil NO emission. This corresponds to our field results, which show a tremendous 'flush' of NO from grassland soil as soon as it experiences the first heavy rain showers at the end of the dry season.

Nomenclature

F = NO flux (ng N s^{-1} m^{-2}) or H$_2$O flux
Q = gas flow rate through the chamber (l min^{-1})
A = chamber area (m^2)
V = gas molar volume (24.265 l mol^{-1}, at 25°C, 1013 hPa)
m_o = NO (H$_2$O) mixing ratio at outlet (ppbv)
m_i = NO (H$_2$O) mixing ratio at inlet (ppbv)
k = NO consumption rate (cm^3 N h^{-1} g^{-1} dw)
m_a = NO ambient mixing ratio (ppbv)
$\theta_1 = \theta$ ($F_{NO} = 0$), for $\theta < \theta_m$ ($\theta_1 \approx 0$)
$\theta_2 = \theta$ ($F_{NO} = 0$), for $\theta > \theta_m$ (θ_2 = saturation moisture of soil)
M = N or H$_2$O molecular weight
M_N = nitrogen molecular weight of N (14 g mol^{-1})
m_c = NO compensation mixing ratio (ppbv)
ρ_s = soil bulk density (cm^3 g^{-1} dry weight)
$\theta_m = \theta$ ($F_{NO} = F_{NO,max}$)

Acknowledgements

The financial support of the Academia Sinica–Max-Planck-Society Agreement for Scientific Cooperation is gratefully acknowledged

References

Davidson, E.A. (1993) Soil water content and the ratio of nitrous oxide to nitric oxide emitted from soil. In: Oremland, R.S. (ed.) *The Biogeochemistry of Global Change*. Chapman and Hall, New York, pp. 369–386.

Fickinger, T., Meixner, F.X. and Nathaus, F.J. (1996) Emission of nitric oxide from Miobao type savanna ecosystems (marondera, Zimbabwe). *EGS Annalles Geophysical (Suppl.II)* 14, 467.

Galbally, I.E. and Johansson, C. (1989) A model relating laboratory measurement of rates of nitric oxide production and field measurement of nitric oxide emission from soils. *Journal of Geophysical Research* 94, 6473–6480.

Grundmann, G.L., Renault, P., Rosso, L. and Bardin, R. (1995) Differential effects of soil water conent and temperature on nitrification and aeration. *Soil Science Society of America Journal* 59, 1342–1349.

IPCC (1994) *Climate Change 1994. Radiative forcing of climate change and an evaluation of the IPCC IS92 emission scenarios*. Cambridge University Press, Cambridge, 325 pp.

Millington, R.J. (1959) Gas diffusion in porous media. *Science* 130, 100–102.

Parsons, D.A.B., Scholes, M.C., Scholes, R.J. and Leaven, J.S. (1996) Biogenic NO emissions from savanna soils as a function of fire regime, soil type, soil nitrogen and water status. *Journal of Geophysical Research* 101, 23,683–23,688.

Remde, A., Ludwig, J., Meixner, F.X. and Conrad, R. (1993) A study to explain the emission of nitric oxide from a marsh soil. *Journal of Atmospheric Chemistry* 17, 249–275.

Williams, E.J., Guenther, A. and Fehsenfeld, F.C. (1992) An inventory of nitric oxide emissions from soils in the United States. *Journal of Geophysical Research* 97, 7511–7519.

Yienger, J.J. and Levy II, H. (1995) Empirical model of global soil-biogenic NO_x emission. *Journal of Geophysical Research* 100, 11,447–11,464.

Denitrification Losses from Urine-affected Soils

T.J. Clough and S.F. Ledgard
AgResearch, New Zealand Pastoral Agricultural Research Institute Ltd, Hamilton, New Zealand

Nitrogen (N) can be a limiting factor to pasture production. Grazing of pasture results in urine patches which contain large quantities of N, often in excess of the pasture's requirements. Measurement of denitrification from urine-affected soil can help to provide information on losses of dinitrogen (N_2) and nitrous oxide (N_2O) from the grazed pasture system. Concentrations of N_2O in the global atmosphere have been rising over the last 20 years: this is of significance, because N_2O contributes to global warming and depletion of stratospheric ozone (O_3) (Duxbury *et al.*, 1993). The objective of this study was to measure the denitrification losses of N from different soils following the application of ^{15}N-labelled synthetic urine.

Intact soil cores were used (9 cm diameter × 50 cm deep), extracted from four soil types (sandy loam, silt loam, clay and peat) which were then watered to field capacity. Synthetic urine (containing urea labelled with 50 atom % [15]N) was applied at a rate of 1000 kg N ha^{-1} to the soil cores (replicated thrice) in winter. Denitrification fluxes of ^{15}N-labelled gases and N_2O-N were monitored on 12 occasions over a 31-day period using a head-space cover technique. After 1-h intervals gas samples were collected for analysis. ^{15}N-labelled gas concentrations were determined using a triple collector mass spectrometer and the ^{15}N ratio technique (Mulvaney and Boast, 1986). Nitrous oxide was measured independently on a gas chromatograph using a [63]Ni electron capture detector. Labelled ^{15}N gas was assumed to consist of N_2 and N_2O. Dinitrogen gas concentrations were calculated by subtracting the N_2O-N concentrations from the ^{15}N labelled gas concentrations. Fluxes of N_2-N and N_2O-N were calculated and the total N gas losses determined by integrating the area under the flux curves (Fig. P3.1).

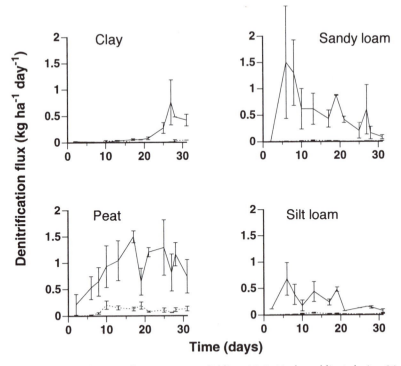

Fig. P3.1. Denitrification fluxes (N_2-N, solid line; N_2O-N, dotted line) during 31 days after urine application. Data points are means of three replicates with error bars ± SEM.

Table P3.1. Denitrification losses from urine applied to soils after 31 days.

	Soil type				Level of significance	
	Clay	Peat	Sandy loam	Silty loam	Peat vs. Mineral	Mineral
Ratio $N_2O : (N_2 + N_2O)$	0.12	0.14	0.03	0.09	NS	NS
Total N loss (kg N ha^{-1})						
N_2-N	2.9	20.2	13.1	5.9	NS	NS
N_2O-N	0.43	3.24	0.37	0.58	< 0.01	NS
Total N loss (%N applied)						
N_2-N	0.29	2.02	1.31	0.59	NS	NS
N_2O-N	0.04	0.33	0.04	0.06	< 0.01	NS

Total N_2O-N losses over the 31-day period in the four soil types ranged from 0.43 to 3.24 kg N ha^{-1}, while total N_2-N loss ranged from 2.9 to 20.2 kg N ha^{-1} (Table P3.1). The magnitude and temporal variation of the denitrification fluxes varied due to soil type (Fig. P3.1). Denitrification was slow to commence in the clay soil, beginning to decline in the sandy and silt loam soils, and remaining high in the peat soil after 31 days.

The slow start in denitrification from the clay soil was thought to be due to slow nitrification rates in the clay soil. Gas losses as N_2 tended to be highest in the peat soil but were not significantly different from the mineral soils due to variability in the magnitude of denitrification fluxes. To reduce variability, a higher degree of replication needed to be used. Loss of N_2O was greatest from the peat soil; however, N_2O comprised only a small fraction of the total denitrification loss (Table P3.1).

References

Duxbury, J.M., Harper, L.A. and Mosier, A.R. (1993) Contributions of agroecosystems to global climate change. In: *Agricultural Ecosystem Effects on Trace Gases and Global Climate Change*. ASA Special Publication Number 55. Madison, WI, USA, pp. 1–18.

Mulvaney, R.L. and Boast, C.W. (1986) Equations for determination of nitrogen-15 labelled dinitrogen and nitrous oxide by mass spectrometry. *Soil Science Society of America Journal* 50, 360–363.

Significance of Gaseous Losses on the Nitrogen Balance in Subalpine Grasslands

F. Rück[1], C. Homevo-Agossa and K. Stahr
Hohenheim University, Institute für Soil Science and Land Evaluation (310), D-70593 Stuttgart, Germany; [1]Present address: Federal Environment Agency, General Affairs of Soil Protection, Postfach 330022, D-14191 Berlin, Germany

The aim of the study was to describe the fate and transformation of nitrogen (N) in grassland ecosystems within the moraine landscape of the 'Württembergisches Allgäu'. The Allgäu, the foothills of the Alps in south-west Germany, is, because of its humidity (1100–1800 mm mean annual precipitation) and cool (6–7°C mean annual temperature) climate predetermined for grassland use. The soil catena includes Regosols-Eutric Cambisols, Phaeozems and Eutric and Stagnic Gleysols as well as Histosols with sandy loam to sandy clay as the dominant texture. The area is characterized by high percolation and particularly high leaching rates in winter. The use of large amounts of farm manure may create an environmental hazard through nutrient losses from this intensively used grassland.

Results of measured components of the nitrogen (N) budget of grasslands and semi-natural meadows in the Allgäu were summarized and combined with yield and N uptake data to provide an inorganic-N balance (Table P4.1). The tested plots cover a wide range from extensively cultivated, non-fertilized natural meadows to intensively fertilized grasslands (up to 550 kg N ha^{-1} year^{-1} N) and showed a distinct differentiation in N mineralization depending on fertilizer level. Soil N mineralization was clearly correlated with intensity of fertilization and cultivation. Net N mineralization

Table P4.1. Nitrogen balances (kg N ha^{-1} year^{-1}) 1991–93 for the experimental sites.

	Site 1		Site 2	
	Plot A	Plot HV	Plot HN	Plot SR
Fertilization:	Conventional	Conventional	Without	Without
Input				
Fertilizer application	313	136	0	0
N deposition (wet)	11	11	11	11
Legume fixation	20	19	40	40
Mineral N at beginning	48	85	38	28
Net N mineralization	444	243	226	50
Sum inputs	836	494	315	129
Output				
N uptake (minus transport losses)	394	286	241	65
Nitrate leaching	158	8	11	1
Mineral N at end	41	74	50	30
Immobilization	53	12	12	17
Gaseous losses				
NH_3-*volatilization*	*97*	*54*	*0*	*0*
N_2O-*losses*	*21*	*6*	*2*	*4*
N_2-*losses*	*8*	*7*	*2*	*3*
Sum outputs	772	447	318	120
N balance	+64	+47	−3	+9

ranged from 50 kg N ha^{-1} year^{-1} in a non-fertilized natural meadow to 444 kg N ha^{-1} year^{-1} in highest intensity level grassland. Nitrogen turnover ranged between 0.8% and 3.7% of total N content of the top soils (Rück and Stahr, 1995). Nitrate (NO$_3^-$) concentrations, collected with suction cups at 40 cm below the surface, were determined as 0.3–0.4 mg NO$_3$-N l^{-1}, on average, whereas at the highest intensity level on a well-drained soil NO$_3^-$ concentrations increased to 11 mg NO$_3$-N l^{-1}. Nitrate leaching ranged from 2 to 77 kg NO$_3$-N ha^{-1} year^{-1}.

Nitrous oxide (N$_2$O) emission rates from between 2 and 5 kg N ha^{-1} year^{-1} and total denitrification losses from 2.2 to 11 kg N ha^{-1} year^{-1} were measured at the non-fertilized natural meadows (mollic Gleysol, c. 11% OM, permanently high water saturation). Grasslands fertilized with slurry and mineral fertilizer had N$_2$O-N emission rates between 6 and 21 kg N ha^{-1} and total denitrification losses from 12 to 29 kg N ha^{-1} year^{-1}. These denitrification losses were 10-fold greater than those from arable land.

Ammonia (NH$_3$) volatilization after slurry application (Braschkat *et al.*, 1993) was calculated to be up to 97 kg N ha^{-1} year^{-1} and was quantitatively the most important gaseous loss.

Nitrogen balances at the field scale of the Allgäu experimental sites give a deficit of 185 kg N ha^{-1} year^{-1}, on average, since the start of the trials in 1988 (Stahr *et al.*, 1988). Inorganic N balances also include input figures for inorganic soil N at the beginning, *in situ* N mineralization, N fertilization, N deposition and N fixation, as well as output figures for N uptake, N immobilization, N leaching, NH$_3$ volatilization and denitrification. From these components, a surplus in grassland plots and a deficit in the native meadow were calculated. This proves that the N deficits in the N balances at the field scale were balanced by net mineralization from the organic N in soil and slurries. Therefore, recommendations for land use and agricultural policies have to take into account the differences in site patterns.

References

Braschkat, J., Mannheim, T., Horlacher, D. and Marschner, H. (1993) Measurement of ammonia emissions after liquid manure application: I. Construction of a wind-tunnel system for measurements under field conditions. *Zeitschrift Für Pflanzenernährung und Bodenkunde* 156, 75–81.

Rück, F. and Stahr, K. (1995) Beitrag der Stickstoffmineralisierung zur N-Bilanz auf Wirtschaftsgrünland und Streuwiesen im Allgäu. *Mitteilungen der Deutschen Bodenkundlichen Gesellschaft* 76, 923–926.

Stahr, K., Fischer, W.R., Rück, F., Fiedler, S., Homevo-Agossa, C., Kleber, M., Mickley, W., Näfe, D., Soeseno, I. and Aquino-Moscoso, O. (1995) TPA2: Landschaftsbezogene Nährstoff- und Wasserhaushaltsuntersuchungen im Allgäu (insbesondere Stickstoff- Phosphor- und Kohlenstoffhausahlt. In: Zeddies, J. *et al.* (1995) *Arbeits- und Ergebnisbericht 1993–1994, Sonderforschungsbereich 183 'Umweltgerechte Nutzung von Agrarlandschaften'*. Hohenheim University.

Carbon and Nitrogen Transformations in Grazed Grassland Systems

C. Mills and C. Watson
Department of Agriculture for Northern Ireland and The Queen's University of Belfast, Newforge Lane, Belfast BT9 5PX, UK

Studies of the influence of fertilizer input and grazing intensity on a number of soil process rates commenced in January 1995 at the Agricultural Research Institute, Hillsborough, Co. Down with the aim of assessing the environmental impact of management strategy.

Grassland plots were established and received 0 (grass/clover), 100, 200, 300, 400 or 500 kg N ha^{-1} year^{-1} as calcium ammonium nitrate (CAN) in six equal applications over the growing season. The plots were grazed by beef steers from April to October to maintain a constant sward height of 7 cm. Soil was collected to a depth of 7.5 cm, roughly sieved and incubated at soil temperature over 2–3-week periods throughout the year using specially adapted Kilner jars. Acetylene was applied to inhibit nitrification and minimize potential loss of nitrate-N (NO$_3$-N) by denitrification. To maintain aerobic conditions during each short-term incubation the jars were aerated every 2–3 days and acetylene re-applied. Nitrous oxide (N$_2$O) and carbon dioxide (CO$_2$) concentrations in the head space were determined prior to aeration with a Perkin Elmer 8500 gas chromatograph with a thermal conductivity detector. Net carbon (C) mineralization was calculated as cumulative CO$_2$ production. Net nitrogen (N) mineralization was calculated as the difference in total mineral N content in the soil at the beginning and end of each incubation period.

Nitrification was calculated as the % gain in soil NO$_3$-N over a 5-week incubation period at 15°C after applying the same rate of granular CAN (equivalent to 38 kg N ha^{-1}) to soil collected from each sward.

Cumulative net N mineralization values (kg ha^{-1}) were calculated over the first 8 months of the experiment. There was a highly significant increase ($P < 0.001$) in N mineralization with increasing fertilizer application rate and hence grazing intensity (Fig. P5.1). The total N mineralized from January to August 1995 was 250, 88, 133, 278, 502 and 407 kg N ha^{-1} for the 0 (grass/clover), 100, 200, 300, 400 and 500 kg N ha^{-1} year^{-1} treatments, respectively. There was no substantial burst in N mineralization in the early spring of 1995; however, from July until the end of August there was an apparent increase in the rate of N mineralization on all plots.

Carbon mineralization was greater than N mineralization, reflecting the C : N ratio of the soil (mean 11.7). The rate of C mineralization over the first 8 months was relatively constant but began to plateau out over the autumn/winter period (Fig. P5.2). Total C mineralized from January 1995 to March 1996 was 6628, 5406, 5283, 6604, 5875 and 5147 kg C ha^{-1} for the swards receiving 0 (grass/clover), 100, 200, 300, 400 and 500 kg N ha^{-1} year^{-1}, respectively. There was a highly significant difference between plots ($P < 0.001$) but, unlike N mineralization, there was no

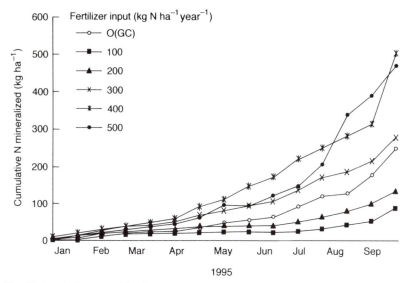

Fig. P5.1. Influence of fertilizer N input on nitrogen mineralization.

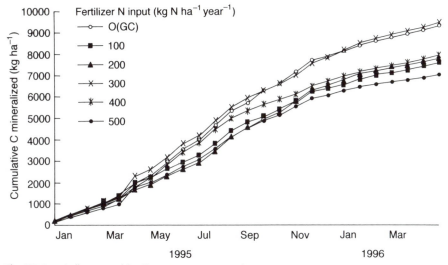

Fig. P5.2. Influence of fertilizer N input on carbon mineralization.

apparent trend with fertilizer N input. Carbon mineralization on the grass/clover sward was similar to the plot receiving 300 kg N ha^{-1} year^{-1}. Regression analysis over the first 8 months of the study indicated that there was no significant relationship between C and N mineralization on any sward. However, there was a significant inverse relationship between soil moisture and temperature (r^2 = 0.56, P < 0.001).

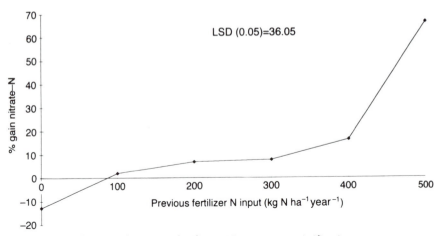

Fig. P5.3. Influence of previous fertilizer N input on net nitrification rate.

There was little net nitrification when the same rate of CAN was applied to swards previously receiving N inputs lower than 300 kg N ha^{-1} year^{-1}. However, on swards with a history of high N inputs, net nitrification rates were significantly greater ($P < 0.05$) (Fig. P5.3).

From these incubation experiments it can be concluded that increasing fertilizer N input, and hence grazing intensity, appeared to have differential effects on soil process rates. The high N mineralization and nitrification rates observed on the swards receiving high N inputs is likely to result in substantial N losses to the environment.

Acknowledgements

Part funding of this work by the Fertilizer Manufacturers Association is gratefully acknowledged.

Components of Ammonia Volatilization from Cattle and Sheep Production

7

S.G. Sommer[1] and N.J. Hutchings[2]

[1]*Department of Soil Science and* [2]*Department of Land Use, Danish Institute of Plant and Soil Science, Research Centre Foulum, PO Box 23, DK-8830 Tjele, Denmark*

Summary. The sources of ammonia (NH_3) volatilization within livestock production systems are reviewed and it is concluded that, in general, the processes driving NH_3 losses are well understood. Exceptions which require particular attention are the losses from solid manure and from arable crops grown for animal feed. Further investigations are needed to explain differences in losses from different species of grazing livestock. To improve the usefulness of future studies, they should be integrated with models for the prediction of emission at the whole farm scale.

Introduction

Livestock production systems are recognized as a major source of atmospheric ammonia (NH_3), which on deposition becomes a threat to the environment (ECETOC, 1994). Legislation encouraging the reduction of these losses is currently under consideration, implemented or under revision, in a number of countries (Hacker and Du, 1993).

Agriculture contributes to atmospheric NH_3 with emissions from livestock housing, manure storage, manure applied to land, urine and dung deposition in grazed pastures and from mineral fertilizers. Agricultural crops may also act as a source or sink of atmospheric NH_3 (Schjørring, 1991). Most of the emission originates from livestock production, less from mineral fertilizers and from crops. Research has identified the sources of NH_3 within livestock systems and the processes driving them (see Chapter 27). It is important that a whole farm approach is adopted because of the interactions

©CAB INTERNATIONAL 1997. *Gaseous Nitrogen Emissions from Grasslands*
(eds S.C. Jarvis and B.F. Pain)

between different sources of NH_3. Models taking this approach can form the basis of decision support tools that enable control measures to be adapted to different livestock systems, soils and climates. Such models can also be used for setting priorities for future research by identifying which components have the greatest influence on NH_3 loss.

Using our knowledge and experience with development of a farm-scale model (Hutchings *et al.*, 1996), the components of NH_3 volatilization from a livestock farm are described and areas where more knowledge or data are needed to improve predictions of NH_3 volatilization, are identified.

Sources of Ammonia

Excretion

The daily production of faeces can be calculated from the intake and apparent digestibility of the feed (Hutching *et al.*, 1996). Urine-N production is then calculated by subtracting the N partitioned to milk, N retained as new animal tissue and faeces from the total daily intake of N. It is often assumed that N in faeces is in the organic form whilst that in urine is urea. This latter assumption is not completely valid, as N in urine is partitioned between urea and other components like hippuric acid, allantoin and uric acid, etc. (Bristow *et al.*, 1992), the partitioning being variable (Smits *et al.*, 1995; Petersen *et al.*, 1996a). Furthermore, both the amount and concentration of urea may vary with the supply of rumen-degradable protein (see Chapter 8). This can cause errors in the predicted loss as only the urea is hydrolysed to NH_4 and contributes to NH_3 volatilization.

Smits *et al.* (see Chapter 9) showed that changing the rumen-degradable protein balance in feed to cattle could reduce NH_3 volatilization by about 40%, because the concentration of urea in urine was reduced. There is considerable scope for increasing the efficiency of utilization of dietary N, and this may be one of the few options for countries like the Netherlands to achieve a significant reduction of NH_3 volatilization in the future (see Chapter 12). Furthermore, formulation of the feed may affect the pH of cattle slurry by as much as 1 unit, i.e. from 6.8 to 7.7, by changing the feed composition from 75% sorghum to 75% barley (Kellems *et al.*, 1979). As pH is an important variable in NH_3 emission, more information is required on how slurry pH changes with the animal's diet.

Animal housing

In animal houses the floor may be solid, partial or fully slatted and the dairy cows tied or free. Below slatted floors faeces, urine, drinking water and straw is collected in pits, the mixture being a viscous liquid or slurry. Solid manure is a mixture of faeces and straw in combination with some urine, which is scraped from the floor behind tied dairy cows whilst liquid manure is a mixture of water, urine and soluble components of faeces separated from the solid fraction by drainage to the store via the gutter. Deep litter originates from animal houses with solid floors and is a mixture of straw or other bedding material, urine and faeces. The temperature in livestock houses is controlled by forced or natural ventilation.

In animal houses with slatted floors NH_3 is lost from the floor or from the storage pit below the floor. It has been shown that equal amounts of NH_3 are lost per area from the floor and from the surface of slurry stored in pits below slatted floors (Voorburg and Kroodsma, 1992). The loss of NH_3 from different types of animal houses has been related to the soiled area of the floor. A recent study (Groenstein, 1993), showing that NH_3 emission is reduced significantly by reducing the soiled area of the floor by tying dairy cows, suggests that this assumption is correct.

The losses vary during the year, partly because an increase in indoor temperature increases NH_3 losses (Smits *et al.*, 1995), and partly because of increased ventilation in the livestock barns, the latter being indirectly affected by indoor temperature (Fig. 7.1). The high losses of NH_3 during summer may therefore be an effect of ventilation, as well as higher temperatures inside the animal houses, as the farmers try to keep the indoor temperature low during summer (Oosthoek *et al.*, 1990). The effect of ventilation may vary with the design of the animal house and the placement of air inlets and outlets. The current data are mainly for houses with forced ventilation and more studies are particularly required for naturally ventilated houses.

Within the first few days after collection of faeces and urine in slurry pits, the pH may fluctuate due to changes in the acid and base components. Thus in a fresh sample, pH may be 8.5, declining to 7.5 within 1 week (Husted, 1995, personal communication). In the surface layer of slurry stored in pits below slats and slurry on bars and floor, where new material is added to the surface, the pH may therefore be higher than the bulk pH. Using the pH of urine seems, therefore, to be a good first choice for modelling the emission, but more data from slurry surfaces and soiled floor surfaces are needed to validate this assumption.

Current studies have concentrated on animal housing in which manure is stored as slurry. There is a need for more studies on emissions from houses where urine is separated from faeces, those with deep bedding systems and on the effect of animal behaviour on emission rates. There can be considerable variation even between systems in which manure is stored in the same form. For example, Groenstein *et al.* (1993) and Thelosen *et al.* (1993) studied deep bedding systems in which the bedding material was harrowed once a week to increase nitrification. In houses such as these, composting processes may change the manure temperature considerably and

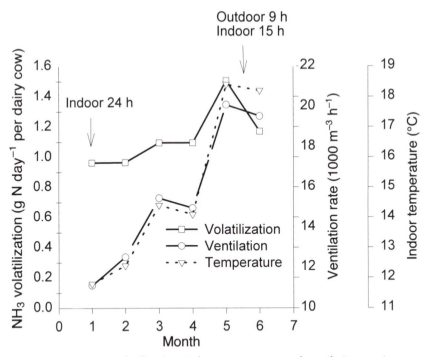

Fig. 7.1. Ammonia volatilization, indoor temperature and ventilation rate in an animal house for 40 dairy cows. (Adapted from Kroodsma *et al.*, 1993.)

losses may be completely different from those with untreated deep litter. Modifications to the design of in-house manure handling can also have an important effect on losses and the models have to be modified, e.g. when new housing systems are developed for the purpose of reducing losses. For example, where a sloping floor causes urine to drain immediately to a frequently scraped channel, volatilization can be considerably reduced (Swierstra *et al.*, 1995).

It is clear that the processes determining NH_3 losses are both complex and interrelated and more research is required before the effect of environment on losses from animal houses can be predicted with confidence.

Manure storage

The loss of NH_3 from stored pig slurry shows a curved relationship to wind speed in a wind tunnel (Olesen and Sommer, 1993), due to a shift from the dominance of the gas phase to liquid phase resistance and to non-linearity in the turbulent transport of NH_3. The NH_3 volatilization from a field scale slurry tank without a surface crust was linearly related to air temperature but no

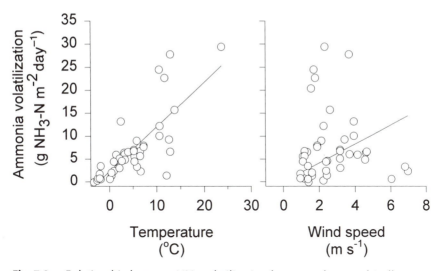

Fig. 7.2. Relationship between NH_3 volatilization from stored, anaerobically fermented animal slurry with no surface crust, air temperature and wind speed. (Adapted from Sommer, 1996.)

relationship with wind speed was established (Fig. 7.2). The store was situated on the plains in western Jutland, which has a windy marine climate, and the wind velocity was always so high that emission was dominated by the chemical resistance which is sensitive to temperature.

A cover of crust, straw, peat or expanded clay particles ('Leca') on the slurry leads to the development of a stagnant air layer through which NH_3 has to be transported by the slow process of diffusion. Furthermore, the covering may reduce natural mixing of the slurry so changing transport of NH_3 to the surface from being convective to diffusive and decreasing NH_3 losses to less than 10% of those from uncovered slurry (Sommer, 1996).

Liquid manures have a high pH and NH_4^+ content and the potential for NH_3 volatilization is therefore high. Uncovered experimental stores containing 1.5 m^3 liquid manure were found to lose up to 50% of total N during 8 months (Iversen, 1925). Volatilization was calculated by a mass balance of total N in the stored liquid manure; there is a need for studies using more modern techniques.

The losses of NH_3 from stored solid manure are variable, depending particularly on the extent of composting. For example, a loss of 20–35% of total-N by NH_3 volatilization occurred from a porous straw pig manure in which composting took place, while losses were 5% from a dense cattle slurry with negligible composting (Karlsson and Jeppson, 1995; Petersen *et al.*, 1996b). More data are needed for the development of models predicting composting processes and NH_3 emission from solid manure.

Applied manure

During the last decade, numerous studies on the emission of NH_3 from livestock slurry applied to land have been reported and there is generally a good understanding of the processes underlying the loss. The initial rate of loss from surface applied slurry is rapid (Fig. 7.3), due to both the initial high concentration of total ammoniacal N (TAN) and a high pH in the surface of newly spread slurry (Sommer and Sherlock, 1996). The TAN in the slurry surface layer decreases rapidly due to volatilization, infiltration and nitrification, so volatilization subsequently decreases. Field studies have shown a direct relationship between NH_3 volatilization and the pH of the applied acidified slurry (Pain et al., 1990; Stevens et al., 1992) and acidifying slurry reduces volatilization significantly (see Chapter 21). Most studies indicate that temperature is the most significant climatic factor affecting NH_3 loss from surface-applied slurry (Bussink et al., 1994; Moal et al., 1995; Sommer, 1996, unpublished results). However, these studies were all from maritime areas in west Europe where the relatively high wind velocities mean that turbulent transport is not the rate limiting factor.

Ammonia losses from surface-applied slurry are inversely related to the rate of infiltration into the soil because sorption of NH_4^+ on to soil colloids reduces the concentration of TAN in the soil solution. Ammonia losses will therefore increase if infiltration is reduced, either due to a high soil water

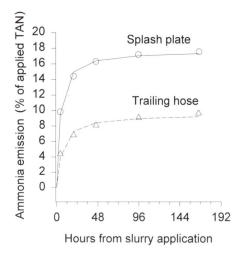

Fig. 7.3.   Ammonia volatilization from slurry applied by trailing hose to soil beneath a wheat canopy and surface spread on the canopy and soil. (Adapted from Sommer et al., 1996, unpublished results)

content or a high slurry viscosity or dry matter content (Jarvis and Pain, 1990). For the same reason, losses are reduced when the slurry is diluted with water or by separating out the coarse solids before application (Brunke *et al.*, 1988; Stevens *et al.*, 1992). Infiltration into the soil is increased and NH_3 losses are reduced by rain or irrigation, but losses may increase again when the water evaporates (Beauchamp *et al.*, 1982). After surface spreading to grass, washing slurry off the canopy may reduce NH_3 volatilization (see Chapter 26). Cultivating the soil surface before surface application of slurry also increases infiltration and reduces NH_3 loss (Horlacher and Marschner, 1990; Bless *et al.*, 1991).

Incorporating the slurry into the soil is a most effective way of reducing NH_3 volatilization. Shallow injection of the slurry into the soil can reduce NH_3 losses by about 70%, whilst deep injection will often stop losses completely (see Chapter 25). Incorporation of slurry by ploughing or by rotary harrow reduces losses of NH_3, the loss being reduced by 80% when slurry is ploughed in rapidly (Pain *et al.*, 1991).

Ammonia loss from slurry applied to the bare soil between rows of plants with trailing hoses is reduced because the surface area of the slurry is reduced, infiltration rate is increased, the wind speed above the slurry is reduced and atmospheric NH_3 concentrations above the slurry surface increased (Bless *et al.*, 1991). The trailing hose application technique does not reduce losses during high wind speed or in a crop with a small leaf area (Sommer, 1996, unpublished results).

Initial slurry pH has not been shown to have a significant effect on NH_3 emission, probably due to the increase in pH that occurs after application of slurry (Sommer and Sherlock, 1996). Hutchings *et al.* (1996) used initial pH of the slurry and van der Molen *et al.* (1990) measured surface pH of slurry amended soil for modelling losses, because the changes of pH are poorly understood. More data are needed before the changes in the pH can be predicted.

Losses from solid manures are poorly understood. The initial loss rate of NH_3 seems to be lower than from slurry, although 50% of the total loss can occur within about 24 h, and significant losses are measured for at least 10 days after surface application (see Chapter 24). The studies of Menzi *et al.* (see Chapter 23) and Chambers *et al.* (see Chapter 24) show that losses are very variable and related to similar factors to those for emissions from slurry. On average, 60–65% of the NH_4 applied in solid manure is lost as NH_3. Ammonia emission from surface-applied liquid manure was in the range 20–86% of the total N applied (see Chapter 10), which is a much higher loss rate than for animal slurry. There is a need for more data on NH_3 volatilization from both liquid and solid manure.

Grazing

Nitrogen in urine excreted during grazing is a significant source of NH_3, whereas the loss from dung pats is insignificant (Kellems *et al.*, 1979; Petersen *et al.*, 1996a). During urination, a large amount of urea is excreted on to a small area and often much of the urine soaks into the soil before the urea is hydrolysed. Consequently, NH_3 emissions tend to be low compared with emissions from animal slurry. It appears that the cation exchange capacity (CEC) and H^+ buffering capacity of soils may have an effect on the emission rate from urine patches. However, the effect may not always be very significant, as Dutch and English studies show similar levels of emission rates from soils with quite different soil characteristics (Fig. 7.4). These studies indicate that NH_3 volatilization from grazing dairy cows are positively related to the N input to the grass or grass clover pasture (Jarvis *et al.*, 1989; Bussink, 1994), whereas NH_3 losses from grazing sheep do not appear to be so (Jarvis *et al.*, 1991). The reason for this difference is unclear and deserves further investigation.

It is striking that emission from urine patches measured with wind tunnels or dynamic chambers is greater than when using a micrometeorological technique (compare Jarvis *et al.*, 1989; Petersen *et al.*, 1996a). One reason may be that NH_3 emitted from urine patches in the open is absorbed by

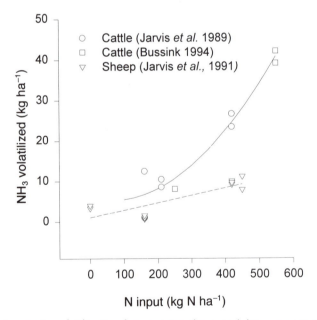

Fig. 7.4.   Ammonia volatilization from grazing sheep and dairy cows in England and the Netherlands. (Adapted from Bussink, 1994; Jarvis *et al.*, 1989, 1991.)

surrounding soil and plants. A similar pattern was seen in a study by Schjør-ring *et al.* (1992), in which 10–50% of the NH_3 emitted from a solution of NH_4^+ in flat beakers placed on the soil was not accounted for when measuring the NH_3 emission. The authors assumed the fraction not accounted for was absorbed by the soil surface. More studies are needed to quantify this absorption of emitted NH_3.

Fertilizers

The typical pattern of NH_3 volatilization from N fertilizers from non-calcareous soils shows a reduction in the following order: urea > diammonium phosphate (DAP) > calcium ammonium nitrate (CAN) (Whitehead and Raistrick, 1990; Sommer and Ersbøll, 1996). It has been shown in several studies that the differences in NH_3 losses from different N fertilizers are caused by the change in soil pH induced by the applied fertilizer. The pH change is often expressed as pH_{max}, the pH in the soil surface 24–48 h after application of the fertilizer. Thus, the NH_3 volatilization from urea, DAP and CAN is exponentially related to pH_{max} (Fig. 7.5), in studies using soils with

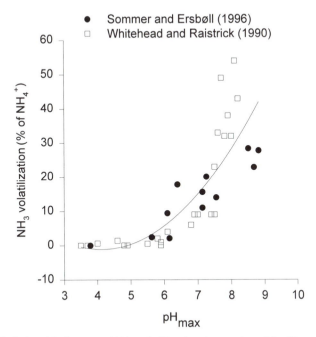

Fig. 7.5. Relationship between NH_3 volatilization from mineral fertilizers applied to a bare soil and maximum pH after application of the fertilizer (pH_{max}). (Adapted from Whitehead and Raistrick, 1990; Sommer and Ersbøll, 1996.)

both high and low clay contents (Lyster *et al.*, 1980; O'Toole *et al.*, 1985; Whitehead and Raistrick, 1990; Sommer and Ersbøll, 1996). Ammonia emission from fertilizers may be predicted by complex mechanistic models (e.g. Kirk and Nye, 1991) or more simply by relating pH_{max} and NH_3 volatilization to soil properties.

Rain reduces volatilization of NH_3 by leaching surface-applied urea and NH_4^+ into the soil. More than 7–9 mm rain within 3 days after application of urea and CAN, therefore, increased the yield of the subsequent cut of grass ley (Bussink and Oenema, 1996).

Crops

Plants may acts as sinks and sources of NH_3 in the atmosphere (Sutton *et al.*, 1994). A diurnal pattern of emission has been shown for many fertilized crops, the emission being highest during the day and lowest during nighttime. In contrast, deposition normally occurs in unfertilized, seminatural ecosystems (Sutton *et al.*, 1991).

It is likely that emission tends to occur when N fertilization provides an excess supply of N to the plant (Schjørring, 1991). During the growing

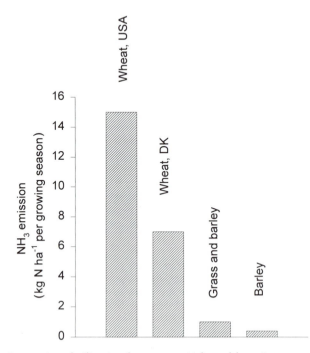

Fig. 7.6. Ammonia volatilization from crops. (Adapted from Sutton *et al.*, 1994.)

season, emission varies due to differences in uptake of fertilizer- or manure-N and to the physiological changes of the plants (Schjørring, 1991). Thus, in a year with adequate water, solar energy and temperatures, emission may occur from the leaves after addition of the fertilizer. Little further loss is seen until a second burst of emission takes place at anthesis and grain filling. Due to climatic differences and fertilization patterns, the NH_3 loss pattern may vary between years, and there may be differences between species and agricultural cultivars. Furthermore, deposition may occur in periods when the atmospheric NH_3 concentration is high, for example, when the crop is in the plume of NH_3 from livestock houses, manure stores or from manure applied to land.

The net NH_3 emission during a growing season measured with micrometeorological techniques is in the range 0.6–15 kg N ha^{-1} (Fig. 7.6). Although this flux of NH_3 is low compared to other agricultural sources, the area under cultivation is so large that crops may be a significant source of NH_3 to the atmosphere. Perhaps more importantly, the figures suggest that crops

Box 7.1. Areas requiring more research.

Animal housing
- Effect of ventilation and temperature
- Changes in manure composition in animal houses (e.g. ammoniacal N and pH)
- NH_3 emission from housing with deep bedding

Manure stores
- The factors influencing surface crust development
- Emission from solid manure
- Emission from composting of manure (and organic farming)

Manure applied in the field
- Effect of slurry viscosity and soil characteristics on infiltration
- Change in pH of slurry surface
- Emission from solid and liquid manure

Grazing
- Effect of variation in urine composition
- Effect of cattle versus sheep

Agricultural crops
- Fertilizer practices
- Climatic conditions
- Crop species and cultivars
- Harvesting practice

Fertilizers
- Field validation using micrometeorological techniques.

---

are not, as previously thought, a sink for atmospheric NH3. The magnitude and pattern of emission from crops deserves further investigation.

Conclusion

The NH3 volatilization sources on livestock farms are clearly identified. In the past, studies have concentrated on losses from applied slurry; knowledge is still lacking concerning other manure types and NH3 sources. Box 7.1 summarizes those areas where more research is needed.

References

Bless, H.-G., Beinhauer, R. and Sattelmacher, B. (1991) Ammonia emission from slurry applied to wheat stubble and rape in North Germany. *Journal of Agricultural Science, Cambridge* 117, 225–231.

Beauchamp, E.G., Kidd, G.E. and Thurtell, G. (1982) Ammonia volatilization from liquid dairy cattle manure in the field. *Canadian Journal of Soil Science* 62, 11–19.

Bristow, A.W., Whitehead, D.C. and Cockburn, J.E. (1992) Nitrogeneous constituents in the urine of cattle, sheep and goats. *Journal of Food and Agriculture* 59, 387–394.

Brunke, R., Alvo, P., Schuepp, P. and Gordon, R. (1988) Effect of meteorological parameters on ammonia loss from manure in the field. *Journal of Environmental Quality* 17, 431–436.

Bussink, D.W. (1994) Relationships between ammonia volatilization and nitrogen fertilizer application rate, intake and excretion of herbage nitrogen by cattle on grazed swards. *Fertilizer Research* 38, 111–121.

Bussink, D.W. and Oenema, O. (1996) Differences in rainfall and temperature define the use of different types of nitrogen fertilizer on managed grassland in UK, NL and Eire. *Netherlands Journal of Agricultural Science* 44, 317–339.

Bussink, D.W., Huijsmans, J.F.M. and Ketelaars, J.J.M.H. (1994) Ammonia volatilization from nitric-acid-treated cattle slurry surface applied to grassland. *Netherlands Journal of Agricultural Science* 42, 293–309.

ECETOC (1994) *Ammonia Emission to Air in Western Europe.* Technical Report No. 62, European Centre for Ecotoxicology and Toxicology of Chemicals. Brussels, Belgium, 195 pp.

Groenstein, C.M. (1993) Animal-waste management and emission of ammonia from livestock housing systems: field studies. In: Collins E. and Boon, C. (eds) *Livestock Environment Fourth International Symposium University of Warwick Coventry, England.* American Society of Agricultural Engineers, St Joseph, MI, pp. 1169–1175.

Groenstein, C.M., Oosthoek, J. and van Faassen, H.G. (1993) Microbial processes in deep-litter systems for fattening pigs and emission of ammonia, nitrous oxide and nitric oxide. In: Verstegen M.W.A., den Hartog L.A., van Kempen G.J.M. and

Metz J.H.M. (eds) *Nitrogen Flow in Pig Production and Environmental Consequences.* Purdoc Scientific Publishers, Wageningen, The Netherlands, pp. 307–312.

Hacker, R.R. and Du, Z. (1993) Livestock pollution and politics. In: Verstegen M.W.A., den Hartog L.A., van Kempen G.J.M. and Metz J.H.M. (eds) *Nitrogen Flow in Pig Production and Environmental Consequences.* Purdoc Scientific Publishers, Wageningen, The Netherlands, pp. 3–21.

Horlacher, D. and Marschner, H. (1990) Schätzrahmen zur beurteilung von ammoniakverlusten nach ausbringung von rinderflüssigmist. *Zeitschrift Pflanzenernährung und Bodenkunde* 153, 107–115.

Hutchings, N., Sommer, S.G. and Jarvis, S.C. (1996) A model of ammonia volatilization from a grazing livestock farm. *Atmospheric Environment* 30, 589–599.

Iversen, K. (1925) Undersøgelser vedrørende ajlens opbevaring. *Tidsskrift for Planteavl* 31, 149–168.

Jarvis, S.C. and Pain, B.F. (1990) Ammonia volatilization from agricultural land. In: *Proceedings No. 298.* The Fertiliser Society, London, 35 pp.

Jarvis, S.C., Hatch, D.J. and Lockyer, D.R. (1989) Ammonia fluxes from grazed grassland: annual losses from cattle production systems and their relation to nitrogen inputs. *Journal of Agricultural Science, Cambridge* 113, 99–108.

Jarvis, S.C., Hatch, D.J., Orr, R.J. and Reynolds, S.E. (1991) Micrometeorological studies of ammonia emission from sheep grazed swards. *Journal of Agricultural Science, Cambridge* 117, 101–109.

Karlsson, S. and Jeppson, K-H. (1995) *Djupströbädd i stall och mellanlager (Deep Litter in Livestock Buildings and Field Storage).* Jordbrukstekniska institutet, S-Uppsala, 120 pp. (summary and legends in English).

Kellems, R.O., Miner, J.R. and Church, D.C. (1979) Effect of ration, waste composition and length of storage on the volatilization of ammonia, hydrogen sulfide and odors from cattle waste. *Journal of Animal Science, Cambridge* 48, 436–445.

Kirk, G.J.D. and Nye, P.H. (1991) A model of ammonia volatilization from applied urea. V. The effect of steady-state drainage and evaporation. *Journal of Soil Science* 42, 103–113.

Kroodsma, W., Huis ingt Veld J.W.H. and Scholtens, R. (1993) Ammonia emission and its reduction from cubicle houses by flushing. *Livestock Production Science* 35, 293–302.

Lyster, S., Morgan, M.A. and O'Toole, P. (1980) Ammonia volatilization from soils fertilized with urea and ammonium nitrate. *Journal of Life Science Royal Dublin Society* 1, 167–175.

Moal, J-F., Martinez, J., Guiziou, F. and Coste, C.-M. (1995) Ammonia volatilization following surface-applied pig and cattle slurry in France. *Journal of Agricultural Science, Cambridge* 125, 245–252.

van der Molen, J., Beljaars, A.C.M., Chardon, W.J., Tury, W.A. and Faassen, H.G. (1990) Ammonia volatilization from arable land after application of cattle slurry. 2. Derivation of a transfer model. *Netherlands Journal of Agricultural Sciences* 38, 239–254.

Olesen, J.E. and Sommer, S.G. (1993) Modelling effects of wind speed and surface cover on ammonia volatilization from stored pig slurry. *Atmospheric Environment* 27A, 2567–2574.

Oosthoek, J., Kroodsma, W. and Hoeksma, P. (1990) Betrieblichemassnahme zur minderung von ammoniakemissionen aus ställen. In: Döhler, H. and van den Weghe, H. (eds) *Ammoniak in der umwelt.* KTBL-Schriften-Vertrieb im Landwirtschaftsverlag GmbH, Münster-Hiltrup, pp. 29.1–29.23.

O'Toole, P., McGarry, S.J. and Morgan, M.A. (1985) Ammonia volatilization from urea treated pasture and tillage soils: effects of soil properties. *Journal of Soil Science* 36, 613–620.

Pain, B.F., Thompson, R.B., Rees, Y.J. and Skinner, J.H. (1990) Reducing gaseous losses of nitrogen from cattle slurry applied to grassland by the use of additives. *Journal of the Science of Food and Agriculture* 50, 141–153.

Pain, B.F., Phillips, V.R., Huisjmans, J.F.M. and Klarenbeek, E.J.V. (1991) Anglo-Dutch experiments on odour and ammonia emission following the spreading of piggery wastes on arable land. Rapport 91-9, IMAG-DLO, Wageningen, 28 pp.

Petersen, S.O., Sommer, S.G. and Aaes, O. (1996a) Ammonia losses from dung and urine of grazing cattle effect of N intake. *Atmospheric Environment* (in press).

Petersen, S.O., Sommer, S.G. and Lind, A. M. (1996b) *Kvælstoftab fra gødningslagre.* Report to the Danish Ministry of Agriculture, DK-Copenhagen, pp. 33.

Schjørring, J.K. (1991) Ammonia emission from the foliage of growing plants. In: Sharkey, T.D., Mooney, H.A. and Holland, E.A. (eds) *Trace Gas Emissions by Plants.* Academic, London.

Schjørring, J.K., Sommer, S.G. and Ferm, M. (1992) A simple passive sampler for measuring ammonia emission in the field. *Water, Air and Soil Pollution* 62, 13–24.

Smits, M.C.J., Valk, H., Elzing, A. and Keen, A. (1995) Effect of protein nutrition on ammonia emission from a cubicle house for dairy cattle. *Livestock Production Science* 44, 147–156.

Sommer, S.G. (1996) Ammonia volatilization from farm tanks containing anaerobically digested animal slurry. *Atmospheric Environment* 31(6), 863–868.

Sommer, S.G. and Ersbøll, A.K. (1996) Effect of air flow rate, lime amendments, and chemical soil properties on the volatilization of ammonia from fertilizers applied to sandy soils. *Biology and Fertility of Soils* 21, 53–60.

Sommer, S.G. and Sherlock, R.R. (1996) pH and buffer component dynamics in the surface layers of animal slurries. *Journal of Agricultural Science, Cambridge* 127, 109–116.

Stevens, R.J., Laughlin, R.J., Frost, J.P. and Anderson, R. (1992) Evaluation of separation plus acidification with nitric acid and separation plus dilution to make cattle slurry a balanced, efficient fertilizer for grass and silage. *Journal of Agricultural Science, Cambridge* 119, 391–399.

Sutton, M.A., Fowler, D., Hargreaves, K.J. and Storeton-West, R.L. (1991) Effects of land-use and environmental conditions on the exchange of ammonia between vegetation and the atmosphere. In: Beilke, S., Slanina, J. and Angeletti, G. (eds) *Field Measurements and Interpretation of Species Related to Photooxidants and Acid Deposition.* Air Pollution Research 39. E. Guyot SA, Brussels, pp. 211–217.

Sutton, M.A., Asman, W.A.H. and Schjørring, J.K. (1994) Dry deposition of reduced nitrogen. *Tellus* 46B, 255–273.

Swierstra, D., Smits, M.C.J. and Kroodsma, W. (1995) Ammonia emission from cubicle houses for cattle with slatted and solid floors. *Journal of Agricultural Engineering Research* 62, 127–132.

Thelosen, J.G.M., Heitlager, B.P. and Voermans, J.A.M. (1993) Nitrogen balances of two deep litter systems for finishing pigs. In: Verstegen, M.W.A., den Hartog, L.A., van Kempen, G.J.M. and Metz, J.H.M. (eds) *Nitrogen Flow in Pig Production and Environmental Consequences.* Purdoc Scientific Publishers, Wageningen, The Netherlands, pp. 318–323.

Voorburg, J.H. and Kroodsma, W. (1992) Volatile emissions of housing systems for cattle. *Livestock Production Science* 31, 57–70.

Whitehead, D.C. and Raistrick, N. (1990) Ammonia volatilization from five nitrogen compounds used as fertilizers following surface application to soils. *Journal of Soil Science* 41, 387–394.

Effect of Nitrogen and Sodium Chloride Intake on Production and Composition of Urine in Dairy Cows

A.M. van Vuuren[1] and M.C.J. Smits[2]

[1]DLO-Institute for Animal Science and Health (ID-DLO), PO Box 65, 8200 AB Lelystad, The Netherlands; [2]DLO-Institute for Agricultural and Environmental Engineering (IMAG-DLO), PO Box 43, 6700 AA Wageningen, The Netherlands

Summary. A study was made of the potential for manipulating urine volume and urea concentration in dairy cows. Four rations were tested. Rations high or low in rumen protein surplus (RPS) (1.0 versus 0.1 kg RPS day^{-1}) were fed with or without 385 g day^{-1} NaCl added. Treatments were tested in four multiparous dairy cows according to a Latin square design. After an 11-day adaptation period, milk, faeces and urine production and composition were measured. Urine production was affected by RPS and NaCl. Differences in RPS were completely recovered as differences in urea excretion. Variation in urine volume resulted in differences in urea concentrations. Thus, a decrease in RPS leads to a small reduction in urea concentrations, unless a reduction in urine volume can be prevented.

Introduction

Of the total acid deposition of 4280 mol H$^+$ ha^{-1} year^{-1}, 24% originates from NH$_3$ emission from cattle husbandry in the Netherlands (van de Ven, 1996). If measures are taken to diminish NH$_3$ emission during storage and manure application, urea excreted in urine on pastures and in cattle buildings becomes the predominant source for NH$_3$ volatilization. Under standardized laboratory conditions, a linear relationship between urinary urea concentrations (UCON) and NH$_3$ volatilization has been observed (Elzing and Kroodsma, 1993). Although a reduction in N intake results in lower urinary N excretion (van Vuuren *et al.*, 1993), urinary urea concentration may be less affected, because low-N diets are often also low in K. The reduced mineral intake may result in a lower urine volume (UVOL) and, consequently, UCON may be similar to those for high-N diets.

In this study the effects of N and mineral intake on mineral balances, UVOL and urinary nitrogenous components in dairy cows were investigated.

Material and Methods

Two levels of rumen protein surplus (RPS) (Tamminga *et al.*, 1994) and two levels of NaCl were tested in a Latin square designed experiment, utilizing four lactating dairy cows in four 21-day periods. The first 11 days of each period were used for adaptation, followed by 10 days for the quantitative collection of feed residues, faeces, urine and milk.

Diets high in RPS contained 50% (on dry matter basis) wilted grass silage, 10% maize silage, 10% maize gluten feed and 30% concentrates. Low-RPS diets contained 20% grass silage, 20% maize silage, 20% sugarbeet pulp, 10% high-moisture ear corn silage with husks and 30% concentrates. Concentrates varied in RPS content and at each RPS level two concentrates were used either with 6 or 66 g NaCl kg^{-1}. Diets were totally mixed and fed in two meals per day. Estimated RPS intakes were 1.0 and 0.1 kg day^{-1} for diets high or low in RPS, respectively. Including more NaCl in the diet resulted in an extra NaCl intake of 0.38 kg day^{-1}.

During weighing, diet components were sampled. Samples were oven-dried at 70°C and analysed for dry matter (DM) by drying at 103°C, Kjeldahl-N (ISO/DIS 5983), sodium (Na), potassium (K) (atomic absorption spectrometry) and chloride (Cl) (ISO 6495). Feed residues and faeces were collected quantitatively per 5-day period. Samples were taken, oven-dried and analysed for DM, ash, N, Na, K and Cl. Water consumption was recorded with micro-oval flowmeters (Type 4550; Brook Instruments BV, Veenendaal, NL), producing 1 pulse per 5 ml. Urine was collected using Foley catheters (24 Fr, 75 ml; Bard Benelux NV, Nieuwegein, NL). Urine was collected in containers with at least 0.5 l of a 10% H_2SO_4-solution. Urine samples were taken and stored at 4°C until analysed for Kjeldahl-N, for urea, creatinine and uric acid (enzymatic methods; Boehringer, Mannheim, Germany) and for allantoin (colorimetric). Milk production was recorded and samples were analysed for N, Na, K and Cl.

Statistical analysis was performed by analysis of variance using Genstat (Genstat 5 Committee, 1993). First, results per 5-day period were analysed, but within each period and cow, no differences between both periods could be detected. Therefore, results were averaged per period. Results were analysed for both treatment factors separately and in combination. Treatment factor interactions could not be analysed since the RPS factor was non-orthogonally distributed.

Results

Dry matter consumption averaged 22.2 kg day^{-1} and was not influenced by diet composition. The increase in N intake ranged from 170 to 180 g day^{-1} due to the higher RPS intake (Table 8.1). The higher N intake was mainly recovered in urine excretion and the level of RPS had no significant effect on milk and faecal N output. Additional NaCl had no effect on the N balance.

As expected, K$^+$ intake was higher for the high-RPS diets (Table 8.2). Extra NaCl increased faecal Na$^+$ excretion but reduced faecal K$^+$ excretion. Eighty per cent of the additional NaCl was excreted via urine. Both RPS and NaCl level influenced water intake and UVOL. Concentrations of N components in the urine are presented in Table 8.3. A higher level of RPS significantly increased the concentrations of urea and allantoin in the urine. Additional NaCl diluted the concentrations of all N components in the urine.

Table 8.1. The effect of rumen protein surplus (RPS) and added NaCl on the nitrogen balance of dairy cows (g day^{-1}; SED = standard error of difference).

	0.1 kg RPS		1.0 kg RPS			P value		
	−NaCl	+NaCl	−NaCl	+NaCl	SED	RPS	NaCl	Treatment
Intake	553	549	742	718	25.9	0.01	0.002	< 0.001
Milk	156	157	161	153	10.9	1.00	0.62	0.90
Faeces	252	241	273	264	13.9	0.27	0.43	0.22
Urine	169	168	308	320	5.6	0.01	0.86	< 0.001

Table 8.2. The effect of rumen protein surplus (RPS) and added NaCl on intake and excretion of Na$^+$ and K$^+$ by dairy cows (SED = standard error of difference).

	0.1 kg RPS		1.0 kg RPS			P value		
	−NaCl	+NaCl	−NaCl	+NaCl	SED	RPS	NaCl	Treatment
Na$^+$ (g day^{-1})								
Intake	36	200	54	200	10.3	0.12	< 0.001	< 0.001
Faeces	13	34	15	28	7.4	0.47	0.02	0.09
Milk	10	11	11	10	1.0	0.58	0.88	0.53
Urine*	12	156	28	168	9.4	0.03	< 0.001	< 0.001
K$^+$ (g day^{-1})								
Intake	402	395	575	540	10.8	< 0.001	0.05	< 0.001
Faeces	64	44	68	46	12.0	0.62	0.04	0.21
Milk	47	47	49	45	2.7	0.93	0.43	0.68
Urine*	292	304	458	463	10.1	< 0.001	0.32	< 0.001
Water (kg day^{-1})								
Intake	97	120	116	130	5.2	0.004	0.003	0.01
Urine	23	42	40	60	1.3	0.02	0.002	< 0.001

*Urine excretion: difference between intake and excretion by milk and faeces.

Table 8.3.   The effect of rumen protein surplus (RPS) and added NaCl on concentrations of nitrogenous components in urine of dairy cows (g N kg^{-1}; SED = standard error of difference).

	0.1 kg RPS		1.0 kg RPS			P value		
	–NaCl	+NaCl	–NaCl	+NaCl	SED	RPS	NaCl	Treatment
Total	7.38	3.87	7.55	5.20	0.37	0.08	< 0.001	< 0.001
Urea	4.91	2.59	6.01	4.04	0.33	0.01	< 0.001	< 0.001
Uric acid	0.06	0.05	0.05	0.03	0.01	0.27	0.03	0.09
Allantoin	1.03	0.55	0.63	0.44	0.04	0.04	0.004	< 0.001
Creatinine	0.26	0.15	0.15	0.11	0.02	0.08	0.03	0.001

Discussion

The reduction in RPS decreased N and K$^+$ intake and consequently their urinary excretion, resulting in a reduction in UVOL by almost 50%. This effect was reduced by NaCl supplementation. In this experiment, a relationship was derived between UVOL and the estimated amounts of Na$^+$ and K$^+$ excreted in urine (intake – {milk + faeces} in g day^{-1}): i.e.

$$UVOL = -8.35 \ (4.62) + 0.136 \ (0.014) \ Na^+ + 0.099 \ (0.012) \ K^+$$

with standard errors in parentheses, $r^2 = 0.93$ and residual standard deviation = 3.79 l day^{-1}.

Due to a lower RPS input, daily urinary N excretion was reduced by 45%. Thus, if NH$_3$ volatilization relates to daily urinary N excretion, a similar reduction in emission would be achieved. However, due to the simultaneous reduction in UVOL, UCON decreased by only 18%. Thus, if NH$_3$ emission relates to UCON, as suggested by the observations of Elzing and Kroodsma (1993), the reduction in emission would be lower than expected from daily urinary N excretion. Including extra minerals in the diet may prevent this reduction in UVOL.

References

Elzing, A. and Kroodsma, W. (1993) *Relatie tussen ammoniakemissie en stikstofcon-centratie in urine van melkvee* (Relationship between ammonia emission and nitrogen concentration in urine of dairy cattle.) IMAG-DLO Rapport 93–3. DLO-Institute for Agricultural and Environmental Engineering, Wageningen, 24 pp.

Genstat 5 Committee (1993) *Genstat Release 3 Reference Manual.* Clarendon Press, Oxford, 796 pp.

Tamminga, S., van Straalen, W.M., Subnel, A.P.J., Meijer, R.G.M., Steg, A., Wever, C.J.G. and Blok, M.C. (1994) The Dutch protein evaluation system: the DVE/OEB system. *Livestock Production Science* 40, 139–155.

van de Ven, G.W.J. (1996) A mathematical approach to comparing environmental and economic goals in dairy farming on sandy soils in the Netherlands. PhD Thesis, Agricultural University, Wageningen, 239 pp.

van Vuuren, A.M., van der Koelen, C.J., Valk, H. and de Visser, H. (1993) Effects of partial replacement of ryegrass by low protein feeds on rumen fermentation and nitrogen loss by dairy cows. *Journal of Dairy Science* 76, 2981–2993.

Effect of Protein Nutrition on Ammonia Emission from Cow Houses

9

M.C.J. Smits[1], H. Valk[2], G.J. Monteny[1] and A.M. van Vuuren[2]

[1]DLO-Institute for Environmental and Agricultural Engineering (IMAG-DLO), Division of Buildings and Environmental Technology, PO Box 43, 6700 AA, Wageningen, The Netherlands; [2]DLO-Institute for Animal Science and Health (ID-DLO), Department of Ruminant Nutrition, PO Box 65, 8200 AB Lelystad, The Netherlands

Summary. The effects of rumen degradable protein and urine production on ammonia (NH_3) emission were studied in two experiments. In the first experiment, the urinary concentration of urea and the NH_3 emission from cows that were fed two diets alternately were measured. Diet L contained 0.1 kg rumen protein surplus day^{-1} and diet H contained 1.0 kg rumen protein surplus day^{-1}. Diet L resulted in a 39% lower emission rate compared with diet H. The urinary concentration of urea was 42% lower when feeding diet L compared with diet H. In the second experiment, three different diets were rotationally fed during seven 3-week periods. The effects of urine volume and urinary concentration of urea on NH_3 emission were studied. A higher urine volume increased not only the number of urinations but also the volume excreted per urination. Both the urinary concentration of urea and the urine volume can be influenced by the diet and do have a significant impact on NH_3 emission. From the experiments it can be concluded that nutrition may substantially contribute to reduction in NH_3 emission.

Introduction

Until recently, the possible effect of nutrition on ammonia (NH_3) emission had not been investigated in housed dairy cattle. Ammonia losses in a building are mainly caused by volatilization following the conversion of urea from fresh urine to NH_3 on the floor and by volatilization of NH_3 from the manure in the pit below the floor. Muck and Steenhuis (1982) indicated that the concentration of urea is one of the main factors involved in the volatilization of NH_3. In a model system of a cow house, Elzing and Kroodsma (1993) found a linear relationship between urinary concentration of urea (UCU) and NH_3 emission.

At a positive rumen protein surplus (RPS) (Tamminga *et al.*, 1994), excess NH_3 will be absorbed, converted into urea in the liver and excreted in the urine (Tamminga, 1992). Urea also originates from intermediary protein metabolism. High nitrogen (N) losses either in the rumen or in intermediary metabolism result in a high excretion of urea in urine (Tamminga, 1992).

The urine volume produced by dairy cows is related to the intake of sodium (Na) and potassium (K) (see Chapter 8). When grass is (partly) replaced by products that are lower in protein, not only N intake but also K intake is often reduced. The volume of urine then will also be reduced. The urea concentration in the urine depends on both the amount of urea which must be excreted in the urine and the volume of urine produced. When both urinary N excretion and urine production are reduced, UCU may be fairly constant. The effect on NH_3 emission then is expected to be limited. A change in urine volume may also result in a change in number of urinations. Model calculations indicated that a lower number of urinations also may decrease NH_3 emission.

The effects of UCU and urine volume on NH_3 emission from cow houses have not been investigated before. In this study, the results of two experiments are presented. In experiment 1, the effect of a reduced UCU, by a lower RPS diet, on NH_3 emission from a cow house was studied. The urine volume was kept at a constant level by adding some salt to the low RPS diet. In a second experiment, the effects of urine volume and N excretion on NH_3 emission were studied by the addition of salt in one diet and the addition of both salt and a higher RPS in another diet. The aim of this paper is to give an impression of the magnitudes of effects on NH_3 emission of the main urine characteristics, obtained by differences in diets.

Experiment 1: Effect of Rumen Protein Surplus

Materials and methods

Animals and treatments

In 1993, two diets were alternately fed to 34 (13 uni- and 21 multiparous) crossbred Dutch Friesian × Holstein Friesian dairy cows (average milk production per cow during the experiment: 31.5 kg 4% fat-corrected milk (FCM) day^{-1}). Cows were housed in a mechanically ventilated two-row freestall barn with a solid concrete floor with an epoxy-mortar covering, sloped 3% to the middle and with a urine drain for quick removal of urine. Every hour, the faeces were removed with a dung scraper. Urine and faeces were stored in the pit underneath the floor. Twice daily the floor was flushed with water via a pipe system with nozzles using approximately 100 l of flushing water per flush.

The diets were offered as totally mixed rations once daily (at 8.30 h). On a dry matter (DM) basis, diet L contained 20% grass silage, 20% maize silage, 10% ground ear maize, 20% sugar beet pulp and 30% compound feed. Diet H contained 50% grass silage, 10% maize silage, 10% maize gluten feed and 30% of a different compound feed. In diet L, additional sodium chloride (NaCl) was added in order to balance the original difference in Na+K content. From the results of Van Vuuren and Smits (see Chapter 8), it followed that the urine production from both diets was approximately the same. Diet L contained an amount of rumen degradable protein which was in balance with the available amount of energy from fermentable organic matter (RPS = 0). In diet H a surplus of over 1000 g RPS per cow per day was available.

The experiment started in January 1993 and lasted for 18 weeks. Diets L and H were offered three times for 3 weeks in a change-over design starting with L and ending with diet H. After the 3-week period, the change-over from one diet to the other took place gradually over 3 days.

Sampling methods

Net intake was measured daily for the whole group of cows. As many urinating cows as possible were randomly sampled (by two persons during the day time and one person during the night), within a 24-h period at the end of the first (n = 106 samples) and second (n = 108 samples) periods. From the third period onwards, samples were taken from at least 50% of the cows, one sample per cow, in four periods (05.30–06.45, 15.30–16.30, 09.00–11.00, 20.00–22.00). Drinking water was supplied in a 60-l reservoir and recorded weekly.

Measurements of NH₃ emission

Ventilation rate was continuously measured with a full-size anemometer in the exhaust chimneys. Ammonia concentration in air samples of exhaust air were automatically analysed every 5 min (Scholtens, 1989). The ventilation rate and the NH_3 concentration were recorded as mean values per hour. The NH_3 emission was calculated by multiplying the hourly ventilation rate by the hourly NH_3 concentration.

Statistical analysis

From the first period, only the last week (days 15–21) was included in the analysis, assuming emission to have reached the normal level related to the diet fed. The average hourly emission per day (E_i; i = 15 . . . 126), expressed in g h⁻¹, was calculated. Natural logarithms of E_i, LOG(E_i), i = 1 . . . 126, were used as the values in the analysis. The model for log (E_i), with the expectation of LOG (E_i) and e_i the deviation from N_i, was:

$$LOG(E_i) = N_i + e_i.$$

In the model for η_i, parameters for the starting level of emission at diet L, the final effect of diet, the speed with which the level of the new diet was approached after a change-over (transition functions), a linear effect of temperature and a linear trend with time were included. The dependency of the data was described by an autoregressive process of order 1 for the errors.

The parameters in the resulting model, a combination of a linear model for the explanatory variables and a time-series (Box and Jenkins, 1970) model for the errors, were simultaneously estimated with GENSTAT (Genstat 5 Committee, 1987).

Results

Feed intake

Both crude protein (CP) and RPS intake were much lower on diet L than on diet H (Table 9.1), which was in agreement with the aim of the experiment. The intake of dry matter, net energy for lactation (NEL) and intestinal digestible protein (IDP) were, respectively, 10, 8 and 6% higher with diet L compared with diet H. The IDP : NEL ratio of the ingested nutrients was slightly lower on diet L compared with diet H.

Urinary urea concentration and ammonia emission

The mean urea concentration in the urine with diet L was 42% (3.5 g l^{-1}) lower compared with diet H. The overall means differed significantly ($P < 0.01$) between the diets. The starting level of NH_3 emission was 35 g h^{-1}. Diet L resulted in a 39% decrease in NH_3 emission compared with diet H. An increase in temperature of 1°C (reference: 15°C) resulted in a 4% increase in emission. Finally there was a trend of 0.2% increase of NH_3 emission per day. More details are given in Smits *et al.* (1995).

Experiment 2: Effects of Urinary N Excretion and Number of Urinations

Materials and methods

Animals and treatments

From February to July 1995, three diets were rotationally fed to 34 (11 uni- and 23 multiparous) crossbred Dutch Friesian and Holstein Friesian dairy cows (average milk production per cow during the experiment: 30 kg FCM day^{-1}). Cows were housed in the same mechanically ventilated two-row freestall barn as in the first experiment, except that a different solid

Table 9.1. Mean intake of dry matter (DM), rumen protein surplus (RPS) and crude protein (CP) of the diets (cow^{-1} day^{-1}).

	DM (kg)	RPS (g)	CP (kg)
L	23.3[a]	40[a]	3.4[a]
H	21.2[b]	1060[b]	4.2[b]

[a,b]Values of diets with different superscripts differ significantly ($P < 0.10$).

Table 9.2. Mean intake of dry matter (DM), rumen protein surplus (RPS), crude protein (CP), sodium (Na), potassium (K) and chloride (Cl) intake of the diets (cow^{-1} day^{-1}). Additional concentrates for high-yielding cows were included.

Diet	DM (kg)	RPS (g)	CP (g)	Na (g)	K (g)	Cl (g)
A	21.3	1060[a]	4310a	180[a]	470[a]	290[a]
B	21.1	250[b]	3320[b]	50[b]	360[b]	112[b]
C	20.8	270[b]	3300[b]	230[c]	360[b]	320[c]

[a,b,c]Values of diets with different superscripts differ significantly ($P < 0.10$).

concrete floor was used. Channels were constructed every 2 m over the full width of the walking alley and the floor sloped towards these.

The diets were offered as totally mixed rations once daily. Diet B contained no special additions. Diet A contained additional N and additional NaCl to increase both N excretion in urine and urine production and to keep UCU at approximately the same level as in diet B (Table 9.2). Diet C contained the same amount of nutrients as diet B and an additional amount of NaCl to increase urine production and to decrease the UCU simultaneously.

Data were collected in the same way as in experiment 1. In addition, the numbers of urinations and defaecations were determined on one day per 3-week period and the Na, K and Cl concentration in urine samples were determined. In the statistical analysis of NH$_3$ emission, only the last 8 days (days 14–21) of each period were included.

Results

In Table 9.3 the urine parameters and the NH$_3$ emission are given per diet. UCU was not significantly different, the number of urinations was significantly higher and subsequently the NH$_3$ emission was also significantly higher (26%; $P < 0.05$) when feeding diet A compared with diet B. UCU was significantly lower, the number of urinations was significantly higher and NH$_3$ emission was slightly, but not significantly, lower when feeding diet C

Table 9.3. Urine parameters and NH_3 emission per diet.

Parameters	Diet		
	A	B	C
UCU-N (g l^{-1})	4.9	4.7	3.3
Urine-N (g l^{-1})	6.3	7.1	4.6
Estimated urinary N excretion (g day^{-1})*	248	158	152
Estimated urine volume (l day^{-1})[†]	40	21	35
Number of urinations per cow during daytime	5.5	4.1	5.2
Relative NH_3 emission (%)	126	100	95

*Assuming a N digestibility of 60% for all diets.
[†]Mean of estimates based on calculated urinary excretion of either chloride, potassium or nitrogen.

compared with diet B. The estimated urine volumes with diets A and C were 90% and 67% higher than with diet B.

Discussion

Diet L resulted in a lowering of the concentration of urea in the urine by 42% compared with diet H, while the NH_3 emission was 39% lower. The lower concentration of urea in the urine resulting from diet L is in accordance with the lower intake of rumen degradable crude protein. When no salt was added in diet L, urine volume with this diet would have been lower and UCU would have been higher.

In the second experiment, the number of urinations was significantly influenced by the diet and had an impact on NH_3 emission. The addition of salt, to lower the UCU by a higher urine volume, does not seem to be an effective way of reducing the NH_3 emission. The water metabolism in the animal and the mineral (and protein) supply in the diet are important factors. Both the UCU and the number of urinations can be influenced by the diet and do have a significant impact on NH_3 emission. To lower the NH_3 emission by lowering the UCU, feeding lower N surpluses is preferred rather than producing higher urine volumes.

References

Box, G.E.P. and Jenkins, G.M. (1970) *Time-series Analysis Forecasting and Control.* Holden-Day, San Francisco, 533 pp.

Elzing, A. and Kroodsma, W. (1993) *Relatie Tussen Ammoniakemissie en Stikstofconcentratie in Urine van Melkvee.* (Relation between Ammonia Emission and

Nitrogen Concentration in Urine of Dairy Cattle.) IMAG-DLO report 93-3, Wageningen, 19 pp.

Genstat 5 Committee (1987) *Genstat 5 Reference Manual.* Clarendon Press, Oxford, 749 pp.

Muck, R.E. and Steenhuis, T.S. (1982) Nitrogen losses from manure storages. *Agricultural Wastes* 4, 41–54

Scholtens, R. (1989) Ammoniakemissionsmessungen in zwangbelüfteten Ställen. In: *Ammoniak in der Umwelt. Proceedings Symposium.* Braunschweig, KTBL, Darmstadt, pp. 20.1–20.9

Smits, M.C.J., Valk, H., Elzing, A. and Keen, A. (1995) Effect of protein nutrition on ammonia emission from a cubicle house for dairy cattle. *Livestock Production Science* 44, 147–156

Tamminga, S. (1992) Nutrition management of dairy cows as a contribution to pollution control. *Journal of Dairy Science* 75, 345–357

Tamminga, S., Van Straalen, W.M., Subnel, A.P.J., Meijer, R.G.M., Steg, A., Wever, C.J.G. and Blok, M.C. (1994) The Dutch protein evaluation system: the DVE/OEBsystem. *Livestock Production Science* 40, 139–155

Spreading of Cattle Urine to Leys: Techniques, Ammonia Emissions and Crop Yields

10

L. Rodhe[1], E. Salomon[2] and S. Johansson[2]

[1]Swedish Institute of Agricultural Engineering, Uppsala, Sweden; [2]Department of Soil Sciences, Swedish University of Agricultural Sciences, Uppsala, Sweden

Summary.   Experiments on the application of cattle urine were carried out on a lucerne dominated ley (year 1) and on a grass ley (year 2). The objective was to determine the effect of application technique (broadcast, trailing hoses and shallow injection) and spreading season on ammonia (NH_3) emission after spreading and on forage dry matter (DM) yield. Urine was applied to both first and second harvest at a rate of 30 kg NH_4-N ha^{-1} on each occasion and the leys were harvested twice. Ammonia emission was measured with a micrometeorological method.

Nitrogen (N) lost as NH_3 for the two years and the two spreading times was in the range of 20–86% of N applied as cattle urine. In both years, N lost as NH_3 was, for all spreading techniques, higher when applied to the second cut than to the first cut. In year 1, there were no significant differences in NH_3 emission for the different techniques, but in year 2, emission was significantly higher for broadcast spreading than for band spreading at both spreading times. Ammonia emission was also significantly higher for broadcast spreading than for shallow injection. There was no significant difference in emission between band spreading and shallow injection in year 2.

No significant differences in DM yield between treatments were established in year 1. In year 2, application of urine and mineral fertilizer resulted in lower DM yields compared with the control for broadcast spreading and shallow injection. Band spreading gave significantly higher DM yield compared with shallow injection. DM yield appeared to be independent of the extent of NH_3 emission.

Introduction

In Sweden, leys are the main crop in cultivated areas outside the flat country-side of the plains and are generally harvested twice a year. Most of the farms with cattle manure handle it as solid manure and urine. Urine from cattle contains half of the total amount of excreted nitrogen (N). Some hours after the urine is excreted, 70–90% of the N is in the form of ammonium (NH_4^+-N). The buffering capacity of urine is weak and stored urine may have a pH of between 8 and 9.

Emission of ammonia (NH_3) in Sweden derives mainly from agriculture. Approximately 50,000 tonnes are lost annually from animal production, with about 26% from spreading of animal manure. The Swedish Parliament has decided to halve NH_3 emission by the year 2000 in the south of Sweden, compared with 1990.

Research results concerning NH_3 emissions from urine are limited in numbers and concern mostly emissions from grassland with grazing animals. In trials with artificial urine the losses of N through NH_3 volatilization were 4–18% (Vertregt and Rutgers, 1988), 6–19% (Vertregt and Rutgers, 1991) and 19–24% (Vallis *et al.*, 1982) of the N in the urine applied to grassland. Trials with animal urine also show losses less than 25% of the N applied with urine (Sherlock and Goh, 1984; Castle and Reid, 1987; Morken,1992).

Factors affecting the magnitude of the emission include spreading season (Vallis *et al.*, 1982; Sherlock and Goh, 1984) and cation exchange capacity (CEC) in soil (Whitehead and Raistrick, 1993). Sherlock and Goh (1984) found that changes in NH_3 gas fluxes were related to measured changes in soil pH and air temperature.

The objective of the present 2-year experiment with urine was to evaluate the influence of spreading techniques and spreading time on NH_3 emission after spreading and on forage dry matter (DM) yield.

Materials and Methods

A 2-year field experiment in Uppsala, Sweden (1993–94) was conducted on the application of stored cattle urine to leys. In the first year, the trial was located at 59°53′ N and 17°42′ E in a field with 70% legumes and 30% grass. In the second year, the trial was moved to a nearby field with 100% grass, located at 59°49′ N and 17°41′ E. For the area, the average temperature for the period 1961–90 during April to August was +11.8°C and, on average, the total precipitation was 256 mm (Alexandersson *et al.*, 1991). The soil texture was classified as a silty clay loam at the first site and as a clay at the second site (FAO, 1990). In spring, each site was fertilized with 40 kg N, 25 kg phosphorus (P) and 100 kg of potassium (K) ha^{-1}, as mineral fertilizers.

Treatments were organized into a randomized block design with four replicates. The treatments included two spreading times, in spring to the first cut and then to the second cut. On each occasion urine was spread with three techniques: broadcast spreading, band spreading with trailing hoses (0.25 m apart) and shallow injection. The working depth of the shallow injector (Jako) varied between 0 and 50 mm. Before spreading, a sample of urine was taken for analyses of fluidity, dry matter (DM), pH, NH_4^+-N, total N, P and K according to Swedish standard methods (Table 10.1). Urine was applied at a rate of approximately 30 kg NH_4^+-N ha^{-1} on each occasion. A treatment receiving the same amount of mineral-N as fertilizer was included as a control.

Ammonia emissions were measured from plots treated with urine. A micrometeorological method of measuring gaseous NH_3 was used which was based on passive diffusion sampling close to the ground (Svensson, 1994). On each plot, measurements were carried out with two chambers to estimate the equilibrium concentration and with one ambient measuring unit to estimate the ambient concentration of NH_3. The NH_3 emissions were measured in four periods with the first two periods directly after spreading. The emissions between measuring periods were calculated from interpolated values of the NH_3 emission corrected for climate conditions prevailing during the interval and the period after the last measuring. Air temperature, soil surface temperature and wind speed were measured for the time period when NH_3 emissions occurred (Table 10.2).

Table 10.1. Analyses of stored cattle urine, 1993-1994.

Year	Spreading	Fluidity*	DM-content (%)	pH	NH_4^+-N	Total-N	P	K
1993	1	6.8	1.6	8.8	1.8	2.5	0.025	4.3
	2	6.7	1.9	8.6	1.9	2.5	0.029	4.7
1994	1	6.5	1.3	8.8	2.1	2.4	0.025	3.1
	2	6.7	1.9	8.8	2.4	3.0	0.027	5.2

The column group "Plant nutrients (kg tonne^{-1})" spans NH_4^+-N, Total-N, P and K.

*Malgeryd, 1994.

Table 10.2. Average air temperature, soil surface temperature and wind speed measured for the time period when NH_3 emissions occurred.

	1993			1994		
	Air temp. (°C)	Soil surface temp. (°C)	Wind speed (m s^{-1})	Air temp. (°C)	Soil surface temp. (°C)	Wind speed (m s^{-1})
Spring	12.5	15.9	2.7	11.1	10.2	1.6
Summer	12.5	18.2	3.0	14.2	13.9	3.1

The plots were harvested twice a year. The fresh crop was weighed, sampled for analysis of botanical, chemical composition and DM content. Within each year, total DM yields were statistically analysed by a General Linear Model (GLM) (SAS/STAT, 1989). Within each year and spreading time, NH_3 emissions after spreading of urine were statistically analysed by a GLM. The difference in NH_3 emission between spreading times within years was also statistically analysed.

Results and Discussion

Nitrogen lost as NH_3 for the two years and the two spreading times was in the range 20–86% of N spread as urine (Table 10.3). These are remarkably high emissions compared with published data. A compact upper soil layer resulting in low infiltration rates, dry and warm weather conditions and lack of high protecting vegetation may have contributed to this. Nitrogen losses as NH_3, when spreading swine urine to growing barley with a porous soil was, in comparison, less than 10% of N applied (Rodhe and Johansson, 1996). In both years, N lost as NH_3 was higher when applied to the second cut than to the first cut for all spreading techniques. One reason could be higher average soil surface temperature when spreading on the second cut compared with the first cut (Table 10.4).

In the first year, no significant differences were established between NH_3 emissions and the different techniques (Table 10.5). In the second year, the NH_3 emissions were significantly higher for broadcast spreading than for band spreading. This applied to both spreading times. There was also a significantly higher NH_3 emission for broadcast spreading than for shallow injection. No significant difference in NH_3 emission was established between band spreading and shallow injection. This could be explained by the

Table 10.3. The proportion of N released in the form of NH_3 as a percentage of the amount of total-N spread in urine with three different spreading techniques and at two different times in 1993 and 1994.

| | Proportion of N released as NH_3 (%) | | | |
| | 1993 | | 1994 | |
Spreading technique	Spreading 1	Spreading 2	Spreading 1	Spreading 2
Broadcasting	33	86	47	83
Band spreading	38	41	20	58
Shallow injection	49	55	28	37

Table 10.4.   Difference in NH₃ release (kg ha⁻¹) between spreading of urine to the second and first cuts. The spreading was done with three different techniques. The values were calculated from measured emissions during the first 6 h after spreading.

| | Difference in loss as NH_3 (kg ha^{-1}) | | | |
| | 1993 | | 1994 | |
Spreading technique	Difference, N loss	P value	Difference, N loss	P value
Broadcasting	7.52	0.02*	13.9	0.0003***
Band spreading	3.20	0.19	9.64	0.0006***
Shallow injection	−0.89	0.64	6.69	0.0031**

Table 10.5.   Loss of NH₃ (kg ha⁻¹) after spreading urine on a ley using three different techniques, 1993 and 1994. The values refer to measured emissions during the 6 h following spreading.

| | Ammonia loss (kg ha^{-1}) | |
Treatment	1993	1994
To 1st cut		
Broadcasting	10.59[a]	16.74[a]
Band spreading	11.29[a]	4.50[b]
Shallow injection	12.32[a]	5.40[b]
To 2nd cut		
Broadcasting	18.11[a]	30.27[c]
Band spreading	13.46[a]	14.15[d]
Shallow injection	14.12[a]	12.09[d]

[a,b]Means with different letters within each column and group are significantly different ($P < 0.05$).
[c,d]Means with different letters within each column and group are significantly different ($P < 0.001$).

shallow working depth of the injector, due to unsuitable design, and the compact soil conditions.

There were no significant differences in total DM yield between treatments in the first experimental year. In the second year, spreading of urine and mineral fertilizer resulted in lower total DM yield compared with the control, although the difference was significant only for broadcast spreading (3669 kg DM ha⁻¹) versus control (3940 kg DM ha⁻¹) and shallow injection (3541 kg DM ha⁻¹) versus control. No significant differences between band spreading (3811 kg DM ha⁻¹) and control were established.

Linear regression analysis of DM yield on NH₃ emission did not reveal any significant relationship except for one spreading time. In that case, the regression coefficient was very small, which implies that there was a negligible influence of the NH_3 emission on the yield. Furthermore, the slope of the regression line was not as expected, and suggested an increased yield with increased NH_3 emission.

References

Alexandersson, H., Carlström, C. and Larsson-McCann (1991) *Temperature and Precipitation in Sweden* 1961–90. Reference normals. SMHI, Norrköping (in Swedish).

Castle, M.E. and Reid, D. (1987) A comparison of the effects of liquid manure (urine and water) and nitrogen fertilizer applied to a grass-clover sward on soils of different pH-value. *Journal of Agricultural Science, Cambridge* 108, 17–25.

FAO (1990) *Guidelines for Soil Description*, 3rd edn (revised). Land and Water Development Division, Rome, 88 pp.

Malgeryd, J. (1994) Manure characterization. *International Agrophysics* Vol. 8, No. 1. Polish Academy of Sciences, Lublin, pp. 93–101.

Morken, J. (1992) Ammonia losses after application of slurry to grassland. The effect of application techniques and type of slurry. *Norsk landbruksforsking* 6, 315–329.

Rodhe, L. and Johansson, S. (1996) Urin – Spridningsteknik, ammoniakavgång och växtnäringsutnyttjande. In: *JTI-rapport Lantbruk & Industri nr 217*. Swedish Institute of Agricultural Engineering, Uppsala.

SAS/STAT (1989) *SAS/STAT User s Guide, Version 6*, Vol. 2, 4th edn. SAS Institute Inc., Cary, NC, 846 pp.

Sherlock, R.R. and Goh, K.M. (1984) Dynamics of ammonia volatilization from simulated urine patches and aqueous urea applied to pasture. I. Field experiments. *Fertilizer Research* 5, 181–195.

Svensson, L. (1994) Ammonia volatilization following application of livestock manure to arable land. *Journal of Agricultural Engineering Research* 58, 241–260.

Vallis, I., Harper, L.A., Catchpoole, V.R. and Weier, K.L. (1982) Volatilization of ammonia from urine patches in a subtropical pasture. *Australian Journal of Agricultural Research* 33, 97–107.

Vertregt, N. and Rutgers, B. (1988) Ammonia volatilization from urine patches in grassland. In: Nielsen, V.C., Voorburg, J.H. and L'Hermite, P. (eds) *Volatile Emissions from Livestock Farming and Sewage Operations.* Proceedings of a Workshop held in Uppsala, Sweden, 10–12 June 1987. Elsevier Applied Science, London, pp. 85–91.

Vertregt, N. and Rutgers, B. (1991) Ammonia emissions from grazing. In: Nielsen, V.C., Voorburg, J.H. and L'Hermite, P. (eds) *Odour and Ammonia Emissions from Livestock Farming.* Elsevier, London, pp. 177–184.

Whitehead, D.C. and Raistrick, N. (1993) The volatilization of ammonia from cattle urine applied to soils as influenced by soil properties. *Plant and Soil* 148, 43–51.

Ammonia Dispersion and Deposition Around Livestock Buildings

11

S. Couling

Silsoe Research Institute, Wrest Park, Silsoe, Bedfordshire MK45 4HS, UK

Summary. The aerial concentration of ammonia (NH_3) close to an idealized live-stock building was measured by means of a photoionization device and found to vary considerably with time. For example, with a point source of 10 l NH_3 min^{-1} on the roof ridge of the building, the concentration of NH_3 at 2.5 m above the ground, 10 m from the lee-side of the building, had peaks of 400 µg (N) m^{-3}. Over a 15-min sampling period, the NH_3 concentration was above 100 µg (N) m^{-3} for 16% of the time. The average concentration measured simultaneously by active denuders was 30 µg (N) m^{-3}.

Short-range NH_3 deposition downwind of the building during rain events was studied. Event rain samples from an area between 10 and 80 m from the building were collected over about 30 min and analysed for ammonium (NH_4^+). On all occasions enhanced NH_4^+ concentrations were measured in the downwind rain samples. The NH_4^+ concentration in the collected rain samples was dependent on the rainfall rates. At higher rates, a decrease in the concentration per millilitre of collected rain was observed. Under drizzle conditions concentrations of 8.0 µg (N) ml^{-1} were measured 12 m from the building, falling to 5.0 µg (N) ml^{-1} 50 m downwind of the building. Dry deposition during collection of the rain samples has been investigated and shown to be small but significant. These measurements demonstrate that there is a significant increase in the concentration of NH_3 in rain deposited downwind of livestock buildings even in the short range (tens of metres).

Introduction

Ammonia (NH_3) emissions from livestock housing are a significant problem. The current UK estimate of NH_3 emissions is about 200 kt NH_3-N per year, of which about 85% comes from agricultural sources, a significant proportion of which originates from livestock housing and from slurry storage systems (ApSimon et al., 1997; Pain et al., 1997). A large proportion of the NH_3 released from livestock buildings may possibly be re-deposited within a short range of the farm but there are very few studies of the magnitude of this deposition or of the processes controlling it. At Silsoe Research Institute, we have the facilities for simulating the emissions of NH_3 from a building source and following the dispersion of the gas downwind. Ammonia emitted from a livestock building can be either dry deposited or washed out of the atmosphere by precipitation. A detailed study is being carried out using an experimental building termed the Silsoe Structures Building to examine these phenomena in the short range downwind. The building has a traverse section equipped with Yorkshire boarding on both sidewalls to simulate a naturally ventilated building such as is used for housing cattle. Ammonia can be released at any desired position and source strength.

Full-scale wind pressure measurements have been made on the Silsoe Structures Building which have been used for comparison with computational fluid dynamics (CFD) solutions (Hoxey et al., 1993; Hoxey and Robertson, 1994; Richards and Hoxey, 1994; Richardson et al., 1995). CFD is a numerical modelling system which calculates the velocity and pressure of a fluid in a given situation, in this case the air flow around the building for any given wind direction and speed. The results of this model can then be combined with a particle tracking model, based on a random-walk principle, to predict the concentration of pollutants downwind of the building. These techniques can potentially give more accurate predictions than a Gaussian model in the near-field (< 1 km), because one is able to include the local features of terrain, building design, fetch, etc., which are significant in determining local concentrations and deposition.

Aerial Ammonia Concentrations Around the Building

A CFD model is being developed to characterize the dispersion of gases around a building. The flow vectors and gaseous concentrations around the Silsoe building are being measured to validate and improve the model. The concentration of aerial NH_3 close to the building varies considerably with time and will depend on the turbulence. The rapid fluctuations in NH_3 concentrations are being measured by means of a rapid photoionization detector (UVIC) (Griffiths, 1993). This instrument is not specific to NH_3 but it detects all species with an ionization potential less than 10.6 eV so cannot be used in a true farm environment. However, it can be used under the simulated conditions around the Silsoe Structures Building where only pure NH_3 is being released. The UVIC has a detection limit of about 100 μg (N) m^{-3} and can sample at 20 Hz. The longer term average concentrations are being measured by means of simple active denuder tubes (Ferm, 1979). Typical sampling times close to the building are about 30 min.

Work is in progress on characterizing the concentration field around the building under various conditions. The results from one particular day illustrate the fluctuations in NH_3 concentration with time. With a point source of 10 l NH_3 min^{-1} on the ridge of the building, the concentration of NH_3 10 m from the lee-side of the building had peaks of 400 μg (N) m^{-3}. Over a 30-min sampling period, the NH_3 concentration was above 100 μg (N) m^{-3} for 16% of the time. The average concentration measured simultaneously by active denuders was 30 μg (N) m^{-3}.

Ammonia Deposition

Wet deposition downwind of the building

A series of wet deposition measurements were made downwind of the Silsoe Structures Building with an NH_3 source within the building. These were based on event-sampling, where clean funnels and bottles were exposed for about 30 min during a rain event, after which the sample was immediately capped and stored in a cold room and analysed within 48 h. For these wet-deposition experiments, the source was a long permeable tube along the ground of the traverse section of the Silsoe Structures Building. For each rain event, a rain-sampler was placed upwind of the building to measure the background concentration of NH_3, while an array of rain-samplers was placed downwind of the building. Ammonia was released into the section at a rate of between 5 and 10 l min^{-1} (~212 g (NH_3) h^{-1} to 424 g (NH_3) h^{-1}), resulting in a concentration inside the section of about 20 ppm. At the end of the rain event, the collected rain samples were analysed for ammonium (NH_4^+) ions and the pH and volume were measured.

The background concentration in rain collected during two summer storm events was extremely high, up to 2.0 μg (N) ml^{-1} for a thunderstorm on 23 August 1995. The typical mean concentration in this area is about 0.7 μg (N) ml^{-1} (RGAR, 1990). The background concentration for the days with steady rainfall ranged from 0.4 to 0.8 μg (N) ml^{-1}.

Three sets of rain samples were collected during one rain event and the variation in concentration with time at a particular site compared. Although the background concentration remained almost constant, the NH_4^+ concentration in the collected downwind rain samples varied with the rainfall rates. At higher rates, a decrease in the concentration per millilitre of collected rain was observed; however, there was an increase in the total amount of NH_4^+ collected per minute. The temperature, wind speed and wind direction remained steady throughout the three runs.

The concentration in the downwind rain samples varied between rain events. For example, under drizzle conditions in September, concentrations of 8.0 μg (N) ml^{-1} were measured 12 m from the building section, falling to 5.0 μg (N) ml^{-1} 50 m downwind from the building (Fig. 11.1). Under similar conditions in December, the concentration varied from 3.0 μg (N) ml^{-1} 15 m from the building to 1.4 μg (N) ml^{-1} at 100 m. The concentration in the collected rain samples declined exponentially with distance from the building. However, the location of the plume was frequently displaced from the expected line of the wind due to the disturbance of the flow over the building.

These measurements demonstrate that there is a significant increase in the concentration of NH$_4^+$ in rain deposited downwind of livestock buildings, even in the short range.

Dry deposition to wet collectors

An array of rain collectors were set up downwind of the building on an overcast day with the aim of estimating the significance of dry deposition to

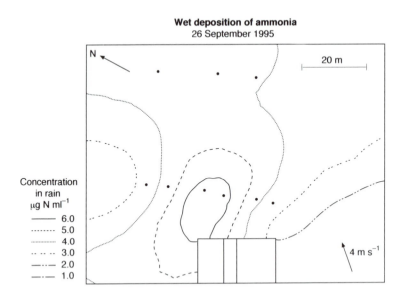

Fig. 11.1. Concentration of NH$_4^+$ in rain water samples collected downwind of a simulated livestock building. The blocked area represents the building, the NH$_3$ was released from a line source within the marked traverse section of the building at a rate of 10 l min^{-1}: the building has an open ridge along the centre perpendicular to the traverse section.

the collectors during a rain event. The funnels of the collectors were washed with ultra-high-purity water and left wet. Ammonia was released from the building by the diffuse source (see above) at $10\,l\,min^{-1}$. After 20 min the funnels were washed again with ultra-high-purity water into the collection bottles. The experiment was repeated several times. Later that day, a rain event was sampled, although a number of the samples were lost due to high winds. The samples were analysed for NH_4^+-N content and results reported as an uptake rate in $\mu g\,(N)\,min^{-1}$, which is the concentration of the collected sample times the volume of the sample divided by the sampling time in minutes. Table 11.1 gives the concentration, volumes and calculated uptake rates for runs 2 and 4. Note that run 2 was a measure of the dry deposition while run 4 was a rain event. For sampling periods of comparable length, the concentration of NH_3 collected by dry deposition ranged from about 5% of that collected from a rain event at the edges of the plume up to 26% close to the building and in the centre of the plume.

Figure 11.2(a) maps the uptake rate for dry deposition around the building. Note the wind direction is a few degrees off north but the higher areas of deposition are out to the south west of the building due to its influence on the wake flow. Figure 11.2(b) compares the uptake measured during a previous rain event: the wind on this occasion was from the south west.

These experiments show that dry deposition to wet collectors contributes a minor but significant portion of the NH_3 collected during rain event sampling. The proportion from dry deposition seems to vary across the plume, but further work is in progress in order to properly characterize the distribution across the plume. The deposition to a *rain*-wetted funnel may be higher, since the pH of rain will be lower than that of pure water. Laboratory work is in progress to characterize the deposition rates to a range of synthetic rain solutions from various aerial NH_3 concentrations.

Table 11.1. Dry and wet deposition of ammonia.

	Run 2: dry deposition rates to wet collectors			Run 4: wet deposition to wet collectors		
Sample point no.	Volume of wash water (ml)	Conc. (ng (N) ml^{-1})	Uptake (ng (N) min^{-1})	Volume of rain (ml)	Conc. (ng (N) ml^{-1})	Uptake (ng (N) min^{-1})
1	19	150	100			
2	13	150	60	14	3000	1050
3	23	120	90			
4	28	450	400			
5	19	290	180			
6	20	800	520	16	6800	2640
7	15	900	440	14	4800	1620
8	19	200	110	10	3400	810

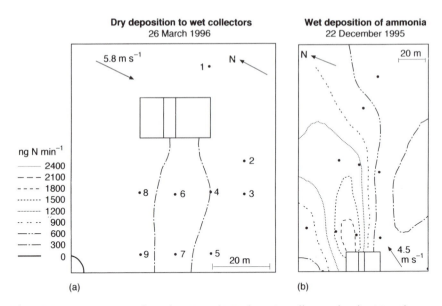

Fig. 11.2. Comparison of uptake rates of NH₃ by rain collectors by dry (a) and wet (b) deposition – see text for explanation of uptake rates. The blocked area represents the building, the NH₃ was released from a line source within the marked traverse section of the building at a rate of 10 l min⁻¹: the building has an open ridge along the centre perpendicular to the traverse section.

Conclusions

Aerial NH₃ concentrations can vary considerably close to a building source over a short time period. Peak concentrations can reach several hundred ppb, depending on the source strength and location. Up to 100 m from the building source, there is a significant enhancement in NH_4^+-N concentration in rain. Further work is required to investigate the extent of this enhancement. Some preliminary work has been carried out on the significance of dry deposition to rain collectors during a rain event. Deposition to funnels wetted with deionized water has been shown to be small but significant.

Acknowledgements

This work was funded by the Office of Science and Technology, and by the Ministry of Agriculture, Fisheries and Food within Open Contract OC 9117.

References

ApSimon, H.M., Couling, S., Cowell, D. and Warren, R.F. (1997) The potential for reducing acidification and eutrophication across Europe by abatement of agricultural emissions of ammonia. *Atmospheric Environment* (in press).

Ferm, M. (1979) Method for determination of atmospheric ammonia. *Atmospheric Environment*, 13, 1385–1393.

Griffiths, R.F. (1993) Emissions and environmental monitoring using energetic UV radiation: a new development in portable ambient monitoring. In: *Conference Proceedings, Monitor 93, Spring Innovations Ltd Manchester, October 1993*, pp. 57–62.

Hoxey, R.P. and Robertson, A.P. (1994) Pressure coefficients for low-rise building envelopes derived from full-scale experiments. *Journal of Wind Engineering and Industrial Aerodynamics* 53, 283–297.

Hoxey, R.P., Robertson, A.P., Basara, B. and Younis, B.A. (1993) Geometric parameters that affect windloads on low-rise buildings: full-scale and CFD experiments. *Journal of Wind Engineering and Industrial Aerodynamics* 50, 243–252.

Pain, B.F., van der Weerden, T.J., Chambers, B.J., Phillips, V.R. and Jarvis, S.C. (1997) A new inventory for ammonia emissions from UK agriculture. *Atmospheric Environment* (in press).

RGAR (1990) *Acid Deposition in the United Kingdom 1986–1988*. The United Kingdom Review Group on Acid Rain, UK Department of the Environment, London.

Richards, P.J. and Hoxey, R.P. (1994) Computational and wind tunnel modelling of mean wind loads on the Silsoe Structures Building. *Journal of Wind Engineering and Industrial Aerodynamics* 41–44, 1641–1652.

Richardson, G.M., Hoxey, R.P., Robertson, A.P. and Short, J.L. (1995) The Silsoe Structures Building: the completed experiment Part 1. In: *Proceedings of the 9th International Conference on Wind Engineering*. New Delhi, India, January 1995.

Ammonia Volatilization from Dairy Farms: Experiments and Model

A.H.J. van der Putten and J.J.M.H. Ketelaars

DLO Research Institute for Agrobiology and Soil Fertility (AB-DLO), PO Box 14, 6700 AA Wageningen, The Netherlands

Summary. In this chapter some developments in research on ammonia (NH_3) volatilization, from different components of dairy farming systems, are briefly presented. The effects of measures to reduce NH_3 emissions are illustrated, using the FARM-MIN model for the average Dutch dairy farm on sandy soils and compared to the situation in the mid-1980s. Calculations indicate that a decrease in total NH_3 volatilization of 70% can be achieved by a large reduction in the rate of N fertilization and, hence, in the excretion of N by the animal and a decrease in the percentage of the excreted N that volatilizes as NH_3. Although encouraging, the figures presented are a long way from the ultimate objective set by the Dutch government. Achieving reductions to the level proposed will severely decrease farm income. As the final targets of the national NH_3 policy are derived from values considered acceptable for the deposition of acid and N on forests and nature reserves, taking legal steps to achieve these targets seems only justified after very careful consideration of deposition targets.

Introduction

The long-term environmental policy of the Dutch government aims to reduce total ammonia (NH_3) volatilization from 251 kt in 1980 to 75 kt in 2010 (Anon., 1995). Since animal excreta are responsible for 90% of total NH_3 volatilization in the Netherlands (van der Meer, 1994), accomplishing this target requires drastic alterations to slurry application techniques, livestock housing systems, slurry stores and grazing systems. Their combined effect should, together with changes in farm structure and management, decrease

NH_3 volatilization from excreted nitrogen (N) from an estimated 31% in the mid-1980s (Fig. 12.1) to less than 5% in 2010 (Anon., 1990).

Realizing such reductions requires knowledge of the processes determining NH_3 volatilization. Furthermore, it is essential to evaluate measures at a farm level to identify side effects and to obtain an indication of the cumulative results achieved by the introduction of a selection of the techniques and measures available.

Some developments in research, conducted at AB-DLO, on NH_3 volatilization from components of dairy farming systems are discussed. Subsequently, the effects of increased knowledge and improved techniques are quantified for an average Dutch dairy farm on sandy soil. Results are compared with the environmental objectives of the government regarding NH_3 volatilization.

Ammonia Volatilization from Applied Slurry

In the mid-1980s, slurry application, housing systems plus slurry stores and nitrogen excreted during grazing accounted for 50%, 30% and 20%, respectively, of the total NH_3 volatilization from dairy farms in the Netherlands (Fig. 12.1). Research, therefore, initially focused on techniques to reduce NH_3 volatilization from applications to land. Immediate incorporation of slurry into the soil proved to be an effective method of minimizing volatilization losses (Loonen et al., 1992). Ammonia volatilization losses, expressed as a percentage of total N in slurry, vary from 1% for deep injection to less than 5% for injection with open slits and 5–15% for trailing hose application (van der Meer, 1994), representing a considerable improvement when compared to the 30–35% which is lost after conventional surface application of slurry. Different types of machinery using these principles have since been developed which are now enforced by legislation and used on commercial farms. A point of concern, however, is the increasing popularity of machinery using the principle of trailing hose application. This requires less investment but does not achieve the maximum reduction in NH_3 volatilization.

Ammonia Volatilization from Grazed Grassland

Urine and dung patches are a source of NH_3 volatilization inherent to grazing. Analysis of a large series of measurements of NH_3 volatilization from urine patches on a sandy soil, using a system of wind tunnels, showed soil surface pH (Fig. 12.2) and soil temperature to be the main variables determining differences in volatilization rates. The high pH values observed after urine application were to a large extent caused by hydrolysis of urea.

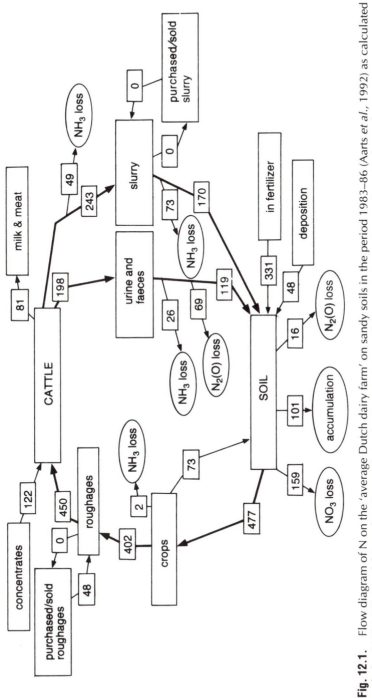

Fig. 12.1. Flow diagram of N on the 'average Dutch dairy farm' on sandy soils in the period 1983–86 (Aarts *et al.*, 1992) as calculated with the FARM-MIN model (van der Meer and van der Putten, 1995). All figures presented are in kg N ha^{-1} year^{-1}.

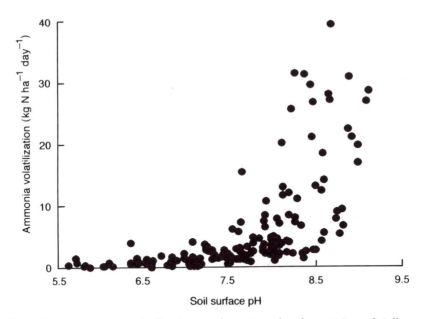

Fig. 12.2. Rate of NH$_3$ volatilization in relation to soil surface pH in artificially created urine patches on sandy soil (Vertregt and Rutgers, 1988).

Dependent on prevailing conditions, the proportion of urinary N that volatilized as NH$_3$ ranged from 3.5 to 34.6% (average 13.0%). Loss percentages were found to be independent of urinary N concentration and N fertilization rate. Therefore, possibilities to reduce NH$_3$ volatilization from urine patches at present appear to be limited to decreasing total N excretion, either by restricting grazing to part of the day or year, or by lowering the daily N intake of the grazing animal.

Ammonia Volatilization from Cattle Housing Systems

The cattle housing system most commonly used in the Netherlands is the cubicle house with slatted floors and a slurry store underneath. Ammonia volatilizes from the slats as well as from the stored slurry. From N budgets, total emission was calculated to be 8.8 kg NH$_3$ per cow per winter season of 190 days (13% of total excreted N). The design of low-emission housing systems – e.g. eligible for the so-called 'Green Label' certificate – aims at a reduction of at least 50% compared to the value mentioned above. To achieve this, the majority of low-emission stables have had the slatted floor replaced by a (semi-)closed, sloping floor equipped with a manure scraper and urine gutter. These floors are thought to decrease volatilization from the floor as well as from the storage underneath. However, measurements have shown a large variation in volatilization rates and the target of 50% reduction is not always achieved.

The floor as a source

Under normal nutritional conditions, fresh faeces and urine contain little NH_3 so volatilization from stable floors must depend on the production of NH_3 from labile nitrogenous excretion products, of which urea is quantitatively the most important. Therefore, research at AB-DLO has focused on the transformation of urea on floors.

Hydrolysis of urea is catalysed by urease. This enzyme is found on the floor, partly in the manure film on the surface, partly encapsulated in a mineral deposit of carbonate and struvite from urinary and faecal origin. Enzyme activity appears to build up over a period of several weeks as the result of fouling of the floor with faeces and urine. Urease activity can easily be removed by dissolving the mineral deposit of carbonate and struvite with dilute acid (Fig. 12.3) and this is accompanied by a decrease in the volatilization rate. Measurements of urease activity have pointed to large differences both within and between floors (Fig. 12.4). Only when urease activity is relatively low is an effect of differences noted: with higher enzyme activities NH_3 volatilization appears to be limited by the supply of urea on the floor. Urea supply is particularly low on sloping floors because only a small quantity of urine is retained on the floor after excretion.

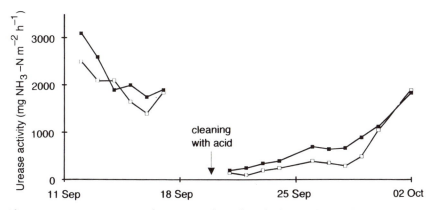

Fig. 12.3. Urease activity of a concrete barn floor before and after cleaning with dilute formic acid. Measurements were made with (solid symbols) and without (open symbols) the presence of a manure film.

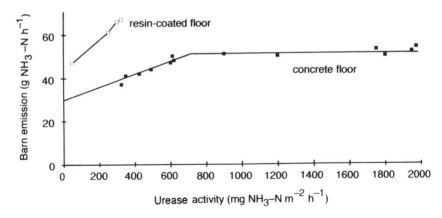

Fig. 12.4. Ammonia volatilization from a barn versus urease activity of the floor and floor type. The high emission at zero activity points to a substantial contribution from the slurry store and lying area in this building.

The slurry store as a source

On reaching the slurry store, urea is hydrolysed within hours. Laboratory experiments demonstrated that after mixing urine and faeces, urease activity builds up fast and is maintained by a continuous supply of new substrate. The fate of NH_3 in a slurry store under the floor is determined to a large extent by the ventilation in the store. Even with an almost completely closed store, emission can be unexpectedly high. This was observed in a cubicle house which had a floor made of 2-m-wide concrete plates separated by narrow slits for urine drainage. Although the total open surface amounted to less than 2% of the passageway area, sealing the slits for short periods was followed by a reduction of emission from the building of between 18 and 43%.

Studies of sources of NH_3 volatilization in cattle housing systems indicate that, apart from floor type, an adequate store construction and suppression of urease activity on the floor can help to reduce NH_3 volatilization. The level to which NH_3 volatilization from naturally ventilated housing systems can be reduced is the subject of current research.

Ammonia Volatilization at a Farm Level

Research has yielded knowledge and techniques which enable a substantial reduction to be made in NH_3 volatilization from dairy farms. Calculations with the FARM-MIN model (van der Meer and van der Putten, 1995) indicate that NH_3 volatilization can be decreased by implementing low-emission techniques for application of slurry and housing, restricting grazing to daytime

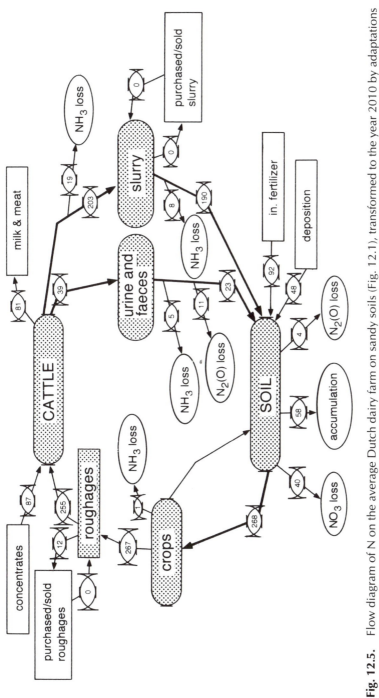

Fig. 12.5. Flow diagram of N on the average Dutch dairy farm on sandy soils (Fig. 12.1), transformed to the year 2010 by adaptations in farm structure and management, aiming to reduce N losses in general and NH_3 volatilization in particular. All figures presented are in kg N ha^{-1} year^{-1}, calculated by the FARM-MIN model.

only and supplementing the animals with silage maize to decrease N intake (and excretion) whilst maintaining current levels of milk production per hectare. Comparison of Figs 12.1 and 12.5 indicates that a decrease in total NH_3 volatilization from 150 to 33 kg N ha^{-1} (78%) can be achieved by a reduction in the amount of excreted N from 490 to 261 kg N ha^{-1} (47%) together with a decrease in the percentage of the excreted N that volatilizes as NH_3 from 31 to 12%.

Preliminary results obtained at the Experimental Farm for Dairy Husbandry and Environment 'De Marke' also suggest that total NH_3 losses from animal manure can be decreased to about 10% of the excreted N whilst maintaining the current level of milk production (Anon., 1995). However, it is obvious that such reductions can only be achieved by extra investments in buildings and machinery, which have a negative effect on farm income.

Although encouraging, the figures presented are a long way off the ultimate objective set by the Dutch government. Attempts to achieve reductions to the level proposed will further reduce farm income. As the final targets of the national NH_3 policy are derived from critical loads of acid and N on forests and nature reserves, taking legal steps to achieve these targets seems only justified after very careful consideration of deposition targets.

References

Aarts, H.F.M., Biewinga, E.E. and van Keulen, H. (1992) Dairy farming systems based on efficient nutrient management. *Netherlands Journal of Agricultural Science* 40, 285–299.

Anon. (1990) *Plan of Action for Reduction of Ammonia Emission from Agriculture.* Regeringsbeslissing, SDU, Den Haag, 97 pp. (in Dutch).

Anon. (1995) Tussenbalans 1992–1994 De Marke. In: *De Marke Rapport nr. 10,* 159 pp. (in Dutch).

Loonen, J., Geurink, J.H., Hoekstra, H., Huijsmans, J.F.M. and Snijders, H. (1992) *Final Report of the Working Group for Tine Injection, Research Results. Rapport No 1.* Werkgroep Mestinjectie Propro Noord Brabant, PR, Lelystad, The Netherlands, 89 pp. (in Dutch).

van der Meer, H.G. (1994) Grassland and society. In: Mannetje, L. 't and Frame, J. (eds) *Grassland and Society: Proceedings of the 15th General Meeting of the European Grassland Federation June 6–9, 1994,* pp. 19–32.

van der Meer, H.G. and van der Putten, A.H.J. (1995) Reduction of nutrient emissions from ruminant livestock farms. In: Pollott, G.E. (ed.) *Grassland into the 21st Century: Challenges and Opportunities: Occasional symposium no. 29 of the British Grassland Society.* Harrogate, December 4–6, 1995, pp. 118–134.

Vertregt, N. and Rutgers, B. (1988) Ammonia volatilization from urine patches in grassland. In: Nielsen, V.C., Voorburg, J.H. and l'Hermite, P. (eds) *Volatile Emissions from Livestock Farming and Sewage Operations.* Elsevier Applied Science, London, pp. 85–91.

Gradients of Atmospheric Ammonia Concentrations and Deposition Downwind of Ammonia Emissions: First Results of the ADEPT Burrington Moor Experiment

13

M.A. Sutton[1], C. Milford[1], U. Dragosits[1], R. Singles[1], D. Fowler[1], C. Ross[2], R. Hill[2], S.C. Jarvis[2], B.F. Pain[2], R. Harrison[3], D. Moss[3], J. Webb[3], S. Espenhahn[4], C. Halliwell[4], D.S. Lee[4], G.P. Wyers[5], J. Hill[6] and H.M. ApSimon[6]

[1]Institute of Terrestrial Ecology (ITE), Edinburgh Research Station, Bush Estate, Penicuik, Midlothian, EH26 0QB, UK; [2]Institute of Grassland and Environmental Research (IGER), North Wyke, Okehampton, North Devon EX20 2SB, UK; [3]ADAS, Boxworth, Cambridge CB3 8NN, UK and Woodthorne, Wergs Road, Wolverhampton WV6 8TQ, UK; [4]AEA Technology, NETCEN, E5 Culham Laboratory, Abingdon OX14 3DB, UK; [5]Netherlands Energy Research Foundation (ECN) Petten, The Netherlands; [6]Imperial College Centre for Environmental Technology (ICCET), 48 Princes Gardens, London SW7 2PE, UK

Summary. A joint field experiment was organized to examine the air concentrations and deposition of ammonia (NH_3) to grassland downwind of NH_3 emissions from slurry spreading. The experiment represents the first intercomparison of NH_3 measurement techniques performed along a substantial gradient in NH_3 concentrations. This provided as reliable estimates as possible of the depletion of NH_3 concentrations with distance. Flux measurements with the grass showed large NH_3 emissions from the sward itself, largely as a consequence of grass cutting, which is a new finding. The result is that in some daytime runs, the grass sward emission was probably only suppressed to permit deposition in the first 70 m from the slurry source, with the remaining slurry-emitted NH_3 travelling further afield.

As well as providing direct measurements of the depletion of NH₃ concentrations with distance, the results are being used to test local scale atmospheric transport models. By providing the models with measured emissions, exchange rates with the grassland and meteorology, it is possible to compare the NH₃ air concentrations predicted by the models with the measured values. This work is ongoing, but a preliminary application of a multilayer model has shown reasonable agreement with the measured NH₃ concentrations. With further development and validation from the measurements, the models will be used to examine local deposition budgets for different scenarios of NH₃ emissions and surrounding land-use.

Introduction

A key uncertainty in assessing the environmental impacts of atmospheric ammonia (NH₃) is the fraction of the emitted NH₃ that is deposited locally, versus that travelling further afield (ApSimon *et al.*, 1994; Sutton *et al.*, 1994). Thus, NH₃ that is deposited back to agricultural land immediately downwind of sources may not represent a net loss from the farming system. Coupled to this is the question of the magnitude of the horizontal gradients in NH₃ concentration and deposition downwind of sources. Information on this is necessary to assess the total local deposition within a particular distance from source, but is also relevant in assessing the local impacts of NH₃ deposition on sensitive ecosystems in source areas (Asman *et al.*, 1989). Thus, although large near-source deposition of NH₃ to agricultural land may be a benefit, deposition to nearby woodlands or semi-natural land may have major ecological impacts.

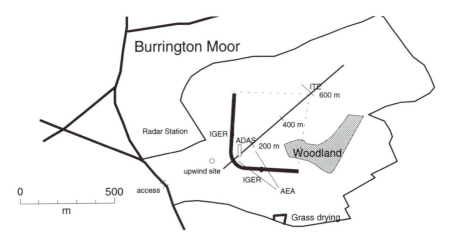

Fig. 13.1.   Simplified map of the field site of the Burrington Moor experiment, showing the slurry strip and principal locations of the different research groups.

To address these issues requires both field measurements and atmospheric models applied to local case studies of NH_3 emissions. As part of an Ammonia Deposition and Effects Project ('ADEPT'), a joint field campaign was organized to provide a major line source of NH_3 emission (minimizing the problems of lateral dispersion) and to measure the downwind gradients in NH_3 concentrations and deposition. The field experiment took place at Burrington Moor, Devon in May 1995 at a site identified by IGER, North Wyke. Initially a single straight line source of slurry was planned, although, because of constraints on wind direction, this was changed to a formation with two arms allowing measurements to be made from a much wider wind sector (Fig. 13.1). The slurry strip area was approximately 600 m long × 15 m wide. The field used was short grassland, managed by cutting (4–5 cuts per year) for grass drying and pelleting, and receiving a fertilizer rate of approximately 300 kg N ha^{-1} year^{-1}.

Approach and Methods

The measurements made (Table 13.1) included NH_3 emissions (micrometeorological mass balance method) (e.g. Pain and Thompson, 1989), concentration measurements upwind and downwind of the source, including a 12-m-high vertical NH_3 profile, and exchange fluxes of NH_3 with the grassland using both the aerodynamic gradient method and dynamic chamber techniques (Sutton et al., 1995).

Apart from providing a direct measure of the concentration depletion with distance from the source, the measurements provide the opportunity to test short-range models of NH_3 dispersion and deposition. By providing the models with the meteorological conditions, emissions and information on surface atmosphere exchange, it is possible to compare the measured with modelled air concentrations. Should the models be shown to be acceptable, it will then be possible to apply these to a much wider range of emission scenarios.

Results

Ammonia emissions and horizontal gradients

Ammonia emissions were determined using passive shuttle samplers at two replicate locations on the slurry strip. Measurements were made with an upwind and a downwind mast at each site, from which emissions were estimated using the mass balance method.

Both gaseous NH_3 and aerosol ammonium (NH_4^+) were measured downwind of the slurry edge. All emission is expected to occur by volatilization

Table 13.1. Summary of measurements made during the ADEPT Burrington Moor experiment.

Measurements	Method	Number of sites	Operating institute
NH₃ emissions	Shuttle samplers – mass balance method (5 heights)*	2	IGER
Horizontal NH₃ concentration gradients	Ferm denuders (paired)[†]	6	ADAS
	Filter packs (paired)[‡]	5	ITE, ICCET
	Batch wet annular denuders[§]	3	ITE, ECN, Univ. of Kiel
	AMANDA continuous annular denuder[‖]	3	AEA, ITE
Vertical NH₃ concentration gradients	Ferm denuders up to 12 m (6 heights)[†]	1	ADAS
	AMANDA denuders up to 2 m (3 heights)[‖]	1	ITE
Horizontal NH₄⁺ concentration gradients	Filter packs[‡]	5	ITE
	Rothero-Mitchell (high volume samplers)	2	AEA
	Post denuder filters	6	ADAS
Acid gases: HNO₃, HCl, HNO₂, SO₂	Batch wet annular denuders[§]	3	ITE, ECN, Univ. of Kiel
	Pulsed fluorescence analyser (Thermo electron, 43S)	1	ITE
NH₃ fluxes with surrounding grassland	AMANDA continuous denuder – aerodynamic gradient method (3 heights)[‖]	1	ITE
	Dynamic cuvette with batch wet denuders	2	ITE, ECN, Univ. of Kiel
	Plant 'biomonitors' IGER	6	IGER
Meteorological information	Wind profiles and direction	2	ITE, ADAS
	Sonic anemometer (Gill instruments)	1	ITE
	Temperature and water vapour fluxes	1	ITE, IGER

See author affiliations for definitions of institute acronyms. *Leuning *et al.* (1985). [†]Ferm (1979). [‡]Harrison and Kitto (1990). [§]Keuken *et al.* (1989). [‖]Wyers *et al.* (1993).

as NH₃, so any enhancement of NH₄⁺ would be attributed to reaction of the emitted NH₃. With the clean conditions at this site (low SO₂, HNO₃ and existing aerosol) slow rates of reaction were expected and no trend in NH₄⁺ concentration was detectable. In contrast, the increase in NH₃ concentrations downwind of the slurry was considerable (Fig. 13.2), as shown by the range of independent measurements. Ammonia is well known to be a difficult gas to measure accurately, and the results show an encouraging agreement in the horizontal gradient away from the source. One caveat on this comparison is that a constant multiplicative correction (factor 0.6) had to be made to the results of the Ferm denuders which, on the basis of comparison with the other methods, were consistently under-reading. This

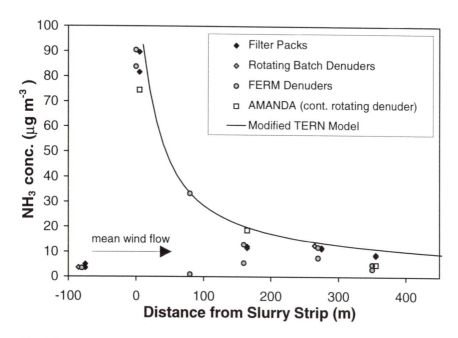

Fig. 13.2. Comparison of NH₃ concentrations measured by a range of independent methods at different distances from the slurry strip for 19 May 1995 (14:00–16:00 h), together with a preliminary model estimate based on the measured emission rates and meteorology.

may have been due to a low collection efficiency in the present implementation of these samplers.

Micrometeorological measurements of ammonia fluxes with the grassland

It was expected that the experiment would show emission of NH₃ (from the slurry) followed by deposition to the downwind grassland. However, micrometeorological measurements of the NH₃ flux using the aerodynamic gradient method, showed that for large periods of time the grassland itself was normally acting as a further source of NH₃ emission.

The field was largely newly cut grass (recently fertilized with ammonium nitrate), but also (near the 600 m site) some less recently cut longer grass. The fluxes over long grass are typical for managed agricultural vegetation (Sutton *et al.*, 1995); however, much larger fluxes were measured over the cut grass (Fig. 13.3). Additional chamber measurements showed that purely mechanical cutting (with removal of the cut vegetation) could explain the increased

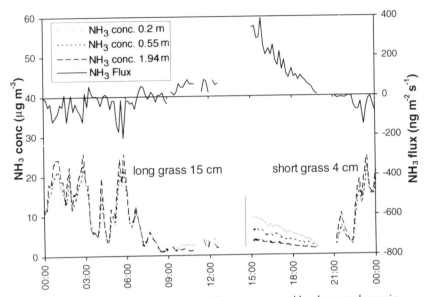

Fig. 13.3. Ammonia concentrations and fluxes measured by the aerodynamic gradient method using the continuous denuder system at 550 m from the slurry strip for 19 May 1995. At 13:00 h the system was moved from long to short grass.

emissions from the short grass. This may be because of a surplus of soil-N coupled to a lack of N 'sink' due to the absence of growing leaves.

Ammonia fluxes with the grassland close to the slurry edge

It is difficult to use the aerodynamic method to quantify fluxes close to a source because of advection errors. Hence a dynamic chamber method was applied to measure fluxes close to the source of emission (Fig. 13.4), using the batch rotating denuder.

Although such chamber techniques are subject to difficulties due to adsorption of NH_3 on chamber walls, these problems are often less at large concentrations, such as here by the slurry edge. Nevertheless, the absolute magnitude of the fluxes should be treated with caution. The chamber measurements confirm the picture of bidirectional fluxes of NH_3 with the grassland, with large daytime emissions from the short grass, but show that this may be suppressed by the high concentrations immediately after slurry spreading, pushing the flux back toward net deposition. These results may be explained on the basis of a compensation point for NH_3 (40 μg m^{-3} in the daytime) that controls the net emission from the short grassland, as well as cuticular adsorption-desorption processes.

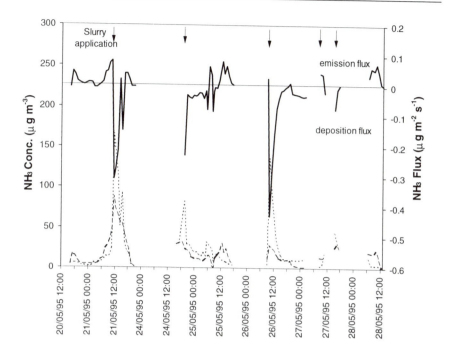

Fig. 13.4. Ammonia concentrations and fluxes determined by a batch wet rotating denuder and the dynamic chamber method at 2 m downwind of the slurry strip.

Discussion

So far, a more detailed analysis of the surface exchange fluxes of NH_3 is needed before this can be included in the modelling, but a first model comparison may already be made. Using a modified version of the TERN model (ApSimon *et al.*, 1994) and taking the meteorology and measured emissions from the slurry as inputs, Fig. 13.2 also shows the predicted concentrations estimated by the model. The slight overestimation of the model particularly at the slurry edge may be due to the need for increased vertical resolution in the model, though the agreement from this preliminary estimate is encouraging and supports the further development of this model. With improved parametrizations of NH_3 surface exchange from the measurements, this will then be used to examine different scenarios of the fraction of the emitted NH_3 that is deposited locally and of the gradients in deposition away from sources.

In addition to the main aim of this experiment in examining local gradients and budgets of NH_3 concentration and deposition, the results also provide several other major scientific contributions. While the range of techniques used here provided a more robust estimate of the NH_3 gradient, this also represents the first NH_3 sampling intercomparison performed along a substantial gradient in NH_3 concentrations. For example, this has shown the filter packs to discriminate NH_4^+ well, even at high NH_3/NH_4^+ ratios.

The result of large NH_3 emissions following grass cutting is also a new result and warrants further investigation. Although emissions from decomposing cut grass have been shown before (Whitehead and Lockyer, 1989), in this case the emissions derived from the remaining sward itself. This result also makes an important point with regard to the deposition budget of NH_3 downwind of sources, which is clearly dependent on land use and management. For the run in Fig. 13.2, where the compensation point was ≈ 40 mg m^{-3} (Figs 13.2, 13.3), deposition would have only occurred for the first 70 m from the slurry edge, after which emission from the sward would have been expected. Had the downwind vegetation been unfertilized semi-natural land or forest, a much more efficient recapture of the emitted NH_3 would have been observed.

Acknowledgements

This work was funded by UK MAFF contract WA0613 (CSA2644) which the authors gratefully acknowledge. Work on ammonia fluxes by ITE was also supported by the UK DOE (EPG 1/3/28). We thank ECN, Petten and the University of Kiel for their contributions to the measurements, even though not funded by ADEPT. Finally, particular thanks are due to Mr B. Fraser-Smith and colleagues of Aylescott Driers for access to the field site and their generous help throughout the experiment.

References

ApSimon, H.M., Barker, B.M. and Kayin, S. (1994) Modelling studies of the atmospheric release and transport of ammonia – applications of the TERN model to an EMEP site in eastern England in anticyclonic episodes. *Atmospheric Environment* 28, 665–678.

Asman, W.A.H., Pinksterboer, E.F., Maas, J.F.M., Erisman, J.-W., Waijers-Ypelaan, A., Slanina, J. and Horst, T.W. (1989) Gradients of ammonia concentration in a nature reserve: model results and measurements. *Atmospheric Environment* 23, 2259–2265.

Ferm, M. (1979) Method for the determination of atmospheric ammonia. *Atmospheric Environment* 13, 1385–1393.

Harrison, R.M. and Kitto, A.-M.N. (1990) Field intercomparison of filter pack and denuder sampling methods for reactive gaseous and particulate pollutants. *Atmospheric Environment* 24A, 2633–2640.

Keuken, M.P., Schoonebeek, C., van Wensveen-Louter, A. and Slanina, J. (1988) Simultaneous sampling of NH_3, HNO_3, HCl, SO_2 and H_2O_2 by a wet annular denuder system. *Atmospheric Environment* 22, 2541–2548.

Leuning, R., Freney, J.R., Denmead, O.T. and Simpson, J.R. (1985) A sampler for measuring atmospheric ammonia flux. *Atmospheric Environment* 19, 1117–1124.

Pain, B.F. and Thompson, R.B. (1989) Ammonia volatilization from livestock slurries applied to land. In: Hansen, J.A. and Henriksen, K. (eds) *Nitrogen in Organic Wastes Applied to Soils*. Academic Press, London, pp. 202–211.

Sutton, M.A., Asman, W.A.H. and Schjørring, J.K. (1994) Dry deposition of reduced nitrogen. *Tellus* 46B, 255–273.

Sutton, M.A., Burkhardt, J.K., Guerin, D. and Fowler, D. (1995) Measurement and modelling of ammonia exchange over arable croplands. In: Heij, G.J. and Erisman, J.W. (eds) *Acid Rain Research: do we have enough answers?* Elsevier, London, pp. 71–80.

Whitehead, D.C. and Lockyer, D.R. (1989) Decomposing grass herbage as a source of ammonia in the atmosphere. *Atmospheric Environment* 23, 1867–1869.

Wyers, G.P., Otjes, R.P. and Slanina, J. (1993) A continuous flow denuder for the measurement of ambient concentrations and surface fluxes of ammonia. *Atmospheric Environment* 27A, 2085–2090.

Posters 6–7

II

Ammonia Concentration and Mass Transfer Coefficient Profiles above an Unfertilized Grass Sward

L. Svensson

Swedish Institute of Agricultural Engineering, PO Box 7033, S-750 07 Uppsala, Sweden

Concentrations and mass transfer coefficients of ammonia (NH_3) were measured over a grass sward at Marsta, near Uppsala in Sweden, in September 1992. The sward consisted of ungrazed rough grass of approximately 0.01 m height, and was unfertilized. Measurements were carried out using passive diffusion samplers carrying filters impregnated with oxalic acid. The samplers were of two types, differing only in the position of the absorption filter, i.e. providing different distances for molecular diffusion (Fig. P6.1). Seven sets of samplers were mounted, with each type in duplicate, on a mast at different heights with spacing according to a logarithmic scale. The samplers were protected against precipitation by means of plastic shields (Fig. P6.2).

Based on the prediction of low NH_3 concentration levels, a sampler exposure time of 20 days was chosen. During the measuring period meteorological conditions were varied so it was not possible to correlate the results with meteorological factors. Analysis of the amount of NH_3 absorbed on the filters was made by the indophenol blue method.

The NH_3 concentration increased from 0.47 µg m^{-3} at the lowest measuring height (0.05 m) to 0.91 µg m^{-3} at the second lowest height (0.55 m), thus indicating a potential net deposition of NH_3. Above this, the concentration

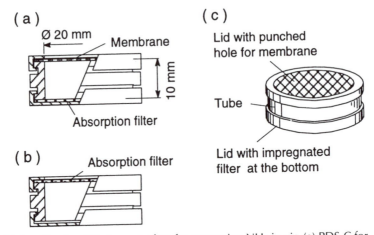

Fig. P6.1. Passive diffusion samplers for measuring NH₃ in air. (a) PDS-C for concentration measurements; (b) PDS-L for measuring laminar boundary layer; (c) schematic representation. (Designed by M. Ferm.)

decreased with increasing height. The decline was initially rapid but slowed down successively, to give a concentration of 0.71 μg m^{-3} at 11.70 m (Fig. P6.3(a)). Thus, concentrations were 10^3–10^4 times lower compared with values normally found for manure-treated fields, and in agreement with data reported by Lemon and van Houtte (1980), Christensen (1988) and Ferm *et al.* (1990).

The mass transfer rate increased with height, following an exponential function, from 9.5 mm s^{-1} at 0.05 m, reaching 13.9 mm s^{-1} at 11.70 m (Fig. P6.3(b)). This is in agreement with numerous earlier ground level measurements and reported by Bouwmeester and Vlek (1981), Ibusuki and Aneja (1984), Svensson and Ferm (1993) and Svensson (1994). The exponential increase with height is also in agreement with what might be expected from generally occurring wind speed profiles.

From the concentration gradient and the transfer rate at the lowest height above the sward, a net deposition of NH₃ corresponding to 13 kg ha^{-1} year^{-1} could be calculated for the 20-day measuring period. This study demonstrates that low concentrations of NH₃ in air and the corresponding transfer coefficients can be estimated by use of a simple kind of passive sampler exposed, in this instance, for 20 days and that NH₃ flux rate and direction (emission or deposition) can be assessed from measured concentrations and transfer coefficients.

References

Bouwmeester, R.J.B. and Vlek, P.L.G. (1981) Rate control of ammonia volatilization from rice paddies. *Atmospheric Environment* 15, 131–140.

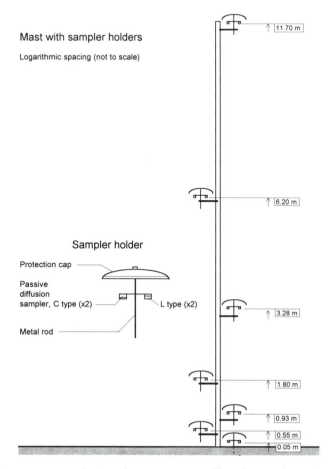

Fig. P6.2. Experimental set-up for measuring profiles of atmospheric NH₃ and meteorological variables at Marsta near Uppsala, September 1992.

Christensen, B.T. (1988) Ammonia loss from surface-applied animal slurry under sustained drying conditions in autumn. In: Nielsen, V.C, Voorburg, J.H. and L'Hermite, P. (eds) *Volatile Emissions from Livestock, Farming and Sewage Operations.* Elsevier Applied Science, London, pp. 92–102.

Ferm, M., Schjørring, J.K., Sommer, S.G. and Nielsen, S.B. (1990) Field investigation of methods to measure ammonia volatilization. In: Nielsen, V.C., Voorburg, J.H. and L'Hermite, P. (eds) *Odour and Ammonia Emissions from Livestock Farming.* Elsevier Applied Science, London, pp. 148–154.

Ibusuki, T. and Aneja, V.P. (1984) Mass transfer of NH₃ into water at environmental concentrations. *Chemical Engineering Science* 39, 1143–1155.

Lemon, E. and van Houtte, R. (1980) Ammonia exchange at the land surface. *Agronomy Journal* 72, 876–883.

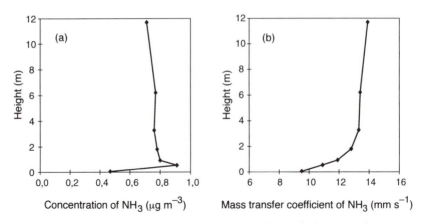

Fig. P6.3. Concentration (a) and mass transfer coefficient (b) for atmospheric NH₃ over a grass sward. Mean values from passive diffusion sampling during 20 days at Marsta near Uppsala, September 1992.

Svensson, L. (1994) A new dynamic chamber technique for measuring ammonia emissions from land-spread manure and fertilizers. *Acta Agriculturae Scandinavica, Section B, Soil and Plant Science* 44(1), 35–46.

Svensson, L. and Ferm, M.(1993) Mass transfer coefficient and equilibrium concentration as key factors in a new approach to estimate ammonia emission from livestock manure. *Journal of Agricultural Engineering Research* 56(1), 1–11.

The Reduction of Ammonia Emissions by Covering Slurry Stores
A.G. Williams
Silsoe Research Institute, Wrest Park, Silsoe, Bedford MK45 4HS, UK

Emissions of ammonia (NH_3) from stored cattle slurry in the UK are estimated to be 37.1 NH_3-N ± 11 kt year^{-1} (ECETOC, 1994). Covering slurry stores reduces the rate of air movement across the slurry surface, hence increasing the head-space partial pressure of NH_3-N and so reducing the potential for mass transfer. The typical daily loss of NH_3-N represents only about 0.1% of total N so that trying to measure these losses by direct chemical analysis of slurry will be prone to considerable error. The effectiveness of a proprietary cover was tested using aqueous solutions of NH_3. This approach quantifies the physical effect of the cover, but does not address any effects that covering has on biological changes in the slurry, such as NH_3 generation from organic N.

Two conventional, epoxy-coated, steel tanks (6.1 m diameter and 2.4 m tall) were used. One had a corrugated, plastic-coated, steel cover, designed originally to prevent algal growth in water tanks, but also sold to help farmers reduce odour problems. The tanks were spaced 30 m apart and aligned at right angles to the prevailing wind to minimize interference between the emission plumes from each tank. The tanks were situated in a field of mown grass with a barley field about 35 m upwind. Each tank had a tractor-driven side stirrer, pH sensor and depth probe.

The tanks were filled to their 50 m^3 working volume with tap water. Ammonia solution was added to bring the total ammoniacal-N (TAN) concentration to approximately 10 g m^{-3} and NaOH was added to adjust the pH to 10. Using a dissociation constant of 1.71×10^{-5}, the NH$_3$-N concentration would then be 8.8 g m^{-3}, equal to that in a slurry of pH 7.2 containing 1000 g m^{-3} TAN. The tanks were stirred and sampled daily over two consecutive periods of 17 and 30 days, with the pH being adjusted to 10 periodically and the NH$_3$ replenished once. Liquid depths, wind speeds, air and slurry temperatures were sampled every 10 s and mean values logged every 10 min. Emissions were quantified from the rate of loss of mass of TAN (dM_a/dt) and a whole system mass transfer coefficient (k_L) using the equation dM_a/dt = $k_L M_a$. The resulting exponential curves were fitted to obtain values for k_L.

The cover reduced NH$_3$ emissions by 82% and 87% during the two periods (Fig. P7.1). Emission rates from the covered and uncovered tanks were in the ranges 0.10–0.36, and 0.21–3.3 g m^{-3} day^{-1}, respectively, during the two measurement periods. Mean wind speeds and ambient temperatures during the periods were 3.0 and 2.9 m s^{-1}, and 13.0 and 15.5°C, respectively. The covered and uncovered liquid temperatures were 16.9 and 16.6°C, and 22.1 and 23°C during the two periods, respectively.

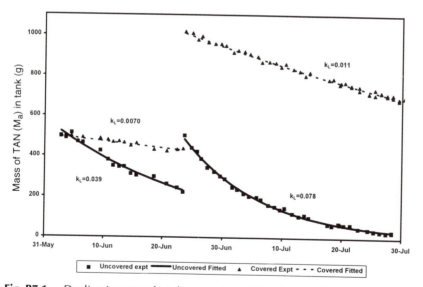

Fig. P7.1.   Decline in mass of total ammoniacal-N in slurry tanks: measurements and fitted curves.

Using covers with an 82% effectiveness could reduce the annual UK NH_3-N emission from slurry stores by 6.7 kt to 30.4 kt. This is potentially valuable as it increases the fertilizer value of slurry, although these nutrients may not be used effectively if the additional retained N is subsequently volatilized or leached before assimilation by grassland or other crops. It was also noted that covering increasingly raised liquid temperatures during storage as weather conditions promoted evaporation. This could promote anaerobic microbial activity and hence increase both NH_3 generation and methanogenesis, which could thus exacerbate harmful methane emissions.

Acknowledgements

This work is funded by the Environmental Protection Division of MAFF, London.

Reference

ECETOC (1994) *Ammonia Emissions to Air in Western Europe. Technical Report Number 62.* European Centre for Ecotoxicology and Toxicology of Chemicals, Brussels.

Emissions of Nitrous Oxide from Grasslands

D. Fowler, U. Skiba and K.J. Hargreaves

Institute of Terrestrial Ecology, Bush Estate, Penicuik, Midlothian EH 26 0QB, UK

Summary. Emissions of nitrous oxide (N_2O) from soil generally represent a very small fraction of the annual input of nitrogen (N) cycling within unmanaged or farmed soils. However, these emissions play a major role in the chemistry of the atmosphere on regional and global scales. Interest in N_2O centres on the direct contribution to radiative forcing of climate through the absorption of long wavelength terrestrial radiation and indirectly through the role of N_2O in the stratospheric chemistry of O_3. Emissions of N_2O from soil increase with temperature and soil N concentrations and show very large spatial variability in rates. This chapter reviews current knowledge of the processes which regulate emissions of N_2O from grassland soils and reports methods used to up-scale fluxes from small chambers to the field scale and extrapolation to provide regional and country emissions of this trace gas.

Introduction

Nitrous oxide (N_2O) is an important contributor to the global greenhouse gas budget because it has a long atmospheric lifetime (~120 years) and, on a per molecule basis, it has a radiative forcing 200 times that of carbon dioxide (CO_2) (IPCC, 1996). Atmospheric concentrations of N_2O increased at about 0.5 ppb year^{-1} in 1993, somewhat smaller than the growth rate during the late 1980s of 0.8 ppb year^{-1}. However, little of the atmospheric budget for N_2O is known with certainty. The atmospheric burden of N_2O (1500 Tg N_2O-N) and the rate of increase are known. The contribution of anthropogenic activity to current atmospheric concentrations can be judged by comparing present day concentrations (310 ppb) with those of 200 years ago from the ice core record (270 ppb), which indicates an increase of about 20%, most of which has occurred during the last 50 years (Khalil and Rasmussen, 1992). The distribution of N_2O in the stratosphere has been estimated to remove between 10.5 Tg N_2O-N (McElroy and Wofsy, 1986) and 12.2 Tg N_2O-N year^{-1} (Minschwaner *et al.*, 1993), so that the current increasing atmospheric burden is due to a 3–4.5 Tg N_2O-N year^{-1} excess of sources over sinks. The steady atmospheric concentrations prior to industrialization may be used to estimate the total 'natural' emissions of N_2O at 8 Tg N_2O-N year^{-1}, which is only between 50 and 60% of the total current source strength of 14–16 Tg N_2O-N year^{-1}.

The various published tables of N_2O sources (Bouwman, 1994; IPCC, 1995) must be regarded as very speculative as the underlying field measurements are quite limited and N_2O emissions from soils, the major global source, have been shown to exhibit huge spatial and temporal variability. However, a common feature of the published global inventories for N_2O is the dominant role of soil emissions, both natural and those resulting from N fertilizer applications. These studies, described in some detail by Bouwman (1994), identify cultivated soils as the major anthropogenic source, with annual emissions in the range 0.3–5.9 Tg N_2O-N. More recent field studies have shown that, especially in northern Europe, grasslands are particularly important sources, with generally a larger N_2O-N release per unit of fertilizer applied than is the case for arable cropland (Velthof *et al.*, 1996). N_2O production and emission is promoted not only by the fertilizer and manure-N applied to grassland soils, but also by the physical properties of a typical grassland soil. Agricultural areas on heavy soils or poorly drained soils and in the higher rainfall areas of north and west Britain are the preferred areas for grassland and stock rearing. By contrast, the land use over the majority of free-draining soils of the drier harvest areas of south-east and eastern Britain is dominated by arable farming, especially cereal production. In Britain, 27% (65,672 km^2) of the total land area is covered by managed grassland and a further 21.8% (52,378 km^2) is used for rough grazing and includes moor grassland (Barr *et al.*, 1993). Of the managed grassland, about 70% is grazed (44,850 km^2) (MAFF, 1993). The fertilizer inputs to all grassland, including both grazed and ungrazed grassland, were found to average 116 kg N ha^{-1}, and range from 60 to 190 kg N ha^{-1}, in 1988 (Chalmers *et al.*, 1990).

The emissions of N_2O from grassland represent an important contribution to the UK N_2O inventory (Skiba *et al.*, 1996). To date the methodology to provide the national emissions inventory estimate of N_2O sources in the UK has been that provided by IPCC and assumes that the percentage of fertilizer-N applied which is subsequently released to the atmosphere as N_2O is 1.25% (Salway, 1995). In this chapter, recent measurements of N_2O emissions from grassland in the UK and at other northern European sites are reviewed. The influence of soil-N, temperature and soil moisture on emissions are briefly considered. The problems of spatial variability of N_2O emission and use of micrometeorological methods to provide field-scale fluxes are described. The chapter concludes with a short section quantifying the magnitudes of UK N_2O emissions from grassland.

Measurements of N_2O

Most flux measurements of N_2O have been made using the static chamber technique, where a small area, usually less than 1 m^2, is enclosed and the increase in N_2O concentration after a time of usually 1 h or less is measured (Smith *et al.*, 1995). This is also the method used in most of the work described in this volume. Gas samples from inside the chamber are collected and stored in syringes, vaccutainers or Tedlar bags until analysis for N_2O. Nitrous oxide concentrations are usually measured by electron capture gas chromatography. The recent development of a photo-acoustic infrared detector (PDA) may offer a good alternative, and has been used successfully by Velthof *et al.* (see Chapter 16) to measure N_2O fluxes from grassland in the UK and the Netherlands. The advantage of the PDA over the GC (gas chromatograph) is its portability, allowing instant analysis of the N_2O flux.

Recent developments in tunable laser spectroscopy (TDL) (Wienhold *et al.*, 1994) and Fourier transform infrared spectroscopy (FTIR) (Galle *et al.*, 1994) and also a modified gas chromatography technique, where repetitive analysis of the same sample improved the resolution to around 1 ppb N_2O (GC_m) (Arah *et al.*, 1994), have made it possible to measure N_2O concentration differences an order of magnitude lower ($\leq$ 1 ppb) than by the conventional GC method ($\geq$ 10 ppb). This permits flux measurements to be made by micrometeorological techniques (Hargreaves *et al.*, 1996) and large and very large chamber (mega chamber) methods (Smith *et al.*, 1994).

Both techniques, the static chambers and micrometeorological/large chamber methods have their advantages and disadvantages. The main advantage of the small static chambers are, they are cheap, versatile and are very useful in process-based studies, for example understanding the affect of soil pH (see Chapter 17) or the contribution of urine patches (see Chapter 18) on the emission of N_2O. However, most chamber methods only measure a flux over a small area and extrapolation to the field scale can only be provided with limited confidence. This is particularly the case for grazed grasslands, which show a high spatial variability. Small chambers have been used to address this spatial variability. The use of many chambers in conjunction with geostatistical analysis can improve the estimate of the field-scale flux; however, these are extremely labour-intensive methods. Micrometeorological techniques are well suited to measure field-scale fluxes. Depending on the technique of choice and the height at which the sample is taken, micrometeorological techniques integrate a flux over 0.1–1 km^2 (Fowler and

Duyzer, 1989). The disadvantages are that the necessary instrumentation is expensive and large uniform areas of land are required. Finding representative fields large and uniform enough for micrometeorological measurements of N_2O can be problematical in some regions and countries. The use of megachambers (> 50 m^2) offers a useful alternative to measure 'large' scale fluxes than the traditional small chambers (< 1 m^2) (Smith *et al.*, 1995). Examples of all three methods to measure N_2O fluxes from grasslands are discussed below. Recent field studies of N_2O emissions from grasslands have addressed the problems of scaling, spatial variability, and identifying the environmental variables, in particular the fractional losses of N_2O from N fertilizer, that determine the magnitude of the flux.

Factors Controlling N_2O Emissions from Grassland Soils

The principal variables that control the emission of N_2O from soil are soil-N, soil temperature and soil moisture.

Soil nitrogen

Applications of N fertilizer to agricultural land make these the largest global sources of N_2O. Bouwman (1994) estimated that overall 1.25% of the mineral N fertilizer input is emitted as N_2O. Studies from some grassland sites compiled in Table 14.1 suggest that the fractional N_2O loss from the fertilizer and excreta input is substantially larger than the 1.25% figure used for the global emission inventory (Bouwman *et al.,* 1995), ranging from 0.3 to 3.7% (median 1.8%) for mowed grassland and from 1.2 to 6.7% (median 2.1%) for grazed grassland. A direct comparison of N_2O emissions from grazed and mowed grassland on three different soil types (Velthof *et al.*, 1996), and areas of a grassland that were grazed the previous year with non-grazed areas of the same grassland (Clayton *et al.*, 1994), suggests that grazing increases the emission of N_2O.

There are several possible reasons for the larger N_2O emissions from grazed than ungrazed managed grassland:

1. Deposition of animal excreta provides localized high inputs of N and carbon (C) in addition to the broad spreading of mineral fertilizers; dung and urine deposition is largely responsible for the inherent patchiness of N and C content of grassland soil and, consequently, N_2O emissions.
2. Soil compaction caused by the animals.
3. Agricultural land used for grazing frequently has wetter soils, is less suitable for crop production and, therefore, dominates the agricultural land use in the

Table 14.1. Emissions of N$_2$O from some managed grasslands in Northern Europe.

Soil type and location	N fertilizer type	Total N input (mineral and/or excreta) (kg N ha^{-1} year^{-1})	N$_2$O emission as % of N input	N$_2$O emission (kg N ha^{-1} year^{-1})	Reference and dates for measuring periods
Mowed grassland					
Peat*	None	0		2, 8	Velthof et al., 1996 (March 1992–March 1994)
	CAN	365, 242	1.9, 3.9	8.8, 18.1	
Clay or sand, Netherlands†	None	0		0.8, 1.1	
	CAN	357, 370	0.9, 1.0	3.8, 4.9	
Clay, Scotland	AN	185	1.7	(153 g N$_2$O-N ha^{-1} day^{-1})	Clayton et al., 1994 (April 1992, 3 weeks)
Sandy loam, Denmark	None	0		1.8–5.1	Christensen, 1983 (April–Aug 1981)
	AN	200‡	8.2§	16.4	
	Slurry	492‡	37	182	
Clay loam, Scotland	AS	360ǁ	0.3, 0.6§	1.1, 2.3¶	McTaggart et al., 1994 (April–Oct 1992 and 1993)
	AN	360ǁ	0.9, 2.5	3.4, 8.9	
	Urea	360ǁ	1.8, 3.2	6.4, 11.6	
Loam, England	AN	250**	1.3		Ryden, 1981 (March–Dec. 1980)
Grazed grassland					
Cattle-grazed peat*	CAN	617, 450	2.1, 6.7	14.6, 38.5	Velthof et al., 1996 (March 1992–March 1994)
Clay or sand, Netherlands†	CAN	644, 735	2.0, 1.2	13.4, 10.3	
Cattle-grazed clay, Scotland††	AN	185	5.1	(557 g N$_2$O-N ha^{-1} day^{-1})	Clayton et al., 1994 (April 1992, 3 weeks)
Sheep-grazed sandy loam, Scotland	AN	160	1.7	2.8	Skiba et al., unpublished (Sept 1995–Aug. 1996)

*The two values in the columns refer to measurements from two different peats. The groundwater table for peat 2 was always lower than for peat 1. †The first value refers to measurements on clay, the second on sandy soil. ‡Total of two split applications. §Calculated from annual totals by authors. ǁThe first value is the annual average for 1992, the second for 1993. **Total of three split applications. ¶Total of four split applications. ††Grazed in previous year. AN = ammonium nitrate; AS = ammonium sulphate; CAN = calcium ammonium nitrate.

high rainfall regions of a country, i.e. the west coast of England, Wales and Scotland. These points are considered further in the following section.

N_2O emission from animal excreta

Nitrous oxide emissions from urine patches tend to be larger than from mineral fertilizer N. Urination results in small areas of the field in which concentrated urine-N is present (30–60 mg urine-N m^{-2}; de Klein and van Logtestijn, 1994). The urea is rapidly hydrolysed to NH_4^+, which can be nitrified to NO_3^- and subsequently denitrified, with N_2O produced by both processes. Nitrous oxide emissions from urine patches peak around 10–15 days after deposition (see Chapter 19; Flessa et al., 1996). The fractional N_2O loss reported in the literature ranges from 2.6% (see Chapter 18) to 16% (de Klein and van Logtestijn, 1994). The magnitude of the emission is partly influenced by the wetness of the soil (see Chapter 18).

Dung patches do not produce the same large increase in N_2O emission as the urine patches, but may increase the N_2O emission indirectly by increasing the C pool and organic N pool of the soil. In the study reported by Flessa et al. (1996), in the 10 days after excretion the N_2O loss from dung was 0.5% of the N input compared to a 3.8% loss from urine patches on the same field. Yamulki and Jarvis (see Chapter 19) observed a peak in N_2O emission from urine patches 15 days after excretion, but from dung patches 4 months after excretion. Flessa et al. (1996) extrapolated from the information given above to the global scale and suggested that N_2O emissions from excreta of pastured animals may be responsible for 12% of the total annual global N_2O emission of 14.5 Tg. For the UK, also, 12% of the total annual N_2O emission has been estimated to originate from livestock wastes (INDITE, 1994).

Organic N input and compaction

Dung deposition and slurry applications provide a C source to the soil, which stimulates denitrification. Microbial degradation of organic matter supplies a readily available energy source for the denitrifiers and reduces the oxygen (O_2) potential of the soil, which may create the necessary anaerobic conditions required for denitrification. The inverse relationship between soil O_2 and N_2O emission is further discussed by Rappoldt and Corre (see Chapter 15). The supply of C in slurry spread onto an ungrazed grassland stimulated N_2O production, and emissions were four times larger when N was applied as NH_4NO_3 followed by slurry rather than NH_4NO_3 only. The total rate of N applied was the same in both treatments (see Chapter 20). Work by Ellis et al. (see Chapter 17) showed that the availability of C to the microbial population is pH dependent, and therefore directly affects N_2O production. For every decrease by 0.1 pH units, the CO_2 and N_2O production decreased by 0.04 and 0.08 mg C and N kg^{-1} h^{-1}, respectively. The stimulating effect of slurry and animal excreta on the N_2O emission is therefore more of a problem on well-managed, limed grasslands, than on unmanaged grasslands, especially on acid upland soils.

The effect of grazing on N_2O emissions has been shown to be long lasting, possibly through a combination of soil, physical and chemical factors. On a grassland, where part of the area was grazed in the previous summer, differences in N_2O emissions were still observed in the following spring: average emissions measured over a 3-week period in April were five times larger from the previously grazed than ungrazed area (Table 14.1) (Clayton et al., 1994). Not only the deposition of excreta, but also the compaction of soil by the cattle contributed to this difference. Slurry spreading and livestock compact the soil, thereby restricting gaseous diffusion within the soil and this influences the rates of N_2O emission. Compaction doubled the emission of N_2O from an imperfectly drained clay loam (see Chapter 20) and increased emissions by 36% on a sandy loam soil (Hansen et al., 1993).

The effect of seasonal changes on N_2O emissions from grassland soils: soil temperature and soil moisture

Typical seasonal patterns of N_2O emissions from agricultural soils show low emissions during the winter months which then increase to maximum emissions in spring, when the soil temperature rises and N fertilizers are applied, with a second peak often occurring in autumn, after ploughing, residue incorporation, etc. During summer the dry soil profile and limited availability of soil-N lead to small emissions of N_2O. An example of the seasonal pattern is shown for a sheep-grazed grassland in SE Scotland, with a stocking density of 12.5 ewes ha^{-1} (Fig. 14.1), and can also be found elsewhere in this volume and the literature (e.g. Ryden, 1981; McTaggart et al., 1994; Velthof et al., 1994, 1996) and accompanying papers. For the sheep-grazed grassland (Fig. 14.1), the first spring mineral N fertilizer application of 31 kg NH_4NO_3-N ha^{-1} coincided with the time the sheep were put back onto the field. It is therefore not easy to determine the individual contribution of sheep-induced and fertilizer-induced emissions of N_2O. Total annual losses of N_2O were estimated at 1.7% of the mineral-N and excreta-N deposited (Table 14.1).

Seasonal patterns in N_2O emission for managed grassland soils in the Netherlands, Scotland and Belgium were very similar. The magnitude of the largest flux was also similar for all sites (around 300–400 g N ha^{-1} day^{-1}). Large differences between the years 1992 and 1993 were observed, indicating the large effect of weather conditions on the N_2O flux. For the mown grassland on a clay loam in Scotland, for example, the annual total N_2O emissions in 1992 and 1993 were 1.6 and 4.1 kg N_2O-N ha^{-1}, respectively, in spite of the same rate of mineral N fertilizer application of 360 kg N ha^{-1} year^{-1}

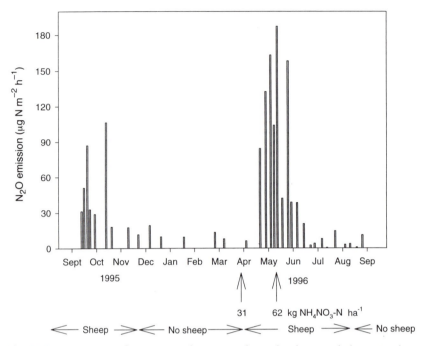

Fig. 14.1. Nitrous oxide emissions from grazed grassland on sandy loam, Bush Estate, SE Scotland. Mean fluxes from six static chambers, always placed onto the same plot of grass prior to flux measurement. Sheep were periodically removed from the site.

(McTaggart *et al.*, 1994; Velthof *et al.*, 1994). Annual differences in N_2O emissions were also observed for a pine forest in Central Scotland. Total annual fluxes were 0.5 kg N_2O-N ha^{-1} year^{-1} in the wetter year 1993 and 0.3 kg N_2O-N ha^{-1} year^{-1} in the drier year 1994 (Skiba *et al.*, 1996). These annual differences show that long-term flux measurements for more than one annual cycle are necessary to obtain a representative flux estimate.

Recent studies have shown that for calculation of annual fluxes the frequency of flux measurements should not be reduced during winter when temperatures are low. For an intensively cultivated sandy loam in Germany, N_2O fluxes for the period December to March contributed to over 50% of the total flux for the period December to July (Kaiser *et al.*, 1996). For a short grass steppe in Colorado, USA, N_2O fluxes for the 4-month period December to February contributed around 30% of the annual total (Mosier *et al.*, 1996). Autumn incorporation of plant residues and daily freeze-thawing cycles are the cause of high winter-time fluxes. The seasonal changes observed depend on changes in soil temperature, soil wetness and, for agricultural soils, the periodic application of N as mineral fertilizer, manure or excreta.

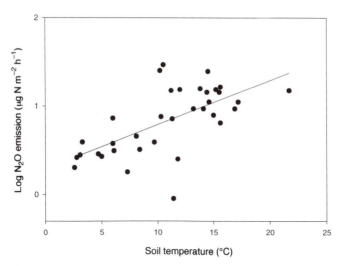

Fig. 14.2. The effect of changes in soil temperature on the logarithm of the emission of N_2O from mowed, unfertilized grassland on a brown forest soil, Glencorse, SE Scotland. The solid line shows the linear regression of this relationship.

Soil temperature

Soil temperature has been shown in many studies to influence N_2O emission fluxes (Conrad *et al.*, 1983). However, in field conditions soil temperature and moisture are often correlated over seasonal timescales. A recent study of N_2O emission fluxes at a mown, ungrazed grassland site in south-east Scotland (Fig. 14.2) shows a marked increase in N_2O flux over the temperature range 5–20°C. The frequency distribution of N_2O fluxes is approximately log-normal and thus the variation of N_2O flux with temperature was plotted as the logarithm of N_2O emission.

Soil moisture

The dependence of N_2O emission from a fertilized ryegrass on clay loam on soil moisture is demonstrated in Fig. 14.3 and discussed further in Skiba *et al.* (1992). For this clay loam, the moisture content needed to exceed 27% (of the dry weight of the soil) before a response in N_2O emission occurred. For the sheep-grazed grassland (Fig. 14.3, Table 14.1) on a sandy loam, a linear increase in N_2O emissions occurred between soil moisture contents of 22 and 40% dry weight, but further increases hardly changed the emission of N_2O. These relationships are, however, not observed in every study. In particular the effect of soil moisture is confounded in some studies by variability in soil temperature. Usually high soil temperatures occur during dry conditions (Velthof and Oenema, 1994), or the application of N fertilizer may overshadow any changes in soil temperature or soil wetness.

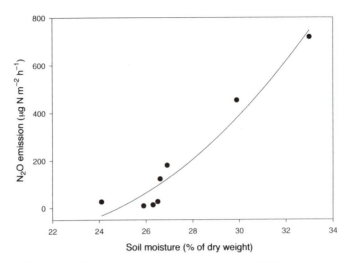

Fig. 14.3. The effect of soil moisture on the emission of N_2O from a ryegrass sward cut for silage on a clay loam soil, Glencorse, SE Scotland.

The spatial variability of N_2O emissions

Spatial variability is an inherent problem with N_2O fluxes, caused by the dependence of N_2O production on areas within the soil of reduced O_2 potential. Microsites of low O_2 potential in soil can develop and disappear rapidly, so that in a field the 'hotspots' of N_2O production may change. This is particularly the case for grazed grasslands caused by the uneven compaction by livestock and deposition of excreted N. For the sheep field, for example, six chambers (0.22 m²) were only placed onto the ground for the duration of the flux measurements. The variation in N_2O emission between the six chambers was large, standard deviations ranged from 6 to 222% of the mean flux. The N_2O emission for the six plots also varied between sequential sampling events. On day 262 (19 September 1995), chamber 2 emitted N_2O at the highest rate, whereas on day 285 (12 October 1995) chamber 6 was the highest emitter (Fig. 14.4). For the ungrazed grassland on clay in Central Scotland (Clayton *et al.*, 1994), however, 'hotspots' persisted over the 3-week measuring period. The rank order of 24 chambers was maintained when fluxes were measured 15 times over a 3-week period. The factors that control the spatial variability of managed grasslands is further outlined by Velthof *et al.* (see Chapter 16).

Micrometeorological and other boundary layer methods

The very large numbers of chambers required to characterize the small-scale (< 1 m) and field-scale variability in N_2O emissions impose an important limitation on the degree to which chamber N_2O fluxes can be safely

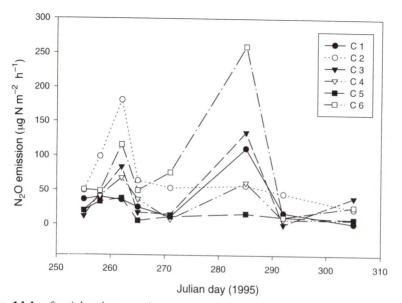

Fig. 14.4. Spatial and temporal variation in N_2O emission between the six chambers (C1–C6) at the sheep-grazed grassland described in Fig. 14.1.

extrapolated. An alternative approach, as briefly mentioned above, is to use methods which integrate fluxes at the field scale using micrometeorological techniques. In principle, the methodology developed originally to study momentum, heat and water fluxes, is well suited to the typical scale of grassland fields (10–20 ha). However, it is important to appreciate that the averaging of the surface-atmosphere flux using micrometeorological methods over the field is not a simple linear process. The measured flux from a single tower is contributed by a 'footprint' of the field, which varies in size according to the height at which the flux is measured and the speed and stability of the airflow over the field. The aerodynamic roughness (z_0) of the vegetation also influences the flux footprint. Typically, for measurement heights of 3–4 m above short grassland in neutral stability, 66% of the flux is contributed by the nearest 200 m to the measurement tower, a further 20% of the flux is contributed by the next 200 m upwind. In unstable conditions, the footprint moves closer to the measurement tower while in stable conditions (generally night or winter conditions with the surface cooler than the air) the footprint extends much further (up to 1 km) into the upwind fetch (Fig. 14.5).

It is difficult, therefore, to design a field study to compare field-scale flux measurements with chamber methods, unless a very large number of chambers are used to characterize all combinations of wind speed, direction and stability. Two recent studies in which field-scale N_2O flux measurements were made using micrometeorological methods (eddy covariance and flux

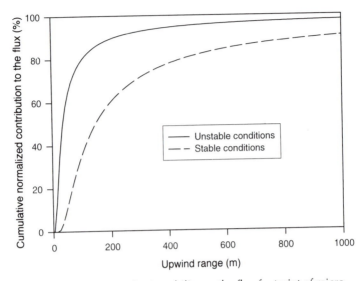

Fig. 14.5. The effect of atmospheric stability on the flux footprint of micro-meteorological measurements.

gradient) have been published (Smith *et al.*, 1994; Hargreaves *et al.*, 1996). In the first of the two comparison experiments over fertilized grassland in Central Scotland, two different micrometeorological methods (eddy co-variance and flux gradient) and three different N_2O detection systems were used, TDL, FTIR and GC. These micrometeorological methods all provided similar fluxes (Table 14.2) within the uncertainties of the methodologies. This, for N_2O, represented an important advance over chamber methods since for the first time the small-scale heterogeneity in N_2O fluxes was overcome by the new methods. However, measurements using chambers in the same field in the Stirling experiment revealed systematically larger fluxes using small chambers (0.13 m^2 and 0.49 m^2). Subsequent analysis of flux footprints for the micrometeorological measurements show that only a small fraction of the footprint for the micrometeorological fluxes was sampled by the chamber measurements, as a consequence of the wind speeds and directions during the successful measurement periods. It transpired therefore that heterogeneity in N_2O emissions across the field prevented a satisfactory comparison of methods, even though for each of the methods (chambers and micrometeorological) the results were internally consistent.

The second of the two comparisons took place over recently harvested cereal stubble in Denmark. Micrometeorological methods used included eddy covariance and flux-gradient, again using TDL and FTIR N_2O detectors (Hargreaves *et al.*, 1996). The different micrometeorological methods yielded different N_2O fluxes (Fig. 14.6). However, the differences in this case were entirely consistent with differences in the footprints of the two measurement systems and known differences in the N₂O emission rate of different sections of the field as measured using an extensive array of chambers throughout the flux footprints. The conclusions drawn from the comparison between chamber and micrometeorological methods was that there was no evidence of bias in either technique (Christensen *et al.*, 1996). However, very large numbers of

Table 14.2. Comparison of N₂O flux measurements from ungrazed grassland on clay, Central Scotland, using chamber and micrometeorological techniques (Hargreaves *et al.*, 1994; Smith *et al.*, 1994).

Method	Dates of measurement (April 1992)	N₂O flux (μg N m⁻² h⁻¹)*
Chambers		
0.13 m² and 0.49 m²	10, 11	29–337 (177), 24–488 (205)
62 m²	10, 11	222–321 (247), 266–384 (295)
Flux gradient†		
GCₘ	11	65
FTIR	10	34–55
TDL	10, 11	20–70, 51–59
Eddy covariance		
TDL	9, 13	75–113, 38–48

*Results given are ranges; figures in parentheses are means.
†Abbreviations are explained in the text.

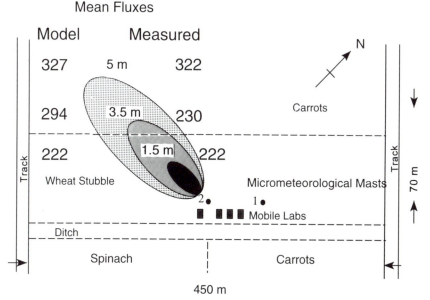

Fig. 14.6. The effect of variation in sensor height on measured N₂O fluxes.

chambers are required to provide field-scale fluxes of N_2O, and the alternative atmospheric methods require substantial quantities of equipment and uniform, extensive field sites for the measurements. The introduction of large (mega) chambers (~60 m^2) provides a useful alternative and may be particularly useful for process studies in which conditions within the chamber may be readily modified. The recent studies of N_2O flux measurements at the field scale are now becoming more widely applied as a consequence of developments with TDL and FTIR instruments. The logical development of these methods will be to monitor seasonal fluxes at sites with well-characterized atmospheric and soil conditions, to provide a check on the extrapolation of fluxes over time.

Extrapolation to UK grasslands

Extrapolation to the regional and global scales relies predominantly on results from small-scale chamber measurements, often made only for short periods during the growing season. In the sections above, it was shown that seasonal and spatial variabilities are very large and extrapolation to annual and large-scale fluxes is very uncertain. With the knowledge to date, the information on fertilizer and livestock N-induced N_2O emissions in Table 14.1 is probably the best option to estimate UK N_2O emission from managed grassland. In a previous estimate (Skiba *et al.*, 1996), the N-induced N_2O emissions for grazed grassland was assumed to be 3.1%, based on the studies by Velthof *et al.* (1994) and Velthof and Oenema (1994). By including more recent measurements from the literature, the average fractional N_2O loss for grazed grasslands has been calculated at 2.1% (Table 14.1).

Applying the fractional loss of 2.1% to calculate annual UK N_2O emissions from grassland, it has been estimated that the grazed area (44,850 km^2) contributes 11 kt N_2O-N to the atmosphere over the UK and the ungrazed grassland contributes a further 4.2 kt N_2O-N (Table 14.3). The annual emission of N_2O from rough grass and moorland is much smaller at 0.6 kt N_2O-N. For this category, an average annual flux of 0.19 kg N_2O-N ha^{-1} $year^{-1}$ was calculated from the data in Table 14.4. These data represent only a small proportion of the environmental variation occurring in this land-cover class and takes no account of atmospheric N inputs to the land. The average fluxes presented in Table 14.4, however, are of the same order of magnitude as those calculated for forest soils (Skiba *et al.*, 1996). No doubt, the emission factors will have to be revised when more flux measurements are published. The growing datasets for UK soils will provide an improved basis to calculate N_2O emissions using UK-specific emission factors, rather than IPCC default values, for example the 1.25% fertilizer-induced emission.

In conclusion, UK grasslands emit approximately 16 kt N_2O-N $year^{-1}$, with grazed grasslands being the largest grassland type and also the largest

Table 14.3. Estimate of annual grassland N_2O emissions for the UK.

Land cover*	Area (km^2)	N_2O emission (kt N $year^{-1}$)
Managed grassland		
Grazed	44,850	11.0[†]
Ungrazed	20,822	4.2[†]
Rough grass/moor grass	32,700	0.6[‡]

*From Barr et al., 1993.
[†]N_2O emission as % of N fertilizer input, median values calculated from Table 14.1: 2.1% for grazed and 1.8% for mowed grassland. Fertilizer input was assumed to be 116 kg N ha^{-1} $year^{-1}$ (Chalmers et al., 1990).
[‡]Assume 0.19 kg N_2O-N ha^{-1} $year^{-1}$ (Skiba et al., 1996 and unpublished data).

Table 14.4. Emissions of N_2O from unmanaged grasslands in Northern Britain.

Vegetation	Soil type	Measuring period	Sample size	N_2O emission (kg N ha^{-1} $year^{-1}$)*
Upland grass[†]	Peat	May 1993, March and June 1995	39	0.31 ± 0.06
Grass/heather[‡]	Peaty podzol	April 1994–Oct 1995	28	0.2 ± 0.05
Grass/sphagnum[§]	Peat	Feb–Sept 1995	7	0.06 ± 0.04

[†]Great Dun Fell, North England, 400, 600 and 800 m above sea level (a.s.l.).
[‡]Dunslair Heights, South East Scotland, 600 m a.s.l.
[§]Auchencorth Moss, 270 m a.s.l.
*Results given are mean $\pm$ SEM.

source of N_2O, accounting for 60% of the total UK soil emissions. This is also the most spatially variable source of N_2O and the certainty of flux estimates needs to be improved by long-term and large-scale measurements.

Acknowledgements

The authors wish to thank the CEC STEP programme and the NERC TIGER programme for financial support.

References

Arah, J.R.M., Crichton, I.J., Smith, K.A., Clayton, H. and Skiba, U. (1994) Automated gas chromatographic analysis system for micrometeorological measurements of trace gas fluxes. *Journal of Geophysical Research* 99, 16,593–16,598.

Barr, C.J., Bunce, R.G.H. and Clarke, R.T. (1993) *Countryside Survey 1990.* Department of the Environment, HMSO, London, 174 pp.

Bouwman, A.F. (1994) *Direct Emissions of Nitrous Oxide from Agricultural Soils*. Report No. 773004004. National Institute of Public Health and Environmental Protection, Bilthoven, The Netherlands.

Bouwman, A.F., van der Hoek, K.W. and Oliver, J.G.J. (1995) Uncertainties in the global source distribution of nitrous oxide. *Journal of Geophysical Research* 100, 2785–2800.

Chalmers, A., Kershaw, C. and Leech, P. (1990) Fertilizer use on farm crops in Great Britain: results from the survey of fertilizer practice, 1969–88. *Outlook on Agriculture* 19, 269–278.

Christensen, S. (1983) Nitrous oxide emission from a soil under permanent grass: seasonal and diurnal fluctuations as influenced by manuring and fertilization. *Soil Biology and Biochemistry* 15, 531–536.

Christensen, S., Ambus, P. and Arah, J.R.M. (1996) Nitrous oxide emissions from an agricultural field: comparison between measurements by flux chamber and micrometeorological techniques. *Atmospheric Environment* 30, 4183–4190.

Clayton, H., Arah, J.R.M. and Smith, K.A. (1994) Measurements of nitrous oxide emissions from fertilized grassland using closed chambers. *Journal of Geophysical Research* 99, 16599–16607.

Conrad, R., Seiler, W. and Bunse, G. (1983) Factors influencing the loss of fertiliser nitrogen into the atmosphere as N_2O. *Journal of Geophysical Research* 88, 6709–6718.

Flessa, H., Dorsch, P. and Beese, F. (1996) N_2O emissions from cattle excrements in pasture land – a global inventory. In: *Transactions of the 9th Nitrogen Workshop, Braunschweig, September 1996*, Technical University of Braunschweig, pp. 141–144.

Fowler, D. and Duyzer, J.H. (1989) Micrometeorological techniques for the measurement of trace gas exchange. In: Andreae, M.O. and Schimel, D.S. (eds) *Exchange of Trace Gases between Terrestrial Ecosystems and the Atmosphere*. Wiley, Chichester, pp. 189–207.

Galle, B., Klemedtsson, L. and Griffith, D.W.T. (1994) Application of an FTIR system for measurement of nitrous oxide fluxes using micrometeorological methods, an ultralarge chamber system and conventional field chambers. *Journal of Geophysical Research* 99, 16,575–16,583.

Hansen, S., Maehlum, J.E. and Bakken, L.R. (1993) N_2O and CH_4 fluxes in soil influenced by fertilization and tractor traffic. *Soil Biology and Biochemisty* 25, 621–630.

Hargreaves, K.J., Skiba, U. and Fowler, D. (1994) Measurements of nitrous oxide emissions from fertilized grassland using micrometeorological techniques. *Journal of Geophysical Research* 99, 16,569–16,574.

Hargreaves, K., Wienhold, F.G. and Klemedtsson, L. (1996) Measurement of nitrous oxide emission from agricultural land using micrometeorological methods. *Atmospheric Environment* 30, 1563–1571.

INDITE (1994) *Impacts of Nitrogen Deposition in Terrestrial Ecosystems*. Report of the United Kingdom Review Group on Impacts of Atmospheric Nitrogen, Department of the Environment, London, 110 pp.

IPCC (1995) In: Houghton, J.T., Meiro Filho, L.G., Callander, B.A., Harris, N., Kattenberg, A. and Maskell, K. (eds) *Climate Change 1995. The Science of Climate Change*. Cambridge University Press, Cambridge, 339 pp.

Kaiser, E-A., Kohrs, K. and Kucke, M. (1996) Nitrous oxide release from cultivated soils: influence of different N-fertilizer types. In: *Transactions of the 9th Nitrogen Workshop, Braunschweig, September 1996.* Technical University of Braunschweig, pp. 149–152.

Khalil, M.A.K. and Rasmussen, R.A. (1992) The global sources of nitrous oxide. *Journal of Geophysical Research* 97, 14,651–14,660.

de Klein, C.A.M. and van Logtestijn, R.S.P. (1994) Denitrification and N_2O emission from urine-affected grassland soil. *Plant and Soil* 163, 235–242.

MAFF (1993) *The Digest of Agricultural Census Statistics, UK, 1992.* HMSO, London, 129 pp.

McElroy, M.B. and Wofsy, S.C. (1986) Tropical forests, interaction with the atmosphere. In: Prance, G.T. (ed.) *Tropical Rain Forests and the World Atmosphere, AAA Selected Symposium 101.* Westview Press, Boulder, Colorado, pp. 33–60.

McTaggart, I., Clayton, H. and Smith, K. (1994) Nitrous oxide flux from fertilized grassland: strategies for reducing emissions. In: van Ham, J., Janssen L.H.M. and Swart, R.J. (eds) *Non-CO_2 Greenhouse Gases.* Kluwer Academic, Dordrecht, pp. 421–426.

Minschwaner, K., Salawich, R.J. and Mc Elroy, M.B. (1993) Absorption of solar radiation by O_2 : implication for O_3 and lifetimes of N_2O, $CFCl_3$ and CF_2Cl_2. *Journal of Geophysical Research* 98, 10,543–10,561.

Mosier, A., Parton, W.J., Valentine, D.W. and Schimel, D.S. (1996) Long-term impact of N-fertilization on N_2O flux and CH_4 uptake in the Colorado shortgrass prairie. In: *Transactions of the 9th Nitrogen Workshop, Braunschweig, September 1996.* Technical University of Braunschweig, pp. 157–160.

Ryden, J.C. (1981) Nitrous oxide exchange between a grassland soil and the atmosphere. *Nature* 292, 235–237.

Salway, A.G. (1995) *UK Greenhouse Gas Emission Inventory, 1990–1993.* AEA Technology, Annual Report for the DOE, AEA/16419178/R/001.

Skiba, U. Hargreaves, K.J., Fowler, D. and Smith, K.A. (1992) Fluxes of nitric and nitrous oxides from agricultural soils in a cool temperate climate. *Atmospheric Environment* 26, 2477–2488.

Skiba, U., McTaggart, I.P., Smith, K.A., Hargreaves, K.J. and Fowler, D. (1996) Estimates of nitrous oxide emissions from soil in the UK. *Energy Conversion Management* 37, 1303–1308.

Smith, K.A., Scott, A., Galle, B. and Klemedtsson, L. (1994) Use of a long-path infrared gas monitor for measurement of nitrous oxide from soil. *Journal of Geophysical Research* 99, 16,585–16,592.

Smith, K.A., Clayton, H., McTaggart, I.P., Thomson, P.E., Arah, J.R.M. and Scott, A. (1995) The measurement of nitrous oxide emissions from soil by using chambers. *Philosophical Transaction of the Royal Society of London A* 351, 327–338.

Velthof, G.L. and Oenema, O. (1994) Nitrous oxide emissions from grassland on sand, clay and peat soils in the Netherlands. In: Van Ham, J., Janssen, L.H.M. and Swart, R.J. (eds) *Non-CO_2 Greenhouse Gases.* Kluwer Academic, Dordrecht, pp. 439–444.

Velthof, G.L., Oenema, O. and McTaggart, I.P. (1994) Nitrous oxide emissions from managed grasslands in the Netherlands, Scotland and Belgium. In: *Proceedings of the 8th Nitrogen Workshop, Ghent, September 1994.* University of Ghent (in press).

Velthof, G.L., Brader, A.B. and Oenema, O. (1996) Seasonal variations in nitrous oxide losses from managed grasslands in the Netherlands. *Plant and Soil* 181, 263–274.

Wienhold, F.G., Frahm, H. and Harris, G.W. (1994) Measurements of N$_2$O fluxes from fertilized grassland using a fats response tunable diode laser spectrometer. *Journal of Geophysical Research* 99, 16,557–16,567.

Spatial Pattern in Soil Oxygen Content and Nitrous Oxide Emission from Drained Grassland

15

C. Rappoldt and W.J. Corré

DLO Research Institute for Agrobiology and Soil Fertility (AB-DLO), PO Box 129, 9750 AC Haren, The Netherlands

Summary.   The presence of drains imposes significant differences in groundwater level between points a few metres apart. This induces a spatial pattern in soil aeration, denitrification rate and nitrous oxide (N_2O) surface flux. During a wet period, 2 days after fertilization, the N_2O surface flux ranged from 0.4 kg N ha^{-1} day^{-1} close to the drains to 4.6 kg N ha^{-1} day^{-1} between two drains, at the wettest place of the field. Throughout this range, the surface flux could not be distinguished from the denitrification rate. Six days after fertilization, the N_2O flux was much lower than the denitrification rate, which confirms, at field scale, the inhibition of N_2O reduction by nitrate.

Introduction

The spatial variability of soil oxygen content and nitrous oxide (N_2O) flux has been studied on relatively homogeneous, drained grassland on a loamy soil. The presence of artificial drains leads to significant differences in groundwater level at places which are just a few metres apart. During wet periods, the groundwater level between two drains can sometimes reach the soil surface. Above the drains, however, the groundwater level always remains below 70 cm. The associated variability in soil aeration also has a largely deterministic nature. The spatial pattern imposed by the drains was made use of in a denitrification experiment. During a wet period fertilizer was applied to the grassland and measurements carried out at different sampling points in the field; these were at different distances from a drain and thus had different groundwater levels and very different N_2O fluxes at the soil surface. Each sampling point was comparable, however, with respect to all other circumstances: soil, weather, vegetation, etc.

Methods

Field site

The field site was situated at the Lovinkhoeve, an AB-DLO experimental farm in the Noordoostpolder (The Netherlands). The soil consisted of a loam topsoil (0–30 cm), an intermediate clay layer (30–39 cm), a thin, very fine sand layer (39–42 cm) and a subsoil consisting of heavy loam and light loam (De Vos *et al.*, 1994). The drain spacing was 12 m. The grassland used for the oxygen measurements and the fertilizer experiment was part of the meteorological station on the farm and was not used for grass production but normally received small amounts of fertilizer. As an extension of the weather station, De Vos *et al.* (1994) installed 13 groundwater tubes between 0.5 m and 6 m from a drain, as well as sensors for measuring soil temperature.

Measurements

Soil oxygen content was monitored by means of 30 open chambers, permanently installed at five different depths and at six different distances from a drain. The volume of each chamber was 28 cm^3. Two metal capillaries ran from each chamber to the surface. A portable measurement system could be connected to these capillaries and the gas in the chamber circulated past an oxygen electrode. During measurement, the soil gas from the chamber was mixed with the gas present in the circuit before the measurement. The volume of the circuit, however, was small enough (8.3 cm^3) to allow for a reliable correction. Note that the oxygen concentration measured in this way is a macropore concentration which is not a direct measure of the degree of anaerobiosis.

During a wet period, on 14 June 1995, 80 kg nitrate N (NO_3^--N) ha^{-1} was applied to the field. The fertilization resulted in a flush of denitrification and N_2O production and observations were made on 16 June and 20 June. Soil oxygen content and groundwater level were measured on other dates as well. For measuring N_2O concentration at different depths, gas samples were taken from the oxygen chambers in the soil.

The surface flux of N_2O was measured by means of a closed chamber with a diameter of 19.1 cm and a height of 9.5 cm. After placing the chamber on the soil surface, the increase in N_2O concentration was determined from repeated measurements with a Brüel Kjær infrared/acoustic gas analyser. The surface flux was assumed to be proportional to the initial concentration increase which was calculated from the data by means of regression analysis. The measurement period used for a flux exceeding 0.1 kg day^{-1} ha^{-1} was 15 min and the relative statistical error in the result was about 5%. Measuring smaller fluxes required more time and the results were less accurate.

Denitrification was measured by incubation of soil samples with acetylene. The samples were taken at three depths, i.e. 0–20, 20–40 and 40–60 cm. The measured denitrification rates were expressed as loss in kg N ha^{-1} day^{-1}. In this paper, denitrification rates for the entire soil profile are given which were calculated as the sum of the results for the three layers. The potential denitrification rate has been measured in a similar way, after incubation in N_2 gas instead of air.

A denitrification model for a soil aggregate

In the discussion, a diffusion-reaction model for denitrification in a partially anoxic soil aggregate is referred to. The model equations and the parameter values of Arah (1990) were applied to a 3-cm-thick, flat soil aggregate. The only difference from this model was that an external source of NO_3^- was not included, but only that initially present in the aggregate was considered. This led to a transient peak of N_2O production just after the aggregate became partially anoxic. Later, denitrification was limited by NO_3^- diffusion towards the anoxic centre and N_2O was reduced to N_2.

Results

Figure 15.1 shows a typical example of the oxygen measurements (data from May 1994). The concentration in the 30 chambers is shown as a function of depth and the horizontal distance to the nearest drain. There were clear effects of both depth and distance and there was little stochastic variability. The example shown does not represent an extreme situation. During a long dry period, the concentrations at 50 and 70 cm below the soil surface may reach 15% and during a wet period concentrations below 15% occur also at a depth of 10 cm.

Figure 15.2(a) shows the daily rainfall (De Vos, 1996, personal communication) and the groundwater level in the period of the fertilization experiment in June 1995. The groundwater level is shown as function of time and the distance to the nearest drain. On 16 June, for instance, the level was 73 cm at 0.5 m from the drain and only 11 cm at 6 m distance, at the wettest place of the field. The three-dimensional grid shows that the reaction of the groundwater table to the rainfall was retarded. The highest groundwater

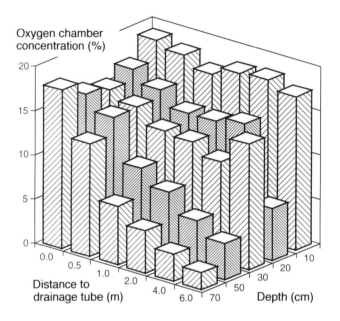

Fig. 15.1. Typical result of chamber oxygen content for five depths and six distances to the drainage tube.

levels were reached towards the end of the wet period. The grids in Fig. 15.2(b) and (c) show how the oxygen content at 10 and 30 cm reacted to the changing groundwater level. The grids have been constructed by means of linear interpolation in both time and distance.

Table 15.1 shows results of the fertilization experiment carried out in the wet period shown in Fig. 15.2. On 16 June, 2 days after fertilizer application, the potential denitrification rate (the denitrification rate without oxygen supply to the soil) was about 10 kg ha^{-1} day^{-1}, independent of the distance to the drain. The actual denitrification rate, however, strongly increased with distance and reached about 50% of the potential rate at 6 m. The N$_2$O surface flux was close to the actual denitrification rate. At 6 m distance, the surface flux reached 4.6 kg ha^{-1} day^{-1} and the N$_2$O concentration in the soil at 10 cm amounted to 230 ppm, more than 700 times the atmospheric concentration.

On 20 June, the denitrification rate had decreased everywhere to less than 0.5 kg ha^{-1} day^{-1}. The N$_2$O surface flux decreased even further and became much lower than the denitrification rate. The potential denitrification rate still reached 10 kg ha^{-1} day^{-1}, however, but only close to the drain. At larger distances, the potential denitrification rate decreased strongly in relation to its value on 16 June.

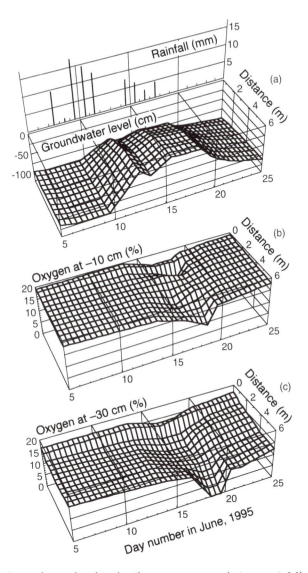

Fig. 15.2. Groundwater level and soil oxygen content during a rainfall period in June 1995. (a) Rainfall as function of day number and groundwater level as function of both day number and distance to drainage tube; (b) chamber oxygen content at 10 cm depth; (c) chamber oxygen content at 30 cm depth. The three-dimensional grids have been constructed by linear interpolation in time and distance.

Table 15.1. Nitrous oxide surface flux, profile denitrification rate and chamber nitrous oxide concentration measured 2 and 6 days after applying 80 kg N ha^{-1} day^{-1} nitrate fertilizer.

Distance to drain (m)	Groundwater level (cm)	N$_2$O surface flux (kg N ha^{-1} day^{-1})	Denitrification rate		Chamber N$_2$O concentration		
			Actual (kg N ha^{-1} day^{-1})	Potential (kg N ha^{-1} day^{-1})	−10 cm (ppm)	−20 cm (ppm)	−30 cm (ppm)
16 June 1995							
6.0	−11	4.6	4.5	11.0	231	66	18
5.0	−18	2.2	1.9	12.5	—	—	—
4.0	−24	1.0	0.59	9.5	120	28.4	6.6
2.0	−42	0.37	0.12	8.1	2.9	3.9	5.7
0.5	−73	0.34	0.64	9.5	—	18	1.7
20 June 1995							
6.0	−52	0.018	0.04	0.89	2.5	1.8	0.74
5.0	−55	0.017	0.10	2.7	—	—	—
4.0	−57	0.009	0.45	2.6	1.9	5.2	4.7
2.0	−74	0.005	0.05	11.2	2.0	5.4	10.7
0.5	−90	0.011	0.20	9.5	1.5	8.4	9.7

Discussion

The results show that the drains impose a spatial pattern in groundwater level and in soil oxygen content, especially under wet conditions. Such a pattern has important consequences for the fraction of anoxic soil, and the associated denitrification rate and N_2O surface flux. Ignoring the regular pattern imposed by drains may lead to severe errors in the measured N_2O flux for a field.

The results for 16 June suggest that a significant part of the soil at greater drain distances was anoxic and almost 50% of the potential denitrification was realized. This happened 2 days after fertilization and all denitrified NO_3^- was converted into N_2O. Six days after fertilization, the N_2O flux was much lower than the denitrification rate. This confirmed, at field scale, the inhibition of N_2O reduction by NO_3^-. Figure 15.3 shows the result of the denitrification model of Arah (1990) for a soil aggregate. Typically, the N_2O flush takes place just after the soil has become partially anoxic. The similarity between the N_2O flux and the denitrification rate could be observed by making use of the drain distance as a single varying factor under otherwise equal circumstances. Separate experiments at different sites would be much more difficult and time consuming.

A N_2O flux of almost 5 kg N ha^{-1} day^{-1} from the soil far from the drain shows once more the potential of the denitrification process on heavy soils. It is comparable to the largest fluxes reported in the literature (Li *et al.*, 1992; Bouwman, 1990; Velthof and Oenema, 1995) and will form a significant part of the yearly emission, although it occurs in part of the field only. A smaller drain distance or extreme care in choosing the fertilization date will decrease

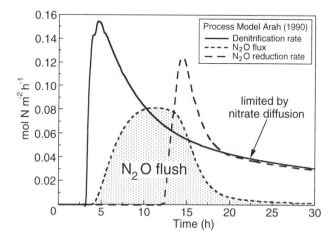

Fig. 15.3. Denitrification in a soil aggregate with a transient peak of N_2O production according to the model of Arah (1990), applied to a 3-cm-thick, flat aggregate.

the risk of N_2O flushes of the observed type and will significantly contribute to a reduction of atmospheric pollution.

References

Arah, J.R.M. (1990) Diffusion-reaction models of denitrification in soil microsites. In: Revsbech, N.P. and Sørensen, J. (eds) *Denitrification in Soil and Sediment.* Plenum Press, New York, pp. 245–258.

Bouwman, A.F. (1990) Exchange of greenhouse gases between terrestrial ecosystems and the atmosphere. In: Bouwman, A.F. (ed.) *Soils and the Greenhouse Effect.* John Wiley & Sons, New York, pp. 61–127.

De Vos, J.A., Raats, P.A.C. and Vos, E.C. (1994) Macroscopic soil physical processes considered within an agronomical and soil biological context. *Agriculture, Ecosystems and Environment* 51, 43–73.

Li, C., Frolking, S. and Frolking, T.A. (1992) A model of nitrous oxide evolution from soil driven by rainfall events: 2. Model applications. *Journal of Geophysical Research* 97, 9777–9783.

Velthof, G.L. and Oenema, O. (1995) Nitrous oxide fluxes from grassland in the Netherlands: I. Statistical analysis of flux-chamber measurements. *European Journal of Soil Science* 46, 533–540.

Spatial Variability of Nitrous Oxide Fluxes from Grasslands: Differences between Soil Types

16

G.L. Velthof[1], S.C. Jarvis[2] and O. Oenema[1,3]

[1]Nutrient Management Institute (NMI), Department of Soil Science and Plant Nutrition, Wageningen Agricultural University, PO Box 8005, 6700 EC Wageningen, The Netherlands; [2]Institute of Grassland and Environmental Research (IGER), North Wyke, Okehampton, Devon EX20 2SB, UK; [3]Research Institute for Agrobiology and Soil Fertility Research (AB-DLO), PO Box 129, 9750 AC Haren, The Netherlands

Summary. Two contrasting grassland sites were investigated to assess the dominant factors controlling spatial variability of nitrous oxide (N_2O) fluxes, namely intensively managed grasslands on peat soil in the Netherlands and on poorly drained clay soil in Devon, UK. Fluxes were measured using 48–144 flux chambers per day. Spatial variability of the N_2O fluxes was large on both sites, with coefficients of variation ranging from 90 to 320%. Multiple linear regression analyses showed weak relations between N_2O flux and contents of nitrate (NO_3^-), ammonium (NH_4^+) and mineralizable carbon (C) and water-filled pore space (WFPS) in the 0–10-cm layer of the clay soil ($R^2_{adj} < 0.15$). By contrast, contents of NO_3^- and NH_4^+ and WFPS in the 0–20-cm layer explained a large part ($R^2_{adj} = 0.37$–0.77) of the variance in N_2O flux on the peat soil. It is suggested that the site of production of N_2O was much shallower in the clay soil than in the peat soil and that soil sampling was inappropriate for these differences.

Introduction

Spatial variability of nitrous oxide (N_2O) fluxes from managed grasslands is large (Ambus and Christensen, 1994) and is generally controlled directly or indirectly by spatial variability in the contents of nitrate (NO_3^-), ammonium (NH_4^+), moisture and mineralizable carbon (C) in the soil. The focus of this study is the effect of soil type on the spatial variability of N_2O flux. Velthof and Oenema (1995b) showed larger denitrification potentials and contents of mineralizable C in peat soils than in mineral soils. It was suggested that the subsoil was a larger source of N_2O in peat soils than in mineral soils. The site of N_2O production in the soil may also affect spatial variability of N_2O flux. Grassland management may attenuate and/or accentuate the spatial variability. For example, it may be expected that homogeneously applied N fertilizer decreases the spatial variability of N_2O fluxes and that grazing increases spatial variability, because of the heterogeneous distribution of urine and dung and effects of treading.

In this study, the dominant factors controlling spatial variability of N_2O fluxes on a field scale on two contrasting grassland sites were compared in intensively managed grasslands on a peat soil in the Netherlands and on a poorly drained clay soil in Devon, UK. Detailed results of the peat soil (Velthof *et al.*, 1996b) and of the clay soil (Velthof *et al.*, 1996a) have been given.

Materials and Methods

Experimental set-up

On the peat soil (Terric Histosol), measurements were carried out on intensively managed grassland on 24 June, 23 September and 9 November 1993. Nitrogen was applied at a total application rate of 244 kg N ha^{-1} in 1993, as both calcium ammonium nitrate (104 kg N ha^{-1}) and cattle slurry (140 kg N ha^{-1}). Grassland was rotationally grazed with dairy cows. In June, 48 flux chambers were placed in a row over the width of the field at a spacing of 0.90 m. In September, 48 flux chambers were placed in a grid of 12 × 4, at a spacing of 1 m and in November, 42 flux chambers were placed in a grid of 6 × 7, at a spacing of 1 m.

On the poorly drained clay soil (Gleyic Cambisol), measurements were carried out on mown and grazed (sheep) grasslands, during 20–23 September 1994. Both plots were fertilized with 125 kg N ha^{-1} as ammonium nitrate on 19 September. Each plot was subdivided into eight square subplots of 40 × 25 m. In each subplot, 18 flux chambers were placed at random in a 3 × 6 m grid. Each day, 8 × 18 flux measurements were made. Fluxes from mown grassland were determined on 20 and 23 September and fluxes from grazed grassland on 21 and 22 September.

Measurements of N_2O fluxes and soil variables

In both studies, fluxes of N_2O were measured using vented closed flux chambers (PVC cylinders: 20 cm internal diameter and 15 cm high) inserted 2 cm into the soil. Changes in the N_2O concentrations in the head space of the chambers were determined in the field, using a photo-acoustic infra-red gas analyser and a multipoint sampler (Velthof and Oenema, 1995a).

After flux measurements, soil in the flux chambers was sampled. In the study on peat soil, two samples of the 0–20-cm layer of all chambers were taken using a 4.7-cm diameter corer and in the study on clay soil, three samples of the 0–10-cm layer were taken from all chambers on the mown plot on 20 September and from the grazed plot on 22 September, using a 2.0-cm diameter corer. Sampling depth was different because a difference was expected between sites of N_2O production within the two soils, due to differences in: (i) time after the latest N application; (ii) drainage; and (iii) denitrification potential in the soil profile. Flux measurements on the poorly drained clay soil and the peat soil were carried out during 1–4 days and 2–4 weeks after fertilizer N application, respectively.

Moisture contents of the soil samples were determined by drying at 105°C for 24 h. Contents of NO_3^- and NH_4^+ were analysed after extraction in 0.01 M $CaCl_2$ in the study on peat soil and in 1 M KCl in the study on clay soil. Water-soluble C in the peat soil was measured in samples taken in November by using a dissolved organic carbon (DOC) analyser. Mineralizable C in the clay soil was determined in samples from the grazed plot taken on 22 September, by measuring the CO_2 evolution during aerobic incubation for 24 h at 20°C.

Results and Discussion

Fluxes of N₂O

Fluxes of N_2O were large, especially on the clay soil (Table 16.1). Apparently, conditions in both soils were favourable for production and emission of N_2O. Spatial variability in N_2O fluxes was also very large, as indicated by the large coefficients of variation (Table 16.1). The larger arithmetic means than medians on all days indicate that the frequency distributions were positively skewed. The similar coefficients of variation in both studies suggest that soil type did not affect overall spatial variability of N_2O fluxes. Clearly, the large pool of organic N and C and the relatively shallow groundwater level in the peat soil did not decrease spatial variability of N_2O fluxes compared with the clay soil with much lower contents of organic N and C and poor subsurface drainage characteristics. The smaller coefficients of variation in grazed grassland than in mown grassland on the clay soil were the opposite of what was expected. This suggests that the intensive grazing buffered changes with respect to N_2O production. It is speculated that the high stocking rate caused an 'unnaturally' uniform distribution of mineralizable C in the grazed grassland, which may have decreased spatial variability of denitrification-derived N_2O fluxes on grazed grassland (Velthof et al., 1996a).

Table 16.1. Summary of results for all measurement days: mean soil temperature at 5 cm depth, water-filled pore space (WFPS), number of flux measurements and arithmetic mean, minimum, median, maximum and coefficient of variation (c.v.) of the N_2O fluxes.

Site and measurement	Soil temp. (°C)	WFPS* (%)	n	N_2O flux (mg N m^{-2} h^{-1})				
				Mean	Min.	Median	Max.	c.v. (%)
Peat soil, NL								
June	14.4	62	48	1.07	< 0.01	0.25	6.66	163
September	14.8	91	48	0.22	< 0.01	0.04	4.62	320
November	7.7	92	42	0.26	< 0.01	0.18	1.28	101
Clay soil, UK								
Mown: 20 Sept.	13.8	85	144	2.59	< 0.01	0.70	20.87	148
Mown: 23 Sept.	14.3	—	144	1.13	0.02	0.54	14.40	155
Grazed: 21 Sept.	13.4	—	144	5.14	0.11	3.76	40.23	105
Grazed: 22 Sept.	13.3	85	144	5.54	0.02	3.95	26.49	90

*WFPS of the 0–20-cm layer of the peat soil and 0–10-cm layer of the clay soil.

Factors controlling N_2O flux

Mineral N content in the 0–20-cm layer was a major factor controlling spatial variability in the peat soil (Fig. 16.1). In contrast, no clear relationship was shown between N_2O flux and mineral N content in the 0–10-cm layer for the clay soil (Fig. 16.1). Multiple linear regression also showed large differences between the sites. Multiple linear regression models with log-transformed contents of NO_3^- and NH_4^+ and WFPS as the independent variables explained 37–77% of the variance in log-transformed N_2O flux on peat soil and less than 13% on the clay soil. Mineralizable C and water-soluble C were not significant variables in the regression models, suggesting that C content was not a major factor controlling spatial variability in N_2O flux from both soils.

Clearly, the soil variables in the samples better represented the interaction of factors controlling N_2O fluxes for the peat soil than for the clay soil. It is suggested that the differences between the soils were due in part to the sampling strategy. Several factors may have played a role in this.

1. The total area of the sampled soil in each flux chamber was larger on the peat soil (34.7 cm^2 or 11% of the chamber area) than that on the clay soil (9.4 cm^2 or 3% of the chamber area). It is suggested that the percentage of variance in N_2O flux accounted for by the soil variables increases when the sampled area of soil within the flux chamber increases.

2. Flux measurements were carried out 1–4 days after N fertilizer application on the clay soil. The major site of N_2O production in the soil was probably that with the highest mineral N contents. It may be speculated that movement of fertilizer N into this poorly drained soil was slow and that after 1 and 3 days

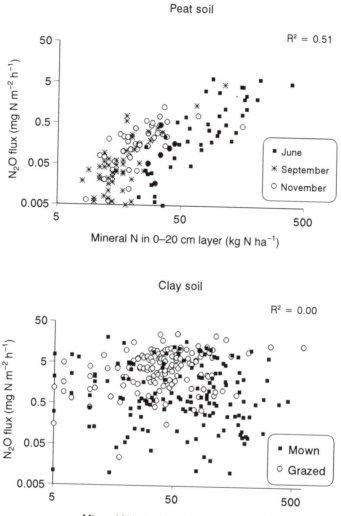

Fig. 16.1. Log-transformed N_2O fluxes versus log-transformed total mineral N contents for both soils. For the peat soil, total mineral N contents are from the 0–20-cm layer, and for the clay soil from the 0–10-cm layer.

only the N contents in the uppermost few centimetres of the clay had been increased and that this was the major site of N_2O production at that time. Moreover, lateral movement of N_2O would also be limited over this period.
3. Flux measurements in the peat soil were carried out 2–4 weeks after N application. In addition, the topsoil of the peat soil was well drained,

suggesting that mineral N would have moved deeper in this peat soil than the clay soil at the time of the measurements.

4. Denitrification potential and mineralizable C contents in the subsoil of peat soils are larger than those of mineral soils (Velthof and Oenema, 1995b), suggesting that the subsoil is of more importance in the production of N_2O in peat soils than in mineral soils and that the sampling depth should be greater in peat.

It is suggested that the major site of N_2O production in the poorly drained clay soil was situated in spots around dissolved fertilizer granules in the uppermost few centimetres during the first days after fertilizer N application. The sites of N_2O production in the peat soil were probably much less restricted and more uniformly distributed.

Conclusions

Characterization of the spatial variability of N_2O fluxes may improve the identification of the factors controlling N_2O fluxes from soils. Flux measurements with chambers and sampling the soil in the chamber after flux measurements can provide insights into these controlling factors. The results of this study suggest, however, that the sampling strategy, i.e. the combination of flux chamber size, soil sample size and sampling depth, is crucial. The choice of the sampling strategy depends on soil type and grassland management. The sampling depth should be larger in peat soils than in mineral soils and the depth of sampling should be increased with increasing time after the latest N fertilizer application.

Acknowledgements

The work was funded by the Dutch National Research Program on Global Air Pollution and Climate Change, the C.T. de Wit Graduate School for Production Ecology of the Wageningen Agricultural University, and the Ministry of Agriculture, Fisheries and Food, London.

References

Ambus, P. and Christensen, S. (1994) Measurement of N_2O emission from a fertilized grassland: an analysis of spatial variability. *Journal of Geophysical Research* 99, 16,549–16,555.

Velthof, G.L. and Oenema, O. (1995a) Nitrous oxide fluxes from grassland in the Netherlands: I. Statistical analysis of flux chamber measurements. *European Journal of Soil Science* 46, 533–540.

Velthof, G.L. and Oenema, O. (1995b) Nitrous oxide fluxes from grassland in the Netherlands: II. Effects of soil type, nitrogen fertilizer application and grazing. *European Journal of Soil Science* 46, 541–549.

Velthof, G.L., Jarvis, S.C., Stein, A., Allen, A.G. and Oenema, O. (1996a) Spatial variability of nitrous oxide fluxes in mown and grazed grasslands on a poorly drained clay soil. *Soil Biology and Biochemistry* 28, 1215–1225.

Velthof, G.L., Koops, J.G., Duyzer, J.H. and Oenema, O. (1996b) Prediction of nitrous oxide fluxes from managed grassland on peat soil using a simple empirical model. *Netherlands Journal of Agricultural Science* 44, 339–356.

Effect of pH on Nitrous Oxide Emissions from Grassland Soil

17

S. Ellis[1], K. Goulding[2] and L. Dendooven[2]

[1]ADAS, Boxworth, Cambridge CB3 8NN, UK; [2]Soil Science Department, IACR-Rothamsted, Harpenden, Hertfordshire AL5 2JQ, UK

Summary. The effect of soil pH (between 3.3 and 6.1) on the emission of nitrous oxide (N_2O) from a permanent grassland soil at IACR-Rothamsted has been studied. Soils were sampled and assayed for denitrification and nitrification potentials, and linked to the size of the microbial biomass. Rates of carbon dioxide (CO_2) and N_2O production decreased significantly with decreasing pH ($P < 0.01$) under anaerobic (denitrifying) conditions. These decreases were equivalent to 0.04 and 0.08 mg carbon (C) and nitrogen (N) kg^{-1} h^{-1} for every decrease of 0.1 pH unit, respectively. Nitrification activity was seen even on the most acid plots of pH < 4 but N_2O production remained < 0.001 mg N kg^{-1} h^{-1}. pH affected N_2O production directly by influencing the biological availability of C substrate to, and thus the size of, the soil microbial biomass. In acid soil, the lag times to NO_2^- reduction and N_2O production under anaerobic conditions increased and thus the kinetics of denitrification were altered.

Introduction

Soil pH is known to affect the rate of denitrification and the ratio of gaseous nitrogen (N) products evolved (Koskinen and Keeney, 1982; Weier and Gilliam, 1986). Acid soil pH favours nitrous oxide (N_2O) production via nitrification in some N-fertilized forest soils, preferentially selecting a dominant population of autotrophic nitrifiers (Martikainen, 1985). It has been suggested that the soil microbial biomass can adapt to changes in pH over time and functions well under stressed conditions induced by an acid environment (Parkin *et al.*, 1985).

There is evidence that the kinetics of denitrification are affected by pH. Thomsen *et al.* (1994) studied the accumulation and destruction of intermediates involved in the denitrification activity of *Paracoccus denitrificans* and found that the complexity of reactions increased as pH decreased. At pH 5.5, nitrate (NO_3^-) was reduced to nitrite (NO_2^-) until all the NO_3^- was depleted; then N_2O reduction to dinitrogen (N_2) began. At pH 5.6–9.5, NO_3^- and NO_2^- reduction appeared to occur simultaneously, resulting in the production of N_2O without a NO_2^- peak.

At Rothamsted, the Park Grass Permanent Hay Experiment provided an opportunity to study, on a managed grassland, the effect of pH on the ratio of gaseous products evolved during denitrification and the contribution of nitrification to the total N loss. On four soils where the pH decreased from 6.1 to 3.3, the following have been studied: (i) denitrification potential and the ratio of the gaseous products evolved; (ii) nitrification potential; and (iii) the size of the microbial biomass.

Materials and Methods

The experiments were carried out with soil from plot 10 of the Park Grass Experiment at IACR-Rothamsted, Hertfordshire, UK. The soil was a silty clay loam (Johnston *et al.*, 1986) and the inorganic fraction contained 23% clay, 58% silt and 19% sand. At sampling, the organic carbon (C) contents, measured by dichromate oxidation (Tinsley, 1950), were 2.4, 2.5, 2.7 and 2.0%, respectively, for plots 10a, 10b, 10c and 10d. The total-N contents were 0.3, 0.2, 0.3 and 0.2%; the gravimetric moisture contents were 31.4, 25.0, 25.6 and 22.6%, and the pH, in 10 mM $CaCl_2$ (soil : solution ratio, 1 : 2.5) was 6.1, 4.9, 3.9 and 3.3, respectively. The soil microbial biomass (SMB) C, measured by automated fumigation-extraction, decreased from 787, to 434, 17 and 35 mg C kg^{-1}, respectively. The moist soil from each of the plots was sieved (< 5 mm) and aerobically conditioned at 25°C for 5 days.

Anaerobic incubation for potential denitrification

Subsamples (15 g) of soil from each of the plots were added to 84 250-ml Erlenmeyer flasks. Twenty millilitres of a solution containing 34 mM KNO_3, equivalent to 500 mg NO_3^--N kg^{-1}, were added to 42 flasks (CON treatment). The flasks were closed air-tight and evacuated for a total of 5 min with the head space replaced with oxygen free N_2 to create an anaerobic atmosphere. From half of the flasks, 30 ml of the head-space gas was removed and replaced with an equivalent volume of acetylene (C_2H_2) (ACE treatment) to prevent the reduction of N_2O to N_2 (Yoshinari and Knowles, 1976).

At times 0, 2, 6, 12, 24, 36 and 48 h, three flasks were chosen at random from each of the CON and ACE treatments of the four plots and the head space analysed for N_2O and carbon dioxide (CO_2) by gas chromatography (Ai 93 chromatogram, Ai Cambridge), using the method reported by Hall and Dowdell (1981). The concentrations of N_2O and CO_2 were corrected for dissolution in water (Weast, 1968; Moraghan and Buresh, 1977, respectively). After gas analysis, the soil was extracted with distilled water, filtered and the NO_2^- concentrations were determined colorimetrically on a Tecator 5010 flow injection analyser.

Aerobic incubation for potential nitrification

Five subsamples of 7.5 g of soil from each of the plots were added to twenty 100-ml conical flasks, sealed, and incubated aerobically on an orbital incubator at 25°C ($\pm$ 3°C) for 48 h. The flask head space was sampled for CO_2 and N_2O at times 0, 2, 6, 12, 24, 36 and 48 h, the atmosphere being renewed with laboratory air after each sampling. After 48 h, the soils were extracted as above and analysed for NO_3^-, NO_2^- and ammonium (NH_4^+).

Statistical analysis

Analysis of variance was used to test for significant differences in gas production under aerobic and anaerobic conditions for each of the four soils and two anaerobic treatments. All analyses were performed using GENSTAT 5 Release 3.1 (1993).

Results

CO₂ production

The CO_2 production rate for the anaerobically incubated soils decreased with decreasing pH, from 1.46 mg N kg^{-1} h^{-1} in plot 10a (pH 6.1), to 1.11 in plot 10b (pH 4.9), 0.49 in plot 10c (pH 3.9) and 0.31 in plot 10d (pH 3.3) (Fig. 17.1(a)). The rate of CO_2 production under aerobic conditions also decreased with decreasing pH, from 2.52 mg N kg^{-1} h^{-1} in plot 10a (pH 6.1) to 0.21 in plot 10d (pH 3.3) (Fig. 17.1(b)).

NO₂⁻ concentration

Nitrite consumption began at approximately 6 h in soils of pH 6.1 (Fig. 17.2). For soils of pH 4.9, NO_2^- behaviour was more complicated and the start of consumption was later, being 24–36 h for plot 10b (pH 4.9) and 24 h for plot 10c (pH 3.9). There was no significant NO_2^- production or consumption in plot 10d (pH 3.3).

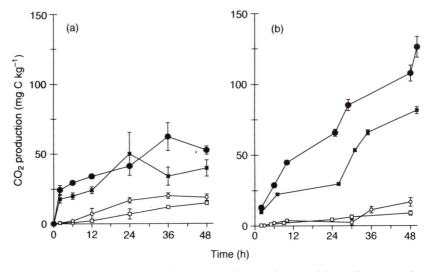

Fig. 17.1. (a) CO_2 production during anaerobic incubation of the Park Grass soils: plot 10a (pH 6.1; —●—), plot 10b (pH 4.9; —■—), plot 10c (pH 3.9; —○—); and plot 10d (pH 3.3; —□—). Error bars are the standard deviation of three replicates. (b) CO_2 production during aerobic incubation of the Park Grass soils. Symbols and error bars as in (a).

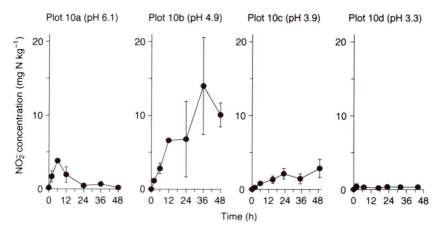

Fig. 17.2. Nitrite production and consumption (—●—) during anaerobic incubation of the Park Grass soils 10a to 10d. Error bars are the standard deviation of three replicates.

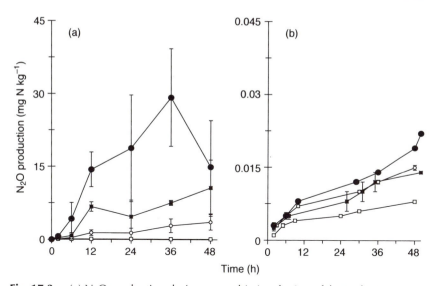

Fig. 17.3. (a) N_2O production during anaerobic incubation of the Park Grass soils: plot 10a (pH 6.1; —●—); plot 10b (pH 4.9; —■—); plot 10c (pH 3.9; —○—); and plot 10d (pH 3.3; —□—). Error bars are the standard deviation of three replicates. (b) N_2O production during aerobic incubation of the Park Grass soils. Symbols and error bars as in (a).

N_2O production

There was no significant difference between the N_2O produced in the soils either un-amended or amended with C_2H_2 ($P < 0.01$). However, the rate of N_2O production decreased with decreasing pH ($P < 0.01$), from an average of 0.45 mg N kg^{-1} h^{-1} in plot 10a (pH 6.1) to 0.001 in plot 10d (pH 3.3) (Fig. 17.3(a)). There was little, if any, N_2 produced at this site. There was no significant difference between the rates of N_2O production for the soils incubated aerobically; they were < 0.001 mg N kg^{-1} h^{-1} (Fig. 17.3(b)).

Mineralization and nitrification activity during the aerobic incubations

Table 17.1 illustrates the soil mineral N concentration at sampling and after 48 h aerobic incubation at 25°C (± 3°C). Soil NH_4^+ concentrations decreased in soils of pH > 4.9 but did not change significantly in soils of pH 4.9. Little NO_2^- was measured in any of the soils, it could only be detected in plot 10a (pH 6.1) after 48 h of incubation. Soil NO_3^- concentrations increased in all of the soils at rates of 0.11, 0.13, 0.04 and 0.05 mg N kg^{-1} h^{-1} for plots 10a to 10d, respectively.

Table 17.1. Soil NO_3^-, NO_2^- and NH_4^+ concentrations (mg N kg^{-1}) (with SD) at sampling and after 48 h of aerobic incubation at 25°C ($\pm$ 3°C).

Plot	NO$_3^-$ At sampling	NO$_3^-$ After 48 h	NO$_2^-$ At sampling	NO$_2^-$ After 48 h	NH$_4^+$ At sampling	NH$_4^+$ After 48 h
10a	0.42	5.57	0.06	0.23	0.43	0.21
(pH 6.1)	< 0.01	(1.52)	(0.00)	< 0.01	< 0.01	(0.07)
10b	0.42	6.53	0.11	nd	0.62	0.92
(pH 4.9)	< 0.01	(0.96)	(0.01)		(0.65)	(0.28)
plot 10c	nd	1.93	0.12	nd	1.30	1.25
(pH 3.9)		(0.24)	(0.02)		(0.20)	(0.12)
plot 10d	nd	2.18	0.07	nd	3.47	3.30
(pH 3.3)		(0.44)	< 0.01		(0.40)	(0.12)

nd = Not determined; SD = standard deviation of mean (n = 3).

Discussion

The Park Grass site with its long-term stable pH values established for over 100 years provided an opportunity to study the potential of the soil microbial biomass (SMB), denitrifiers and nitrifiers in particular, to adapt to changes in soil pH. The slurried soils used to determine denitrification potential ensured that no microsites of neutral pH existed (Parkin *et al.*, 1985) and there appeared to be no adaptation of the SMB *per se* to an acidic environment: SMB decreased significantly with decreasing pH through the influence of pH on the biological availability of substrate. There was no significant difference between the organic C contents of the soils ($P< 0.01$), therefore pH must have affected the biological availability of substrate to the SMB. Koskinen and Keeney (1982) also suggested that pH did not directly control denitrification but exerted an influence in controlling the availability of C to denitrifying microorganisms, which would exert an important control over the size of the SMB.

The kinetics of denitrification were affected by pH. The lag to the start of NO_2^- consumption and N_2O production increased with decreasing pH from approximately 6 h in plot 10a (pH 6.1), to 36 h in plot 10b (pH 4.9), and to 24–48 h in plot 10c (pH 3.9), with no significant difference between the slow rates of N_2O production witnessed in the aerobic soils and those of the anaerobically incubated soil of plot 10d (pH 3.3) (Fig. 17.2). Thomsen *et al.* (1994) found that the complexity of the production and consumption of intermediates in denitrification increased as pH decreased. In the soil of near neutral pH (plot 10a), there appeared to be a synchronous pattern of NO_3^- reduction to NO_2^- and NO_2^- reduction to N_2O. However, in plot 10b (pH 4.9),

NO_2^- appeared to accumulate in the soil whilst N_2O production increased and the synchronous reduction pathway of denitrification appeared to have broken down.

There was no discernible production of N_2 from any of the Park Grass soils. This may have resulted from the high concentrations of added NO_3^-; Blackmer and Bremner (1978) demonstrated that readily available soil NO_3^- was preferentially reduced over N_2O. Alternatively, the kinetics of denitrification may have been such that the incubation time was not long enough to see N_2 production. Where anaerobic events may be infrequent, *de novo* synthesis of N_2O reductase by denitrifying microorganisms may take > 70 h (Dendooven and Anderson, 1995). Indeed, anaerobic events would be infrequent in the Park Grass soils, as they are well-drained sandy loams and Harpenden receives only *c.* 700 mm of rainfall per annum. Emissions of N_2O from denitrification decreased with decreasing pH and very little appeared to be produced by nitrification (Fig 17.1(b)). However, nitrification activity was seen (Table 17.1) and may be important, even in the acid soils, in supplying NO_3^- to denitrifying microorganisms. pH exerted controls over N_2O production by influencing the biological availability of decomposable organic matter inputs to the SMB. The SMB appeared to be adapted *per se* to the inputs of organic matter, in terms of their numbers and their rates of denitrification and nitrification, but not so that they functioned well under low pH soil conditions (Parkin *et al.*, 1985).

This research suggests that N_2O produced by an ungrazed acid grassland comes from denitrification rather than nitrification, and that emission rates will decrease with pH. However, acid grassland tends to be in the wetter west and north of the UK. The reduction in emission rates with acidity is likely to be (more than) offset by the heavier rainfall and greater frequency of anaerobic conditions. These results also suggest that, at pH values of 4 and 5, the NO_2^- produced as an intermediate is not rapidly reduced and could be leached out by sudden, heavy rain. This may provide an additional explanation of the relatively large concentrations of NO_2^- in soil and flushes of NO_2^- in surface waters observed in Northern Ireland by Burns *et al.* (1995a, b) and Smith *et al.* (1995).

References

Blackmer, A.M. and Bremner, J.M. (1978) Inhibitory effect of nitrate on nitrous oxide reduction to nitrogen by microorganisms. *Soil Biology and Biochemistry* 10, 187–191.

Burns, L.C., Stevens, R.J., Smith, R.V. and Cooper, J.E. (1995a) The occurrence and possible sources of nitrite in a grazed, fertilized, grassland soil. *Soil Biology and Biochemistry* 27, 47–59.

Burns, L.C., Stevens, R.J. and Laughlin, R.J. (1995b) Determination of the simultaneous production and consumption of soil nitrite using ^{15}N. *Soil Biology and Biochemistry* 27, 839–844.

Dendooven, L. and Anderson, J.M. (1995) Maintenance of denitrification in pasture soil following anaerobic events. *Soil Biology and Biochemistry* 27, 1251–1260.

GENSTAT 5 Release 3.1 (1993) Lawes Agricultural Trust (Rothamsted Experimental Station), 356 pp.

Hall, K.C. and Dowdell, R.J. (1981) An isothermal gas chromatographic method for the simultaneous measurement of oxygen, nitrous oxide and carbon dioxide contents of gas in soil. *Journal of Chromatographic Science* 19, 107–111.

Johnston, A.E., Goulding, K.W.T. and Poulton, P.R. (1986) Soil acidification during more than 100 years under permanent grassland and woodland at Rothamsted. *Soil Use and Management* 2, 3–10.

Koskinen, W.C. and Keeney, D.R. (1982) Effect of pH on the rate of gaseous products of denitrification in a silt loam soil. *Soil Science Society of America Journal* 46, 1165–1167.

Martikainen, P.J. (1985) Nitrous oxide emission associated with autotrophic ammonium oxidation in acid coniferous forest soil. *Applied and Environmental Microbiology* 50, 1519–1525.

Moraghan, J.T. and Buresh, R.J. (1977) Correction for dissolved nitrous oxide in nitrogen studies. *Soil Science Society of America Journal* 41, 1201–1202.

Parkin, T.B., Sexstone, A.J. and Tiedje, J.M. (1985) Adaptation of denitrifying populations to low soil pH. *Applied and Environmental Microbiology* 49, 1053–1056.

Smith, R.V., Foy, R.H., Lennox, S.D., Jordan, C., Burns, L.C., Cooper, J.E. and Stevens, R.J. (1995) Occurrence of nitrite in the Lough Neagh river system. *Journal of Environmental Quality* 24, 952–959.

Thomsen, J.K., Geest, T. and Cox, R.P. (1994) Mass spectrometric studies of the effect of pH on the accumulation of intermediates in denitrification by *Paracocccus denitrificans*. *Applied and Environmental Microbiology* 60, 536–541.

Tinsley, J. (1950) The determination of organic carbon by dichromate mixtures. *Transcriptions IV International Congress on Soil Science* 1, 161–164.

Weast, A.C. (1968) *Handbook of Chemistry and Physics*, 49th edn. The Chemical Rubber Company, Ohio.

Weier, K.L and Gilliam, J.W. (1986) Effect of acidity on denitrification and nitrous oxide evolution from Atlantic Coastal Plain soils. *Soil Science Society of America Journal* 50, 1202–1205.

Yoshinari, T. and Knowles, R.J. (1976) Acetylene inhibition of nitrous oxide reduction by denitrifying bacteria. *Biochemical and Biophysical Research Communications* 693, 711–715.

Contribution of Urine Patches to the Emission of Nitrous Oxide

18

A. Vermoesen, O. Van Cleemput and G. Hofman

University of Ghent, Faculty of Agricultural and Applied Biological Sciences, Coupure 653, 9000 Ghent, Belgium

Summary. Nitrous oxide (N_2O) fluxes were measured from a grassland after application of fresh urine. On a plot of 4.5 m^2, 5 l of urine m^{-2} were spread. Another plot of 4.5 m^2 was used as a blank. To measure the N_2O flux, six closed boxes were used on each plot. At the same time, the $N_2O:N_2O + N_2$ ratio was determined in the laboratory by using the soil core method, with and without the addition of acetylene. During the measuring days, soil temperature and moisture content were also noted. The upper 10 cm soil was analysed for ammonium (NH_4^+) and nitrate (NO_3^-). As well as nitrous oxide losses, ammonia (NH_3) volatilization and leaching were quantified and the N uptake by the grass was determined. Nitrous oxide emissions of 2.4%, 1.23% and 0.71% of the applied urine-N were measured.

Introduction

Permanent and temporary grassland occupies an important part of the agricultural area. In Belgium it represents 46.6% of the agricultural land (NIS, 1992). In many cases, intensive mineral fertilization is applied. In the last ten decades nitrogen (N) application has increased from 100 to 400 kg N ha^{-1} $year^{-1}$ with peak doses of up to 500 kg N ha^{-1} $year^{-1}$ (Van der Meer, 1991). In addition, on grazed pastures a substantial amount of N returns to the soil through urine and faeces. The main N constituent of urine is urea (80–90%) which hydrolyses readily to ammonium (NH_4^+). This can be taken up by the grass but part of it will be lost as a result of different processes such as nitrification, denitrification and ammonia (NH_3) volatilization. In this field study, it was the aim to determine the percentage of the applied urine-N which can be lost as nitrous oxide (N_2O). In addition, some other losses were determined to establish a N balance.

Materials and Methods

A field experiment was carried out on a grassland on a sandy-loam soil, with a pH of 5.8. It was repeated three times with different climatic conditions. The experimental area of 2×4.5 m^2 was part of a pasture: 4.5 m^2 was treated with urine and the other 4.5 m^2 acted as blank. For each experiment, $5 \, l \, m^{-2}$ of fresh urine were spread. Nitrous oxide emission was measured by using the closed cover box method. In the laboratory, a soil core method was used to estimate the $N_2O : N_2O + N_2$ ratio. Nitrous oxide was measured by gas chromatography.

As well as N_2O losses, NH_3 volatilization was also measured during the first days of the experiment. An open box system was used and NH_3 was captured in boric acid (1%). The mineral N content of the upper 90 cm soil was measured at the end of the experiment as well as the uptake by the grass. Soil moisture content (upper 10 cm), soil temperature (upper 5 cm) and mineral N content (upper 10 cm) were determined on each day of measurement.

Results and Discussion

Experiment 1: 3 May to 7 June 1995 (35 days) (soil temperature: min. 8°C, max. 16°C, mean, 12.1°C; soil moisture content: min. 25.7%, max. 35%, mean, 30.6%)

The evolution of the N_2O emission out of the grassland with and without application of urine is shown in Fig. 18.1. It is clear that upon addition of urine, N_2O emission was enhanced by up to 100 times in comparison with the control. During the first 2 days, almost no difference was noted between the treated and untreated plot. From the third day, the N_2O emission from the urine-treated plot increased. From the analysis of the upper 10 cm soil, it can be concluded that the N_2O was mainly formed during denitrification. At the beginning of the experiment, the NO_3^- concentration was still low while the NH_4^+ content was already high (i.e. 177 kg N ha^{-1}): during this period N_2O emission was limited. From day 6 onwards, the NO_3^- concentration increased, followed by a higher N_2O emission. At day 10, the NO_3^- content was still high but the moisture content and the soil temperature were decreasing. After day 13, moisture content and temperature were increasing again. Why the N_2O emission decreased at days 25 and 28 was not clear. There was still abundant

NO_3^- and NH_4^+, and the temperature was 12°C and the moisture content 33%, i.e. not limiting to N transformation processes. Statistical analysis showed that temperature and NO_3^- were the most influential variables.

By integration of the emission curves, the total N_2O emission was calculated. For the urine-treated plot, a total emission of 7.7 kg N_2O-N ha^{-1} was found over 35 days. The $N_2O : N_2O + N_2$ ratio was determined in the laboratory as 0.42, thus a total $(N_2O + N_2)$-N loss of 18.25 kg N ha^{-1} over 35 days could be calculated. On the blank plot, only 0.08 kg N_2O-N and 0.28 kg $(N_2O + N_2)$-N ha^{-1} was emitted. This means that 5.62% of the applied urine-N was lost as N_2O and N_2 and *c.* 2.4% was due to N_2O emission.

At the end of the first experiment, the N uptake by the grass, as well as the mineral N content in the soil layers 0–30 cm, 30–60 cm and 60–90 cm, was

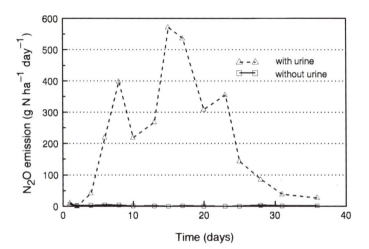

Fig. 18.1. Nitrous oxide emission from the urine-treated and untreated grassland: experiment 1.

Table 18.1. Summary of experiments 1 (35 days), 2 (27 days) and 3 (19 days): N balances as percentage of the applied urine-N (320, 350 and 555 kg N ha^{-1}, respectively).

	N balances (% of the applied urine-N)		
	Experiment 1	Experiment 2	Experiment 3
N_2O emission	2.4	1.11	0.13
$(N_2O + N_2)$ emission	5.6	1.23	2.24
NH_3 volatilization	10.8	—	4.52
N uptake by the grass	30.7	13.6	7.48
Mineral N in the soil (0–60 cm)	19.0	13.0	56.07
Leached mineral N (60–90 cm)	—	1.8	2.6
Total recovered N	66.1	29.6	72.9

measured to allow a N balance to be determined: the results are given in Table 18.1. The losses measured on the blank plot were subtracted from those from the urine-treated plot. The mineral N content of the 60–90-cm layer was the same for the treated and the untreated plot indicating that over the 35 days no NO_3^- leaching occurred. In this first experiment, about 34% of the applied urine-N was not recovered. Other components of the budget which were not measured are NO emission and immobilization of N in the roots of the grass and in soil microorganisms.

Experiment 2: 4–31 July 1995 (27 days) (soil temperature: min. 16°C, max. 26°C, mean, 18.6°C; soil moisture content: min. 12.2%, max. 31.5%, mean, 22%)

The evolution of the N_2O losses of the urine-affected and blank treatments are shown in Fig. 18.2. Here also the highest N_2O emissions were measured only after a few days, which can be explained by the fact that during the first 5 days the NO_3^- content was still low. The N_2O emission stopped after 15 days of the urine application because the soil moisture content decreased to values lower than 20%. Table 18.1 gives the results of the N-balance of experiment 2.

In this experiment, about 70% of the applied urine-N was not recovered. The measurement of the NH_3 volatilization was not successful: this may have been relatively high because of the high temperatures at that time ($\pm$ 20°C).

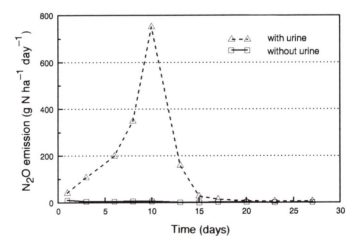

Fig. 18.2. Evolution of the N_2O emission from the grassland with and without addition of urine: experiment 2.

As in experiment 1, NO-losses and immobilization may be important components of the N balance.

Experiment 3: 17 April to 6 May 1996 (19 days) (soil temperature: min. 7°C, max. 13°C, mean, 9.5°C; soil moisture content: min. 17.5%, max. 30%, mean, 25.3%)

In this experiment, the untreated plot was moistened with $5 \, l \, H_2O \, m^{-2}$. The N_2O emissions of the treated and the blank plot are given in Fig. 18.3. The influence of the urine application on the N_2O emission lasted only a few days. The total N_2O emission during the 19 days was again calculated by integration of the curves. Ammonia volatilization was measured during the first 2 days. The results of experiment 3 are summarized in Table 18.1. During this experiment, almost no N_2O loss was noted after 2 days after the urine application. This could be due to the extremely low temperatures and low soil moisture content.

Conclusions

In the three experiments, the values for the N_2O emissions from the urine-treated grassland were low in comparison with other studies: e.g. de Klein (1994) found that 18% of the applied urine was lost through ($N_2O + N_2$)

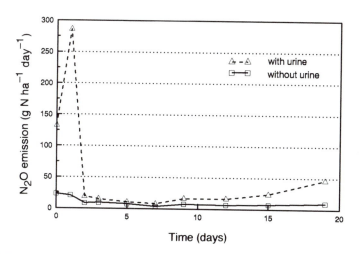

Fig. 18.3. Evolution of the N_2O emission from the grassland with and without addition of urine: experiment 3.

emission. By contrast, Velthof and Oenema (1994) also measured very low N_2O emissions: only 0.5% of the applied urine N was lost as N_2O. These low emissions were attributed to low temperatures and dry weather during the measuring period, as well as to fast immobilization and NH_3 volatilization. Nevertheless, these low amounts of N_2O emissions are important in a total N-budget and are partly responsible for the annual increase of the N_2O concentration in the atmosphere, leading to the enhancement of the green-house effect and the destruction of the stratospheric ozone layer.

References

de Klein, C.A.M. (1994) Denitrification and N_2O emission from urine-affected grass-land soil. PhD thesis, Wageningen, The Netherlands, 120 pp.

van der Meer, H.G. (1991) *Stikstofbenutting en -verliezen van gras- en maïsland.* Ministry of Agriculture, Nature Conservation and Fishery, Wageningen, The Netherlands, 134 pp.

NIS (1992) *Landbouwstatistiek nr. 1.* Nationaal Instituut voor de Statistiek, Koninkrijk België, Ministerie van Ekonomische zaken, 214 pp.

Velthof, G.L. and Oenema, O. (1994) Effect of nitrogen fertilizer type and urine on nitrous oxide flux from grassland in early spring. In: 't Mannetje, L. and Frame, J. (eds) *Proceedings of the 15th General Meeting of the European Grassland Federation.* Wageningen Pers, Wageningen, pp. 458–462.

Nitrous Oxide Emissions from Excreta from a Simulated Grazing Pattern and Fertilizer Application to Grassland

19

S. Yamulki and S.C. Jarvis

Institute of Grassland and Environmental Research, North Wyke, Okehampton, Devon EX20 2SB, UK

Summary. Nitrous oxide (N_2O) emissions from excreta (dung or urine) or fertilizer application were measured from six experiments conducted on six separate plots between September 1994 and November 1995. Applications of treatments were made at different times of the year in order to represent distinct components of a grazing season. Preliminary results showed that N_2O emission rates from excreta were large and could be compared, per unit area, to that from fertilizer application. Nitrous oxide emission rates from the fertilizer treatment were significantly dependent on soil mineral N ($r > 0.7$), whereas those from excreta were not. The mean percentage loss of dung, urine or fertilizer N as N_2O-N from all the experiments was in the order fertilizer (0.82%) > urine (0.68%) > dung (0.26%). The total cumulative N_2O flux varied substantially with the grazing season and the timing of fertilizer application.

Introduction

It is estimated that livestock waste can account for more than 12% of the total nitrous oxide-nitrogen (N_2O-N) emissions from terrestrial sources in the UK (INDITE, 1994). About 250 kt of ammonium (NH_4^+)-N are applied to land each year from livestock waste (INDITE, 1994), which ultimately may contribute to this emission. As well as this source, excreta deposited directly by grazing animals to pastures also contribute to the N_2O flux to the atmosphere. Data on N_2O emissions from animal excreta deposited in grazed grassland are few and their interaction with environmental conditions and their contribution to global N_2O flux are still uncertain.

©CAB INTERNATIONAL 1997. *Gaseous Nitrogen Emissions from Grasslands*
(eds S.C. Jarvis and B.F. Pain)

Earlier studies (Allen *et al.*, 1996) showed that there were important differences in N_2O emission from dung and urine depending on when these were deposited and on which type of soil. This research has been extended, and in this study N_2O emission from cattle dung and urine application in a simulated pattern through a grazing season was measured. The results were also compared to N_2O emissions from applications of mineral N as NH_4NO_3 fertilizer made at the same time.

Materials and Methods

The flux measurements were made on a silty clay loam soil (Halstow series) at the Institute of Grassland and Environmental Research, in Devon, southwest England. The sward, comprising perennial ryegrass (*Lolium perenne* L.) and other grasses, had not been fertilized or grazed for at least 12 months prior to the experiment.

In order to measure fluxes, a plot (4 × 40m) was divided into four subplots, three were used for the dung, urine and fertilizer treatments and the other left untreated as a blank. Within each of the appropriate subplots, 50–60 samples of dung or urine were applied to precisely located 20-cm-diameter areas. Fresh samples were collected on each occasion from dairy cows, and 1.2 kg dung or 200 ml of urine (i.e. representative of typical animal excreta deposition rates on a mass per area basis) were applied to the 20 cm diameter areas. Nitrous oxide emission was also measured from applications of mineral fertilizer (120 kg N ha^{-1} as NH_4NO_3) at the same time. These applications were repeated six times through the grazing season on separate plots in order to represent typical components of a grazing season and were monitored over 15 months. Soil available nutrients (NH_4^+ and NO_3^-), moisture content and meteorological data were also measured.

Nitrous oxide emissions were measured using a closed chamber technique. Between three and six cylindrical chambers (approximately 31 l) per treatment were inserted into the soil: each covered an area of 0.126 m^2, equivalent to twice the area of excreta patches, as indicated by Allen *et al.* (1996). However, for fertilizer treatments, the whole area within the chamber received fertilizer. During the flux measurements, each chamber was sealed with a lid for 40 min and fluxes were determined from the N_2O concentration increase within the chamber above that of ambient. N_2O samples taken by syringe from the chambers were analysed in the laboratory using a GC with an ECD detector connected to a head-space sampler (Hewlett-Packard type 19395A) which analyses up to 24 samples automatically. This paper presents some of the detail of emissions from one application time, and a preliminary synthesis of overall emissions over time from all the application times.

Results and Discussion

A typical example of changes in the soil available NH_4^+ + NO_3^- measured from the dung, urine and fertilizer treatments after application on 31 May 1995, plotted against time, is shown in Fig. 19.1. The results generally showed a similar pattern of changes in available N during all the experimental runs. High peaks of available N were observed from the fertilizer and urine treatments immediately after application, followed by sharp decreases. This was in contrast to the effects of dung which maintained a similar concentration throughout the measurement period, but remained higher than that from the urine which returned to the background levels after 20 weeks. Ammonium concentrations were initially higher than those of NO_3^- in all the treatments and remained higher throughout the whole experimental period.

Nitrous oxide emission rates (Fig. 19.2) followed a similar pattern to that of the mineral N, although a significant correlation was not observed. The patterns of N_2O emissions from the dung treatments were complex. In most of the experimental runs, N_2O emissions did not peak immediately after the application of dung except on one occasion. The peak of N_2O emission in one experiment (on 17 July 1995) occurred immediately after application, even before that from the fertilizer and urine. In contrast, in another experiment N_2O emissions from the dung treatment did not peak until 4 months after the application on 16 September 1994.

Preliminary results from all the experiments showed that the total cumulative N_2O flux (Fig. 19.3) from excreta was large and similar, per unit of

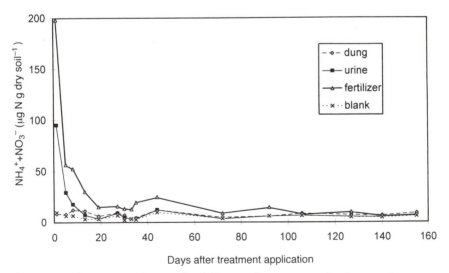

Fig. 19.1. Changes in soil mineral N following dung, urine or fertilizer application on 31 May 1995.

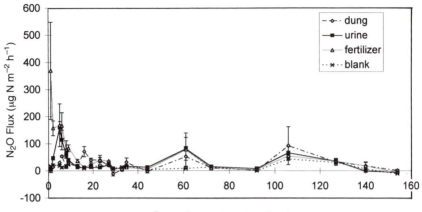

Fig. 19.2.   Effects of dung, urine or fertilizer application on 31 May 1995 on N_2O emission.

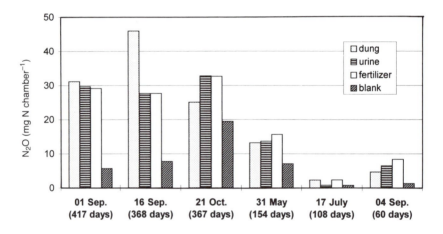

Fig. 19.3.   Cumulative fluxes of N_2O-N after application of dung, urine or fertilizer at various times of the year.

affected area, to that from fertilizer application. The total cumulative N_2O flux varied with application time from trace amounts to more than 45 mg N per treated area. These variations must be attributed to the effect of environmental conditions on the processes controlling N_2O fluxes, as the experimental protocols and application rates were similar for all the experiments. However, the cumulative fluxes clearly demonstrate that the greatest losses occurred as the result of additions made in autumn and early spring. As well as the fluxes of N_2O occurring as a result of additional N being added in excreta or fertilizer, significant amounts were also being lost from the untreated soil, especially in spring. Presumably this is linked with a flush of NO_3^- derived from mineralization but the extent of this should be considered when full accounts of inventories of N_2O emissions are being developed.

Acknowledgements

The research was funded by MAFF, London. IGER is supported by the BBSRC.

References

Allen, A.G., Jarvis, S.C. and Headon, D.M. (1996) Nitrous oxide emissions from soils due to inputs of nitrogen from excreta return by livestock on grazed grassland in the UK. *Soil Biology and Biochemistry* 28, 597–607.

INDITE (1994) Impacts of nitrogen deposition in terrestrial ecosystems, October 1994. In: *Report of the United Kingdom Review Group on Impacts of Atmospheric Nitrogen.* Department of the Environment, London, 17 pp.

Nitrous Oxide Emission from Slurry and Mineral Nitrogen Fertilizer Applied to Grassland

20

I.P. McTaggart[1], J.T. Douglas[1], H. Clayton[2] and K.A. Smith[2]

[1]Soils Department, SAC and [2]IERM, University of Edinburgh, School of Agriculture, West Mains Road, Edinburgh EH9 3JG, UK

Summary. The disposal of livestock slurry on grassland can have significant implications for nitrous oxide (N_2O) emissions, because of increased soil compaction through additional traffic, and increased wetness and soluble carbon. Measurements of N_2O were made on a compacted or uncompacted, imperfectly drained clay loam soil, with a combination of ammonium nitrate (AN) closely followed by slurry N (SN). Emissions were also measured from uncompacted plots with AN only. Nitrous oxide emissions increased dramatically to > 1 kg N ha^{-1} day^{-1} immediately with AN + SN, but decreased quickly after 2–3 days. In April, total emissions from compacted plots were > 5 kg N ha^{-1}, more than double those from uncompacted plots. Soil moisture and temperature had substantial effects at different times of the year. The cause of increased emissions from AN + SN was further investigated. Treatments were established on compacted plots with two applications, 7 days apart, of SN or AN only, AN followed by SN and vice versa, or with an equivalent volume of water instead of slurry. Emissions of N_2O from the SN only treatment remained low throughout, but when AN was applied after SN, emissions increased immediately and followed a similar trend after each AN application, peaking at 0.4 kg N ha^{-1} day^{-1} within 2 days. Ammonium nitrate with added water did not alter emission rates, possibly due to the very wet soil conditions in April. Applying the SN prior to the AN significantly reduced emissions.

Introduction

Microbial processes in soils are the dominant source of atmospheric nitrous oxide (N_2O) (Prather *et al.*, 1995). In the UK it is estimated that intensively managed grassland soils are the major agricultural source of N_2O emissions

(Skiba *et al.*, 1996). A major contributory factor is the large inputs of available nitrogen (N) in the form of mineral fertilizers and/or slurry. Soil physical factors also play a part; Hansen *et al.* (1993) reported that N_2O emissions following application of mineral N fertilizer were higher on soils with increased compaction. The presence of available organic carbon (C) in applied slurry can lead to greater emissions of N_2O from arable soils of low organic matter content than when inorganic fertilizer N is applied (Christensen, 1985; Van Cleemput *et al.*, 1992). However, on grassland, where organic C availability is unlikely to limit denitrification (Elliot *et al.*, 1991), emissions of N_2O following the application of mineral N fertilizers are higher than from slurry N applications (Egginton and Smith, 1986; Hansen *et al.*, 1993).

The aims of this research were to compare emissions of N_2O following no, partial, or the entire replacement of ammonium nitrate fertilizer with livestock slurry applied to grassland. In addition, the effect of soil compaction was studied, following the application of a mixture of both forms of nitrogen.

Materials and Methods

The experiment was located 10 km south of Edinburgh on an imperfectly drained clay loam soil sown to perennial ryegrass for silage cutting. Plots measuring 28 m × 2.4 m were sown, fertilized and harvested without the intrusion of wheel traffic. Uniformly compacted plots were established in March 1992 by successive adjacent passes by a tractor weighing 5 t. In 1993 and 1994 granular ammonium nitrate (AN) was spread from a seed-drill, then an approximately equal amount of low dry matter (DM) (~2%) slurry N (SN) was spread from a tanker with a 2.4 m wide dribble bar, to three replicates each of compacted and uncompacted plots, at the rates shown in Table 20.1. An additional treatment of AN only was applied to a further three replicate uncompacted plots at a rate of 120 kg N ha^{-1} on the same dates. In 1995, the experiment was restricted to 7.2 m^2 areas within compacted plots. Slurry was applied from a cylinder with a 2.4-m-wide dribble curtain moved along a 3-m-long frame. Granular AN was hand-broadcast. Two alternating applications of 75 kg N ha^{-1} each were applied on 17 April and 24 April as AN or SN

Table 20.1. Amounts of nitrogen applied as granular ammonium nitrate (AN) or slurry (SN) in April and June 1993 and 1994.

N applied (kg N ha^{-1})	April 1993		June 1993		April 1994		June 1994	
	AN + SN	AN only	AN + SN	AN only	AN + SN	AN only	AN + SN	AN only
AN	60	120	50	120	60	120	50	120
SN	59	0	51	0	56	0	44	0
Total	119	120	101	120	116	120	94	120

Table 20.2.   Source of nitrogen applied at 75 kg N ha^{-1} to four replicate plots starting in April and June 1995.

Day of application	Treatments applied					
Day 1	SN	AN	ANw	AN	SN	AN
	+	+	+	+	+	+
Day 8	SN	AN	AN	ANw	AN	SN

AN = ammonium nitrate; SN = slurry; ANw = ammonium nitrate with water equal to moisture in slurry.

to four replicate plots (Table 20.2). Water, in the same quantity as that applied in slurry, was applied to AN plots using the slurry applicator described above. The treatments were repeated on 26 June and 3 July.

Nitrous oxide fluxes were measured by means of 40-cm-diameter chambers (Clayton *et al.*, 1994; Smith *et al.*, 1995). In 1993 and 1994 there were two chambers in each of the three replicate plots, while in 1995 there were four replicate plots with one chamber in each. Samples of the enclosed atmosphere taken 1 h after closure were analysed by ECD gas chromatography. Soil moisture, dry bulk density and air-filled porosity were measured on 0–10-cm cores, with the air-filled pore space expressed as a percentage of the void volume.

Results and Discussion

Following application of N in 1993, emissions of N_2O increased in all treatments, but the rise was sharper for ammonium nitrate followed by slurry treatments (AN + SN) than for AN only (Fig. 20.1). Maximum emissions in April were higher for compacted plots than for non-compacted plots, with a maximum flux of 1170 g N ha^{-1} day^{-1}. The duration of high emissions was shorter from AN + SN plots than from AN only plots. In June, the warmer conditions resulted in the highest emission rates, illustrated by the maximum flux from the non-compacted AN + SN plots of 1709 g N ha^{-1} day^{-1}, 2 days after N application (Fig. 20.1). The shortest duration of high fluxes also occurred in June, with the maximum flux of 1500 g N ha^{-1} day^{-1} from the compacted AN + SN plots falling to < 30 g N ha^{-1} day^{-1} within 4 days.

Cumulative emissions of N_2O from compacted AN + SN plots were greater than from non-compacted plots following April N applications in 1993 and 1994 (Fig. 20.2). Following N applications in June 1993, however, cumulative emissions were greater from non-compacted AN + SN plots, nearly double those from the compacted AN + SN plots. In April, the soil temperature was 7°C (5 cm depth) at the time of application, but by June the soil temperature had risen to c. 15°C, while soil air-filled pore space (AFPS) increased only slightly compared with that in April (Fig. 20.3). In June, denitrification would have been the likely main source of N_2O emissions, due to the low percentage of AFPS. As the higher temperature would be expected to increase the overall denitrification rate, the reduced flux from the

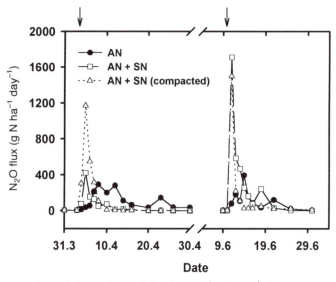

Fig. 20.1. Daily emissions of N_2O following application of nitrogen as ammonium nitrate (AN) or slurry (SN) to compacted or non-compacted grassland soil in April and June 1993. ↓= time of nitrogen application.

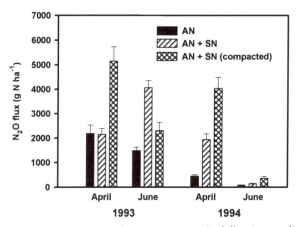

Fig. 20.2. Cumulative emissions of N_2O over 9 weeks following application of nitrogen as ammonium nitrate (AN) or slurry (SN) to compacted or non-compacted grassland soil in April and June 1993 and 1994.

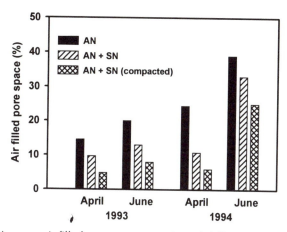

Fig. 20.3. Average air filled pore space over 1 week following application of nitrogen as ammonium nitrate (AN) or slurry (SN) to compacted or non-compacted grassland soil in April and June 1993 and 1994.

compacted AN + SN treatment, compared with emissions following April N applications, could have been due to a proportionately larger increase in the $N_2 : N_2O$ ratio, as greater microbial activity produced more anaerobic sites. The higher emissions from the non-compacted AN + SN treatment in June, compared with April, suggests that here the balance lay in the other direction. Over the same period, emissions from non-compacted plots treated with urea were similar to the non-compacted AN + SN treatment (McTaggart *et al.*, 1994). Hansen *et al.* (1993) found soil N_2O concentrations more than seven times higher in compacted compared with non-compacted soil, although surface emissions were only one and a half times greater, suggesting either low gas diffusion within the soil profile, or significant reduction of N_2O to N_2.

In June 1993 and April 1994, emissions from non-compacted AN + SN plots were up to four times greater than from AN (Fig. 20.2). The higher emissions from AN + SN could have been due to an overall increase in soil moisture, washing in of applied N to sites of denitrification, or organic C derived from the applied slurry enhancing denitrification. Egginton and Smith (1986) reported low emissions when slurry alone was applied to grassland, with much greater emission from applied calcium nitrate, under wet conditions. However, in the drier conditions in June 1994, emissions from all treatments were very low.

The purpose of the experiments in 1995 was to examine whether the higher emissions from AN + SN were caused by increased soil moisture due to slurry addition and whether AN or SN was the main source of the emitted N_2O. In April, emissions from the SN only treatment (SN + SN) remained at < 26 g N ha^{-1} day^{-1} throughout (Fig. 20.4(a)). However, when AN was applied following SN (SN + AN), emissions increased to 300 g N ha^{-1} day^{-1}.

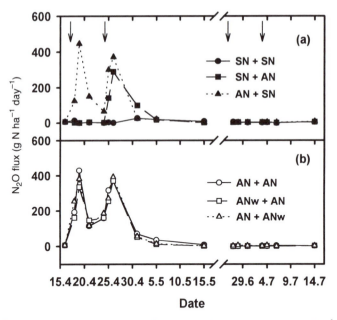

Fig. 20.4. Daily emissions of N_2O following application of 150 kg N ha^{-1} divided in two equal applications 7 days apart, as ammonium nitrate (AN), slurry (SN) or ammonium nitrate with added water (ANw) to compacted grassland soil in April and June 1995. ↓ = time of nitrogen application.

Emissions always increased rapidly following each application of AN, peaking at around 400 g N ha^{-1} day^{-1} within 2 days (Fig. 20.4(a),(b)). Cumulative emissions over 30 days (Fig. 20.5) showed that emissions from a combination of AN and SN were not greater than from AN only. In the two previous years, the comparison had been made on non-compacted soil, where the greater hydraulic conductivity of the soil could have allowed the added water from slurry to wash more of the AN into the soil to sites of denitrification. Ammonium nitrate, with added water to simulate the water in the slurry (ANw), did not increase emission rates in April, possibly due to the wet soil conditions already prevailing, resulting in low values of AFPS (Fig. 20.6). In the 1995 experiment, the compacted soil and wet soil conditions, enhanced by two significant rainfall events (Fig. 20.6), minimized any impact of added water. However, the experiment did show that the main source of emissions was from AN rather than from SN, and that applying SN a few days prior to AN resulted in reduced emissions, compared with the reversed treatment (Fig. 20.5). In the AN + SN treatment, emissions immediately following the SN application were similar to those following the earlier AN application, whereas where the SN was applied first with no previous AN application, emissions remained very low. When the experiment was

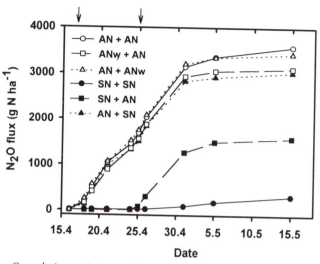

Fig. 20.5. Cumulative emissions of N_2O following application of 150 kg N ha^{-1} divided into two equal applications 7 days apart, as ammonium nitrate (AN), slurry (SN) or ammonium nitrate with added water (ANw) to compacted grassland soil in April 1995. ↓= time of nitrogen application.

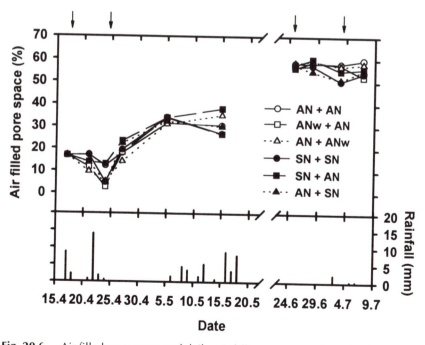

Fig. 20.6. Air filled pore space and daily rainfall on compacted grassland soil following application of nitrogen in April and June 1995. ↓= time of nitrogen application.

repeated under warmer conditions in June, emissions from all treatments remained extremely low (Fig. 20.4(a),(b)). This appeared to be caused by the drier soil conditions at this time (Fig. 20.6). The addition of moisture, either in the slurry or the applied water substitute, did not have any effect.

Conclusions

Emissions of N_2O were greater from compacted soils than from non-compacted soils, except during moist warm conditions in June 1993. On non-compacted soils, a combination of AN followed by SN increased emissions compared with AN only. However, this was not repeated on compacted soils in 1995. The main source of emissions was from applied AN, rather than from applied SN. The application of SN a few days prior to AN gave lower emissions, compared with those from the reversed treatment.

Acknowledgements

The authors would like to thank the Department of Agriculture and Fisheries for Scotland, the Commission of the European Communities (STEP Programme) and the NERC Terrestrial Initiative in Global Environmental Research (TIGER) for financial support.

References

Christensen, S. (1985) Denitrification in an acid soil. Effects of slurry and potassium nitrate on the evolution of nitrous oxide and on nitrate reducing bacteria. *Soil Biology and Biochemistry* 17, 757-764.

Clayton, H., Arah, J.R.M. and Smith, K.A. (1994) Measurement of nitrous oxide emissions from fertilized grassland using closed chambers. *Journal of Geophysical Research* 99, No. D8, 16,541–16,548.

Egginton, G.M. and Smith, K.A. (1986) Nitrous oxide emission from a grassland soil fertilized with slurry and calcium nitrate. *Journal of Soil Science* 37, 59-67.

Elliot, P.W., Knight, D. and Anderson, J.M. (1991) Variables controlling denitrification from earthworm casts and soil in permanent pastures. *Biology and Fertility of Soils* 11, 24–29.

Hansen, S., Mæhlum, J.E. and Bakken, L.R. (1993) N_2O and CH_4 fluxes in soil influenced by fertilization and tractor traffic. *Soil Biology and Biochemistry* 25, 621–630.

McTaggart, I.P., Clayton, H. and Smith, K.A. (1994) Nitrous oxide flux from fertilized grassland: strategies for reducing emissions. In: Van Ham, J., Janssen, L.J.H.M. and Swart, R.J. (eds) *Non-CO2 Greenhouse Gases*. Kluwer Academic, The Netherlands, pp. 421–426.

Prather, M., Derwent, R., Ehhalt, D., Fraser, P., Sanhueza, E. and Zhou, X. (1995) Other trace gases and atmospheric chemistry. In: Houghton, J.T., Meira Filho, L.G., Bruce, J., Hoesung Lee, Callander, B.A., Haites, E., Harris, N. and Maskell, K. (eds) *Climate Change 1994: Radiative Forcing of Climate Change and An Evaluation of the IPCC IS92 Emission Scenarios.* Cambridge University Press, New York, pp. 77–126.

Skiba, U., McTaggart, I.P., Smith, K.A., Hargreaves, K.J. and Fowler, D. (1996) Estimates of nitrous oxide emissions from soil in the UK. *Energy Conversion and Management* 37, 1303–1308.

Smith, K.A., Clayton, H., McTaggart, I.P., Thomson, P.E. and Arah, J.R.M. (1995) The measurement of nitrous oxide emissions from soil using chambers. *Philosophical Transactions of the Royal Society of London A* 351, 327–338.

Van Cleemput, O., Van Hoorde, J. and Vermoesen, A. (1992) Emission of N_2O under different cropping systems. In: François, E., Pithan, K. and Bartiaux-Thill, N. (eds) *Proceedings of COST Workshop 814: Nitrogen Cycling and Leaching in Cool and Wet Regions of Europe.* Gembloux, pp. 20–21.

Posters 8–13

III

Use of a Specially Configured Gas Chromatographic System for the Simultaneous Determination of Methane, Nitrous Oxide and Carbon Dioxide in Ambient Air and Soil Atmosphere

H.J. Segschneider, I. Sich and R. Russow

UFZ-Umweltforschungszentrum Leipzig-Halle GmbH, Sektion Bodenforschung (Leipzig) PF 2, D – 04301 Leipzig, Germany

In recent years investigations on the trace gas turnover of methane (CH_4), nitrous oxide (N_2O) and carbon dioxide (CO_2) in the soil–plant–atmosphere system have become of increasing interest (Granli and Brockman, 1994). Studies on the influence of different agricultural management strategies on the flux of these trace gases, as well as the high spatial and temporal variability of the N_2O emission from soils, require a large number of gas sample analyses to develop reasonable estimations of annual fluxes (Duxbury *et al.*, 1993). Furthermore, in the range of ambient concentrations relatively high volumes of gas samples have to be injected for chromatographic analysis (GC) (Heinemeyer *et al.*, 1991). Nevertheless, GC analysis of air samples drawn from soil covers (Heinemeyer *et al.*, 1995) or from the soil atmosphere (Magnusson, 1989) is still the most commonly applied analytical procedure used to determine these compounds.

An 'easy-to-handle' system comprising an auto-sampler for head-space analysis connected to a gas chromatograph equipped with a flame ionization detector (FID) to measure CH_4, and an electron capture detector (ECD) for N_2O and CO_2 determination, has been developed.

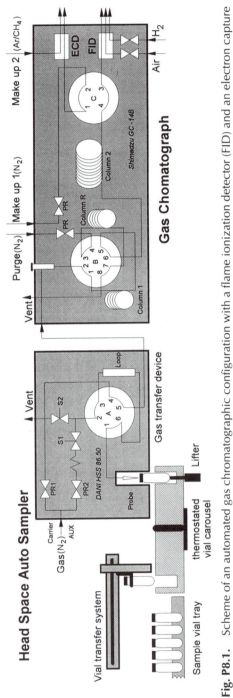

Fig. P8.1. Scheme of an automated gas chromatographic configuration with a flame ionization detector (FID) and an electron capture detector (ECD) for the simultaneous determination of methane (CH_4), carbon dioxide (CO_2) and nitrous oxide (N_2O) in ambient air and soil atmosphere gas samples.

The gas transfer device of the head-space auto-sampler used is based on the 'valve and loop' principle, which allows the injection of a precise and known amount of a gas sample into the GC. To determine CH_4 and N_2O in the range of ambient concentrations, a sample loop with a volume of 3 ml was used. The GC operational conditions were as follows:

1. ECD: temperature, 290°C; current, 1 nA; range, 1.
2. FID: hydrogen flow rate, 50 ml min^{-1}; air flow rate, 500 ml min^{-1}; range, 1.
3. Columns 1 and R: 1 m × 1/8″ i.d.; Column 2, 3 m × 1/8″ i.d. stainless-steel column; column packing, *PORAPAK* Q 80/100 mesh; temperature, 30°C; injector temperature, 100°C; carrier gas, N_2 (5.0); flow rate, 35 ml min^{-1}.
4. Make up gas line 1: N_2 (5.0); flow rate, 35 ml min^{-1}.
5. Make up gas line 2: Ar/CH_4 mixture (P10); flow rate, 5 ml min^{-1}.
6. Purge gas: N_2 (5.0); flow rate, 35 ml min^{-1}.

Column 1 is used as a pre-column to separate water vapour from the injected gas sample. The main separation of the gas compounds takes place in column 2. Column R is installed to provide a constant pressure regime during backflush mode. A PC controlled valve switching program operates the gas flow to both detectors, which protects the ECD from contamination by the front-flushing oxygen. During phase 1 of a GC run, the FID is fed by the carrier gas stream via port 4–3 of valve C (Fig. P8.1). The ECD is supplied from the Make-up gas line 1 during this phase. Immediately after CH_4 has passed the FID, valves B and C are activated. During phase 2, Column 1 is back-flushed by the purge gas stream and water vapour is vented from the system.

Table P8.1. Response and precision of the flame ionization detector (FID) to CH_4 and the electron capture detector (ECD) to CO_2 and N_2O in different gas samples depending on the supply with nitrogen (5.0) or an Ar/CH_4 mixture (P10)* to the Make up gas line 2 (see Fig. P8.1).

Make up gas line 2 supply Sample type	Nitrogen (5.0)				Ar/CH_4 mixture (P10)*			
	Ambient air[†]			N_2O gas standard[‡]	Ambient air[†]			N_2O gas standard[‡]
n		22		22		22		22
Detector	FID	ECD		ECD	FID	ECD		ECD
Compound	CH_4	CO_2	N_2O	N_2O	CH_4	CO_2	N_2O	N_2O
	25,052	20,263	5796	15,581	25,486	21,667	18,341	47,921
±SD (counts)	125	203	284	561	77	325	275	431
±RSD (%)	0.5	1.0	4.9	3.6	0.3	1.5	1.5	0.9

*Ar/CH_4 mixture (P10): 90 vol% argon/10 vol% methane.
[†]Pressurized air from a steel cylinder used as laboratory standard.
[‡]N_2O gas standard: 1.5 ppm N_2O in synthetic air.

The ECD is now served by the carrier stream for CO_2 and N_2O detection. The analysis of one gas sample takes 6 min. Figure P8.2 shows a parallel chromatogram obtained from the separation of an ambient air gas sample . The additional supply of the ECD with an Ar/CH_4-mixture via the Make-up gas line 2 (Fig. P8.1) increases the response to N_2O, which enhances the precision of trace gas determination. Under the chosen auto-sampler GC conditions the detection limit for N_2O was $c.$ 0.03 ppm. In the range of ambient N_2O

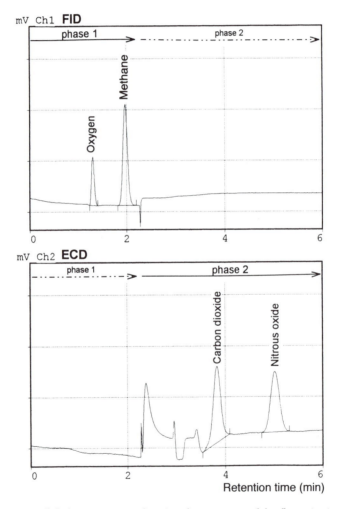

Fig. P8.2. Parallel chromatogram showing the response of the flame ionization detector (FID) and the electron capture detector (ECD) during separation of an ambient air gas sample.

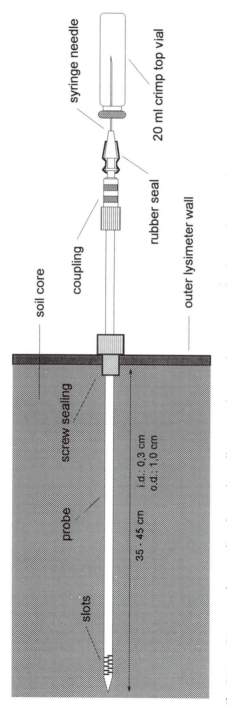

Fig. P8.3. Diagram of a sampling device for collecting soil atmosphere gas samples from soil cores in lysimeters.

concentrations, a precision of ± 1.5% relative standard deviations was achieved (Table P8.1).

A method for soil atmosphere collection based on the partial pressure gradient between hydrogen and ambient air (the diffusion principle) is presented. At depths of 0.5, 1.5 and 2.5 m, stainless-steel probes with slots at the end were inserted through the outer wall of lysimeters up to a length of 35–45 cm into the soil core (Fig. P8.3). Sample vials flushed with hydrogen or helium were fixed at the opposite end of the probe outside the lysimeter. The vials act as diffusion sinks and the air from the soil pores is forced to

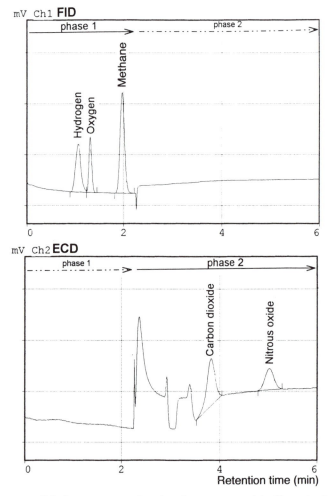

Fig. P8.4. Parallel chromatogram showing the response of the flame ionization detector (FID) and the electron capture detector (ECD) during separation of a soil atmosphere gas sample with incomplete hydrogen–soil air equilibration.

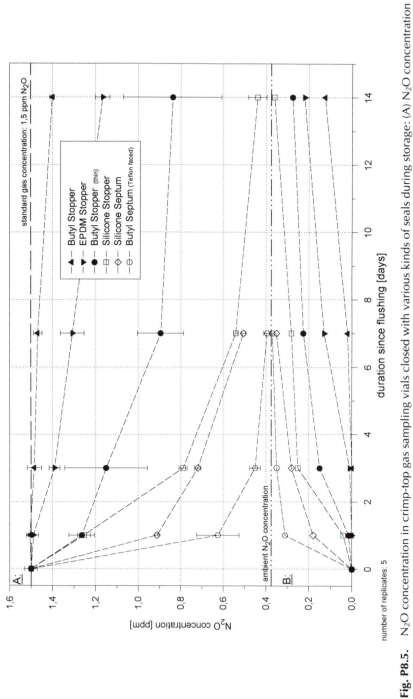

Fig. P8.5.   N₂O concentration in crimp-top gas sampling vials closed with various kinds of seals during storage: (A) N₂O concentration after flushing with a N₂O standard gas mixture. (B) N₂O concentration after flushing the vials with pure hydrogen.

equilibrate rapidly with vessel contents. After a defined time (i.e. 1 week), the vials were removed and replaced with freshly filled sampling containers. The response of the FID to the presence of remaining hydrogen gas sampling vials (Fig. P8.4) was used to check if the equilibration between soil atmosphere and vessel contents had been completed and a representative gas sample obtained. Any poor exchange was normally caused by a blockage of the probe slots with small soil particles.

Figure P8.5 shows the average N_2O concentrations in the 20-ml crimp-top head-space vials closed with different seals during gas sample storage over 14 days. Of the various vial seals tested, only butyl rubber stoppers provided a reliable seal for the 20-ml sampling vessels for gas sample storage but only over 7 days. An additional sealing with fluid silicon glue on the top of the seals did not prevent the degassing of the N_2O (data not shown).

References

Duxbury, J.M., Harper, L.A. and Mosier, A.R. (1993) Contribution of agroecosystems to global climate change. In: Rolston, D.E., Harper, L.A., Mosier, A.R. and Duxbury, J.M. (eds) *Agricultural Ecosystems Effects on Trace Gases and Global Climate Change*. ASA Special Publication Number 55, Soil Science Society of America, Madison, Wisconsin.

Granli, T. and Bøckman, O.C. (1994) Nitrous oxide from agriculture. *Norwegian Journal of Agricultural Science* 12, 7–128.

Heinemeyer, O., Walenzik, G. and Kaiser, E-A. (1991) Zur Methodik der Bestimmung gasförmiger N-Abgaben in Freilandexperimenten. *Mitteilungen der Deutschen Bodenkundlichen Gesellschaft* 66, 499–502.

Heinemeyer, O., Munch, J.C. and Kaiser E-A. (1995) Variabilität von N_2O-Emissionen – Bedeutung der Gasauffangsysteme. *Mitteilungen der Deutschen Bodenkundlichen Gesellschaft* 76, 543–546.

Magnusson, T. (1989) A method for equilibration chamber sampling and gas chromatographic analysis of the soil atmosphere. *Plant and Soil* 120, 39–47.

Nitrous Oxide Exchange in Wet Soil after Ploughing for Grass Establishment: Evidence for Uptake of Atmospheric Nitrous Oxide?

B.C. Ball[1] and H. Clayton[2]*
[1]*Soils Department, SAC and [2]University of Edinburgh, West Mains Road, Edinburgh EH9 3JG, UK; *Present address: Division of Biological Sciences, Lancaster University, Lancaster LA1 4YQ, UK*

An earlier study on an arable site on poorly drained loam soil in South-east Scotland revealed negligible emissions of nitrous oxide (N_2O) under conditions conducive to denitrification (Arah *et al.*, 1991). This was attributed partly to trapping of N_2O in the soil because of low gas diffusivity and partly to the reduction of N_2O to dinitrogen (N_2). After the site was sown to grass in mid September 1992, the situation was re-examined. N_2O fluxes were measured and attempts made to relate them to soil N_2O concentrations and soil gas diffusivities and permeabilities.

The experimental site contained a gleysol (Winton series) and had been the site for a long-term tillage experiment under continuous cereals for the previous 26 years (Ball et al., 1994a). Fluxes were measured in the field using closed chambers of 0.4 m diameter (Clayton et al., 1994) located at the intersections of a 4 × 4 grid of interval 0.8 m, and on cores in the laboratory. In situ and core methods (Ball et al., 1981, 1994b) were used to determine the gas diffusivities. Cores were taken for diffusivity, air permeability and chemical measurements, either near or within the chamber locations. Soil N_2O concentrations were measured by sampling copper tubes (probes) buried in the soil.

The weather was unusually warm and wet for a November study period: 62 mm of rain fell in the week before measurements began and soil temperatures at 200 mm depth were mostly between 4 and 8°C during the sampling period. No N fertilizer was applied on sowing, and soil N contents (0–100 mm) were low (1.9 and 2.4 $\mu g\ g^{-1}$ nitrate and ammonium N, respectively). Most chamber locations showed net uptake rather than net emission of N_2O, and the mean flux on the four measurement days (Fig. P9.1(a)) was always negative, though small, in spite of substantial soil N_2O concentrations (Fig. P9.1(b)). Soil N_2O concentrations were highest (up to 57 $\mu l\ l^{-1}$) around 170–240 mm depth, corresponding to the ploughed-in straw layer. Core incubations with and without acetylene gave zero or small positive N_2O fluxes, two-fold or more greater below 140 mm than above 140 mm depth; only up to 10% of the total dinitrogen (N_2) + N_2O loss was as N_2O.

Diffusivity was measured at two chamber locations, one showing net emission, the other net uptake of N_2O (Fig. P9.2). The core but not the in situ measurements revealed that the 50–100-mm soil layer at the location showing net emission had higher gas diffusivity and air permeability than did the corresponding layer at the chamber where N_2O was taken up. The in situ measurements did not reveal these differences because the trace gas source could not be sealed effectively above 180 mm depth. Soil N_2O at 250 mm depth was also higher under the chamber emitting N_2O. However, over the measurement area as a whole N_2O flux did not correlate with soil N_2O; negative fluxes were sometimes detected at locations with high soil N_2O at depth. Over the first three measurement occasions there was some indication that an increasingly negative mean flux corresponded with decreasing mean soil N_2O, which might in turn have been related to a decline in soil temperature. These results are in agreement with those of Hojberg et al. (1994), who found that nitrate depletion from organic 'hot spots' of denitrifying activity (such as within incorporated residues) could lead to sufficiently great N_2O reduction capacity for such spots to serve as sinks for N_2O.

At this site, under the relatively warm and wet conditions prevailing, mineralization in the straw layer provided inorganic N substrate for nitrifica-

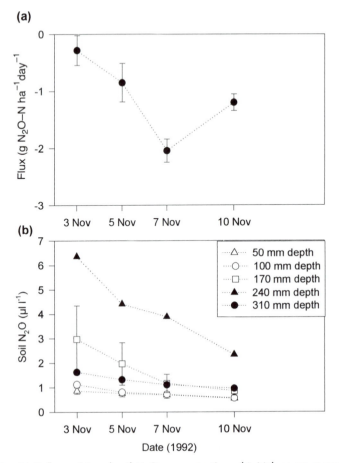

Fig. P9.1. N₂O fluxes (a) and soil N₂O concentrations (b). Values are means of all chambers and sampling probes, respectively. Error bars represent 2× standard error of the mean.

tion and denitrification to N_2O and N_2. The gas diffusivity and air permeability in the soil above the straw, influenced by rainfall, affected (a) the rate at which the N_2O could escape to the atmosphere and (b) the degree of anaerobicity of the upper soil layers. Greater anaerobicity in particularly wet or dense regions is likely to have facilitated greater reduction of N_2O from the lower layers and of N_2O from the atmosphere, leading to a negative flux.

References

Arah, J.R.M., Smith, K.A., Crichton, I.J. and Li, H.S. (1991) Nitrous oxide production and denitrification in Scottish arable soils. *Journal of Soil Science* 42, 351–367.

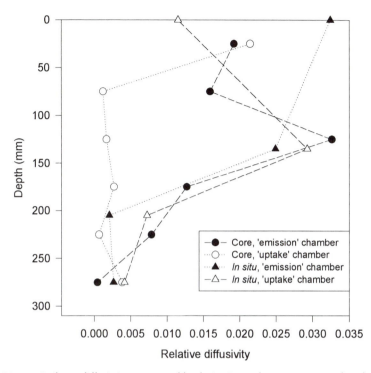

Fig. P9.2. Soil gas diffusivity measured both *in situ* and on cores near chambers showing net emission or uptake of N_2O.

Ball, B.C., Harris, W. and Burford, J.R. (1981) A laboratory method to measure gas diffusion and flow in soil cores and other porous materials. *Journal of Soil Science* 32, 323–333.

Ball, B.C., Lang, R.W., Robertson, E.A.G. and Franklin, M.F. (1994a) Crop performance and soil conditions on imperfectly drained loams after 20 to 25 years of conventional tillage or direct drilling. *Soil and Tillage Research* 31, 97–118.

Ball, B.C., Glasbey, C.A. and Robertson, E.A.G. (1994b) *In situ* measurement of gas diffusivity. *European Journal of Soil Science* 45, 3–13.

Clayton, H., Arah, J.R.M. and Smith, K.A. (1994) Measurements of nitrous oxide emissions from fertilized grassland using closed chambers. *Journal of Geophysical Research* 99, 16,599–16,607.

Hojberg, O., Revsbech, N.P. and Tiedje, J.M. (1994) Denitrification in soil aggregates analyzed with microsensors for nitrous oxide and nitrogen. *Soil Science Society of America Journal* 58, 1691–1698.

Distinguishing between Routes of Nitrous Oxide Flux from Soil in the Presence of Plants

L. Anderson and D.W. Hopkins

Department of Biological Sciences, University of Dundee, Dundee DD1 4HN, UK

Nitrous oxide (N_2O) is produced in soils predominantly during the processes of denitrification and nitrification (Blackmer *et al.*, 1980; Firestone *et al.*, 1980; Robertson and Tiedje, 1987; Webster and Hopkins, 1996). Several laboratory studies have shown that N_2O emission is increased by the presence of plants (Cribbs and Mills, 1979; Klemedtsson *et al.*, 1987). This effect is often attributed, at least in part, to increased microbial activity in the rhizosphere compared with non-rhizosphere soil linked to the supply of organic carbon (C) from roots (Klemedtsson *et al.*, 1987; Christensen *et al.*, 1990). However, reduced soil oxygen (O_2) concentration due to root respiration (Klemedtsson *et al.*, 1987) and localized compaction caused by fine roots (Barber, 1974) are also likely contributory factors. In this study the possibility that *Lolium perenne* L. mediates N_2O emission from the soil and the processes of N_2O production in the presence of *L. perenne* have been investigated.

Laboratory experiments were conducted in which N_2O accumulation in the sealed head spaces of Kilner-type jars containing 180 g (d.w.) soil (sandy-loam texture, pH 6.5, 5.3% organic matter) at field capacity (0.3 g H_2O g^{-1} soil) supporting either 12 or 0 *L. perenne* plants has been determined. When the plants were 12 days old (two-leaf stage and about 4 cm tall), the soil surface around the plants was sealed with wax and an inner chamber (total volume 36 cm^3 covering an area of 4.9 cm^3) was fitted over the plants. This allowed plant-mediated N_2O emission to be determined separately from N_2O emission at the soil surface of the outer chamber (total volume 285 cm^3 covering an area of 76 cm^2). Corresponding jars without plants were assembled with inner chambers over wax-sealed soil to act as controls. The rates of N_2O accumulation in the inner and outer chambers were determined from daily measurements (gas chromatograph fitted with ECD) over a 96-h period. In addition, the effects of 10 Pa C_2H_2 (to inhibit nitrification), 100 kPa O_2 (to suppress denitrification) and 10 kPa C_2H_2 (to inhibit N_2O reduction to N_2 during denitrification and also nitrification) (Robertson and Tiedje, 1987) added to the head space of the outer chamber were determined.

The total N_2O accumulation rate was significantly greater in the presence of plants (2.1 nmol N_2O g^{-1} soil day^{-1}) than in their absence (1.1 nmol N_2O g^{-1} soil day^{-1}). Although N_2O accumulated in the inner chambers in both the presence and absence of plants, the rate of accumulation in the inner chamber without plants was small (Table P10.1) and can probably be attributed to slow leakage across the wax seal. In the absence of plants there was a 5.6-fold difference in the rate of N_2O accumulation between the inner and outer chambers, compared with a difference of only 1.8 times with plants present. This preferential N_2O accumulation in the chambers containing plants suggests that inward leakage from the outer chamber made only a small contribution to N_2O accumulation. If it is assumed that the N_2O emission rate into the outer chamber is representative of that which would have been emitted from the soil surface between the plants had the soil surface not been sealed, then of the additional N_2O produced in the presence of plants, approximately 50% appeared to be conducted into the head space by the plant.

In the presence of plants, the effect of both 10 Pa C_2H_2 and 100 kPa O_2 was to reduce N_2O accumulation in both the inner and outer chambers (Table P10.2). The inhibitory effect of 10 Pa C_2H_2 was far greater than that of 100 kPa O_2, which suggests that nitrification was the process limiting N_2O emission, either directly or indirectly through the supply of NO_3^- to denitrifiers. The fact that the 100 kPa O_2 treatment led to a greater reduction in the N_2O accumulation rate in the inner chamber than in the outer chamber

Table P10.1. N_2O accumulation rates over 96 h in the inner and outer chambers of jars with and without *L. perenne*. Each value is the mean of duplicate jars and the standard deviations are shown in parentheses. Where *L. perenne* plants were present, they were sealed into the inner chambers and the soil surface in the outer chamber was not sealed. For the purposes of comparison all rates are expressed per unit of soil surface covered by the chamber.

Treatment	N_2O accumulation rate (nmol N_2O cm^2 soil surface day^{-1})
L. perenne present	
Outer chamber	4.9 (1.50)
Inner chamber	2.6 (0.053)
L. perenne absent	
Outer chamber	2.6 (0.29)
Inner chamber	0.46 (0.010)

Table P10.2.   Effects of 10 Pa C_2H_2, 100 kPa O_2 and 10 kPa C_2H_2 on the accumulation of N_2O in the inner and outer chambers of jars with *L. perenne*. Each treatment was conducted in triplicate and the standard deviations are shown in parentheses. Note: *L. perenne* plants were sealed into the inner chambers and the soil surface in the outer chamber was not sealed.

Treatment	N_2O accumulation rate (as a percentage of control)	
	Inner chamber	Outer chamber
Control	100	100
10 Pa C_2H_2	19 (9.5)	29 (4.5)
100 kPa O_2	52 (7.5)	71 (7.5)
10 kPa C_2H_2	133 (23.8)	165 (36.5)

suggests that denitrification made a larger contribution to the production of that N_2O conducted from the soil by *L. perenne*. This suggestion is supported by the greater effect of 10 kPa C_2H_2 on N_2O accumulation in the inner chamber compared with the outer chamber. These observations are consistent with denitrification making a larger contribution to N_2O production in the rhizosphere than in non-rhizosphere soil and with *L. perenne* acting as a conduit for N_2O from the soil.

The proposed mechanism for plant-mediated N_2O emission arises from preliminary experiments and there may be other explanations. For example, plant (Dean and Harper, 1986; Chen *et al.*, 1990) and phylloplane sources of N_2O have not been excluded, and it is not known whether N_2O movement may occur in the gas or liquid phase. Mosier *et al.* (1990) have reported significant N_2O emission from waterlogged soils mediated by rice plants, which have a large amount of aerenchyma tissue that may facilitate gas phase transport. The mechanism proposed here for plant-mediated N_2O emission has a parallel in methane (CH_4) emission, whereby methanogenesis is driven by organic exudation from roots and the CH_4 escapes via the plants without the mitigation possible by oxidation during diffusion through the soil (Thomas *et al.*, 1996).

Acknowledgements

This work was undertaken with support from a postgraduate studentship (L.A.) from the UK Natural Environment Research Council

References

Blackmer, A.M., Bremner, J.M. and Schmidt, E.L. (1980) Production of nitrous oxide by ammonium-oxidising chemoautotrophic microorganisms in soil. *Applied Environmental Microbiology* 40, 1060–1066.

Barber, S.A. (1974) Influence of the plant root on ion movement. In: Carson, W.P. (ed.) *The Plant Root and its Environment.* University Press, Charlottesville, Virginia, pp. 525–564.

Chen, G., Shang, S., Yu, K., Yu, A., Wu, J. and Wang, Y. (1990) Investigation on the emission of nitrous oxide by plant. *Chinese Journal of Applied Ecology* 1, 94–96.

Christensen, S., Groffman, P., Mosier, A. and Zak, D.R. (1990) Rhizosphere denitrification: a minor process but indicator of decomposition activity. In: Revsbech, N.P. and Sorensen, J. (eds) *Denitrification in Soil and Sediment.* Plenum Press, New York, pp. 199–211.

Cribbs, W.H. and Mills, H.A. (1979) Influence of nitrapyrin on the evolution of nitrogen oxide (N_2O) from an organic medium with and without plants. *Communications in Soil Science and Plant Analysis* 10, 785–794.

Dean, J.V. and Harper, J.E. (1986) Nitric oxide and nitrous oxide production by soybean and winged bean during the *in vivo* nitrate reductase assay. *Plant Physiology* 82, 718–723.

Firestone, M.K., Firestone, R.B. and Tiedje, J.M. (1980) Nitrous oxide from soil denitrification: factors controlling its biological production. *Science* 208, 749–751.

Klemedtsson, L., Svensson, B.H. and Rosswall, T. (1987) Dinitrogen and nitrous oxide produced by denitrification and nitrification in soil with and without barley plants. *Plant and Soil* 99, 303–319.

Mosier, A.R., Mohanty, S.K., Bhadrachalam, A. and Chakravorti, S.P. (1990) Evolution of dinitrogen and nitrous oxide from the soil to the atmosphere through rice plants. *Biology and Fertility of Soils* 9, 61–67.

Robertson, G.P. and Tiedje, J.M. (1987) Nitrous oxide sources in aerobic soils: nitrification, denitrification and other biological processes. *Soil Biology and Biochemistry* 19, 187–193.

Thomas, K.L., Benstead, J., Davies, K.L. and Lloyd, D. (1996) Role of wetland plants in the diurnal control of CH_4 and CO_2 fluxes in peat. *Soil Biology and Biochemistry* 28, 17–23.

Webster, E.A. and Hopkins, D.W. (1996) Contributions from different microbial processes to N_2O emission from soil under different water regimes. *Biology and Fertility of Soils* 22, 331–335.

Application of Uncertainty Analysis in Modelling of Nitrous Oxide Emission from Grassland

O. Oenema[1,2], C. Wierda[2] and G.L. Velthof[2]
[1]AB-DLO, PO Box 129, 9750 AC Haren, The Netherlands; [2]NMI, Department of Soil Science and Plant Nutrition, Wageningen Agricultural University, PO Box 8005, 6700 EC Wageningen, The Netherlands

A large number of soil and climate factors, and grassland management factors, affect nitrous oxide (N_2O) emissions from managed grasslands. These interacting factors make N_2O emissions highly variable because soil and climate conditions vary tremendously, both in space and time. Consequently, estimates of N_2O emissions from grassland based on field measurements have a wide confidence interval due to random variations and with uncertainty due to bias associated with inappropriate sampling. The width of the confidence interval of N_2O emissions estimates from managed grassland was explored, using a simple mechanistic/empirical simulation model combined with Monte Carlo simulations.

The N-risk model (Wierda and Oenema, 1996) has been written in a Lotus 1-2-3 spreadsheet, which contains as add-in @RISK, a risk analysis and Monte Carlo simulation tool. Using general soil, grassland management and climate data as inputs, the model calculates consecutively the soil hydrology, dry matter production, N uptake, N mineralization-immobilization turnover, ammonia volatilization, nitrification, denitrification and leaching. Dry matter production is calculated on a daily basis, all other processes and variables are calculated per decade. Nitrous oxide is released from nitrification and denitrification reactions, using formulations as described by Bril *et al.* (1994).

Briefly, the amount of N_2O released from nitrification depends on water filled pore space (WFPS), temperature, ammonium (NH_4^+) and N_2O concentrations and a number of rate constants. The amount of N_2O released from denitrification depends on WFPS, amount of mineralizable carbon, temperature, the relative affinity of microorganisms for nitrate (NO_3^-) and N_2O, and a number of rate constants.

All soil, climate and management input variables are assigned a mean value, a standard deviation and a type of frequency distribution, based on literature data. Also the rate constants in the model were assigned a mean and standard deviation, using literature data and best estimate values. These standard deviations mimic the range of variations. Simulations were made for different grassland managements on sandy soil, using 15-year-average climate data. Mean output was obtained from, on average, 1000 simulations. Calibration of the model was based on literature data (e.g. Bril *et al.*, 1994) and results of an extensive monitoring study that was carried out on four sites for two consecutive years (Velthof *et al.*, 1996).

The results showed that mean annual losses from grassland on sandy soil ranged from 5 to 15 kg N_2O-N ha^{-1}, depending on N input via slurry and fertilizer, and on grazing versus mowing. Urine and dung from grazing animals were large sources of N_2O. Preliminary results of a grassland management regime with two grazing cycles, four mowing cuts, and 300 kg fertilizer N and 60 m^3 cattle slurry ha^{-1} $year^{-1}$ are shown in Fig. P11.1. The probability

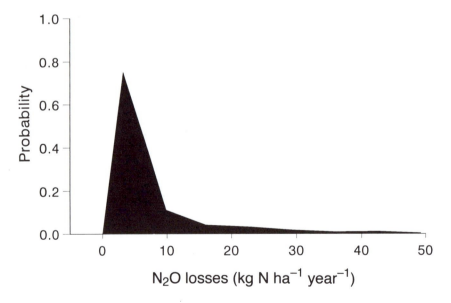

Fig. P11.1. Probability distribution of total annual N_2O losses from managed grassland on sandy soil.

(*P*) distribution of total annual N_2O losses is positively skewed. This holds for N_2O from both denitrification and nitrification. Approximately one-third of the total loss came from nitrification. Mean N_2O loss was about 9 kg ha^{-1} year^{-1}, with a *P* value of 0.7. The probability of extreme high values of > 40 kg ha^{-1} year^{-1} was about 0.01. Similar skewed distributions were obtained for NO_3^- leaching and total N losses.

Quantification of possible variations in N_2O losses from managed grassland and understanding the factors that contribute to these variations are essential steps in the development of nutrient management strategies for reducing these emissions. The present modelling approach provides a tool for exploring effects of improved nutrient management on N_2O losses from grassland, and a tool for decision analysis at the farm level. It also clearly shows the stochastic nature of N_2O emissions. Based on environmental legislation and personal preference of the farmer, utility functions can be assigned to herbage yield and annual N_2O losses, but also to, for example, NO_3^- leaching and NH_3 volatilization. The probabilities of the output factors combined with the utility functions of these factors provide an optimal and rational strategy for nutrient management planning (Anderson *et al.*, 1977). Current research is focused on elaborating the possibility for improving the N-risk model and decision analysis.

References

Anderson, J.R., Dillon, J.L. and Hardaker, J.B. (1977) *Agricultural Decision Analysis.* Iowa State University Press, 344 pp.

Bril, J., Van Faassen, H.G. and Klein Gunnewiek, H. (1994) *Modeling N2O Emission from Grazed grassland. AB-DLO Report* 24, Haren, The Netherlands, 45 pp.

Velthof, G.L., Brader, A. and Oenema, O. (1996) Seasonal variations in nitrous oxide losses from managed grasslands in the Netherlands. *Plant and Soil* 181, 263–274.

Wierda, C. and Oenema, O. (1996) Uncertainty and risk analysis of nutrient management: a model survey for nitrogen fertilization of grassland. *Mestotten* 1996, 30–37.

Nitrous Oxide Emissions from Grazed Grassland at Different Levels of Nitrogen Fertilization

S. Poggemann[1], F. Weißbach[1], U. Küntzel[1] and E.A. Kaiser[2]

[1]*Institute of Grassland and Forage Research and* [2]*Institute of Soil Biology, Federal Research Center of Agriculture (FAL), Bundesallee 50, D-38116 Braunschweig, Germany*

Grazed pastures appear to be one of the major sources of nitrous oxide (N_2O) emitted from agriculture. This N_2O originates from the nitrogen (N) of excreta distributed extremely unevenly by grazing animals as well as from applied N fertilizer during the grazing season. The total flux from grassland into the atmosphere is not yet well quantified, partly because of the large variation in

N_2O fluxes. As expected, the N_2O emissions increase with increasing N input via fertilizer or excreta. However, the combined effects of changes in soil water content and temperature, and their effects on the chemical soil properties under the areas affected by urine or faeces, make it difficult to generalize about quantitative estimations of N_2O emissions from grazed grassland.

The individual and combined effects of N fertilization and cattle excreta on the N_2O emission rate were studied in a field experiment on a sandy loam soil in Germany. The seasonal fluctuation in N_2O emissions was investigated for the following treatments.

The conditions of grazed pasture were simulated by the application of fresh urine and faeces to small plots. The experiment was designed in randomized blocks, with three levels of N fertilization, 0, 130 and 260 kg N ha^{-1} year^{-1} as ammonium nitrate in three replicates. Field measurements were carried out on 0.28 m^2 plots after application of 20 g N 0.18 m^{-2} urine or 10 g faeces 0.09 m^{-2}. A control treatment, without any excreta, was generally included to facilitate the calculation of N_2O emissions from urine and faeces *per se*. The application of excreta took place in spring, summer and autumn.

The closed soil cover box technique in conjunction with electron capture gas chromatography was used for the estimation of field loss rates. Rates were estimated at least twice a week for more than 6 months and the information obtained was compared with chemical properties of the soil as well as climatic data.

The N_2O emission rates were closely related to the total N input (Table P12.1). High seasonal variations in N_2O emission rates were found for all treatments. Irrespective of the kind of excreta applied, the total N_2O emissions were highest in spring compared to the other seasons, corresponding with the higher soil moisture in spring. With regard to the pattern of the N_2O emissions, most was observed shortly after deposition, especially of urine in the field (Fig. P12.1), but significant amounts were still released from urine

Table P12.1. Mean N_2O emission rates (μg N m^{-2} h^{-1}) from grassland (1994/95)

| N_{min} fertilizer (kg N ha^{-1} year^{-1}) | Excreta | Date of excreta application | | |
		Spring	Summer	Autumn
0	—	11	6	4
130	—	18	8	6
260	—	33	16	13
0	Faeces	20	14	30
130	Faeces	37	19	31
260	Faeces	77	19	59
0	Urine	90	78	21
130	Urine	123	86	26
260	Urine	171	106	41

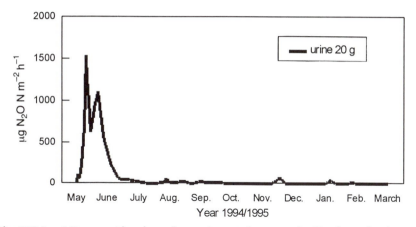

Fig. P12.1. Nitrous oxide release from urine patches on unfertilized grassland (1994/95) after application of excreta in spring.

and faeces even after several weeks following deposition. On this basis, the calculated N_2O-N losses ranged from 0.1 to 1.1% of the applied N depending on season, form and amount applied.

Nitrous Oxide Emissions from Fodder Crop Systems

O. Heinemeyer[1], M. Kucke[2], K. Kohrs[2], E. Schnug[2], J.C. Munch[1] and E.A. Kaiser[1]

[1]*Institute of Soil Biology and* [2]*Institute of Plant Nutrition and Soil Science, Federal Research Centre of Agriculture (FAL), Bundesallee 50, D-38116, Braunschweig, Germany*

Productive grasslands or fodder crop systems are the basis for milk and meat production with ruminant animals. High applications of nitrogen (N) fertilizers in order to maximize fodder production is thus a common practice in grassland and fodder crop system managements. High nitrous oxide (N_2O) emissions can be expected from such systems.

In order to determine N_2O losses, experimental plots for grass, grass/clover and clover fodder production were established at the FAL, Braunschweig, Northern Germany (52° 17′35 N, 10° 26′55 E) on a sandy loam soil (42% sand, 53% silt, 5% clay) in close vicinity. Nitrogen fertilization was adjusted to expected crop demand so that grass received 350, grass/clover received 170 and clover received 0 kg N ha^{-1} year^{-1} in split doses. Either calcium ammonium nitrate or a urea ammonium nitrate solution was used. On all three plots very large closed soil cover boxes (5.8 m^2, see Fig. P13.1) were installed in order to integrate spatial flux variations (Kaiser *et al.*, 1996). Five times a week, the N_2O flux rates were estimated by using electron capture gas chromatography (Heinemeyer and Kaiser, 1996). For a season-based comparison, the respective N_2O-N losses from June to October 1994 (season I), November 1994 to March 1995 (winter) and March 1995 to October 1995 (season II) were integrated in Fig. P13.2. A simple N input versus N output summary for the first complete year, November 1994 to October 1995 is presented in Table P13.1.

High seasonal variations were found for N_2O emission measurements (n = 320 per plot, c.v. 200–350) from all three experimental plots. While N_2O losses in seasons I and II clearly were ranked clover < grass/clover < grass, the picture for the winter period indicated similar losses for all three plots. Due to severe frosts and failing equipment, one important emission peak was probably missed on the grass/clover plot.

The measurements showed that clover was the most effective fodder crop, with highest N yields but lowest N_2O losses. Approximately 1% of the applied fertilizer-N was lost as N_2O-N. This was comparable to the figure for losses from arable crops on this soil (Kaiser and Heinemeyer, 1996). Signifi-

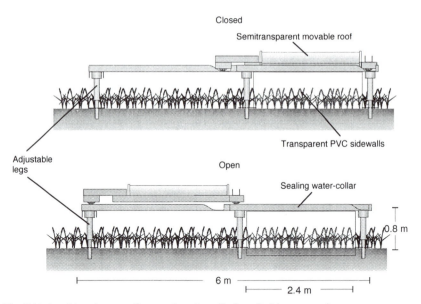

Fig. P13.1. Very large soil cover box installed on fodder grass plot.

Table P13.1. Summary of N-fertilizer input, N_2O emissions and N yields for three fodder crop production plots (November 1994 to October 1995).

	Grass	Grass/clover	Clover
N fertilization (kg ha^{-1})	350	170	0
N yield in harvest (kg ha^{-1})	193	243	311
N_2O emitted (kg ha^{-1})	3.53	2.31	2.24

cant N₂O emissions occurred during winter. For proper budgeting this finding
needs attention.

Acknowledgements
This work received funding from the German Federal Ministry of Agriculture
(BML) and the Ministry of Education and Science (BMBF).

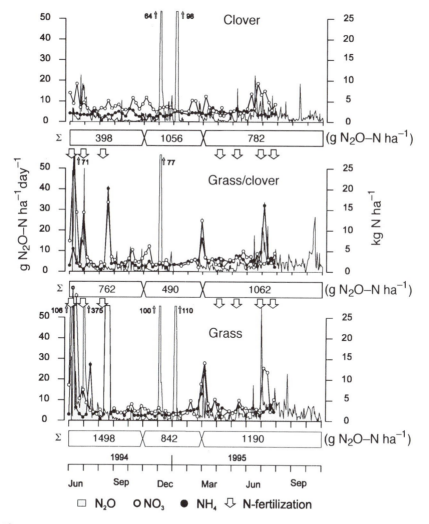

Fig. P13.2. N₂O emissions and mineral N contents of three fodder crop systems.

References

Heinemeyer, O. and Kaiser, E.A. (1996) An automated gas injector system for gas chromatography – atmospheric N_2O analysis. *Soil Science Society American Journal* 60, 808–811.

Kaiser, E.A. and Heinemeyer, O. (1996) Temporal changes in N_2O-losses from two arable soils. *Plant and Soil* 181, 57–63.

Kaiser, E.A., Munch, J.C. and Heinemeyer, O. (1996) Importance of soil cover box area on the determination of N_2O emissions from arable soils. *Plant and Soil* 181, 185–192.

The Impact of Cattle Slurries and their Management on Ammonia and Nitrous Oxide Emissions from Grassland

21

R.J. Stevens[1,2] and R.J. Laughlin[1]

[1]Agricultural and Environmental Science Division, Department of Agriculture for Northern Ireland; [2]Department of Agricultural and Environmental Science, The Queen's University of Belfast, Newforge Lane, Belfast BT9 5PX, Northern Ireland

Summary.   The fluxes of ammonia (NH_3) and nitrous oxide (N_2O) from grassland following slurry application are highly variable and depend on slurry properties, soil properties and environmental conditions. Ammonia losses following slurry application are typically 50% of the ammonium (NH_4^+) applied, whereas N_2O-N losses are likely to be < 5% of the NH_4^+-N applied.

Management strategies are reviewed for their potential to lower NH_3 volatilization and N_2O emissions from cattle slurry applied to grassland. More research has been done on strategies to control NH_3 volatilization than on strategies to lower N_2O emissions. Few strategies have been evaluated for their ability to lower NH_3 and N_2O fluxes simultaneously. This is felt to be an important omission, as effective control of one N-cycle process has the potential to enhance another loss process. No strategy has been evaluated for its ability to control all the N-gas fluxes.

Strategies for controlling NH_3 volatilization affect either slurry total solids (TS), slurry–air contact or the NH_4^+/NH_3 equilibrium. Lowering TS content in slurry by mechanical separation or dilution can halve NH_3 volatilization compared with raw slurry. Decreasing the slurry–air contact time by injection or by shifting the NH_4^+/NH_3 equilibrium to NH_4^+ by acidification can lower NH_3 volatilization by > 90% compared with raw slurry.

Factors that influence N_2O flux after slurry application are timing of application, rate of application and soil type. Strategies with the potential to lower N_2O flux are those that minimize the accumulation of NO_3^- outside the growing season. Good management practices and the use of nitrification inhibitors can minimize N_2O losses. Control of NH_3 volatilization by injection or acidification has the potential to enhance N_2O emission, as this creates conditions conducive to denitrification.

Measuring N_2O emission without simultaneously measuring N_2 emission can lead to erroneous conclusions about the extent of denitrification. Little is known about the factors which affect the $N_2 : N_2O$ ratio when slurry is surface spread or the coupling of nitrification and denitrification processes.

Introduction

Under conditions of intensive farming, 77% of the nitrogen (N) ingested by cattle is voided as excreta (Blaxter, 1977). At least half of the excrement of housed livestock is collected and applied to the land as slurry (Royal Society, 1983). As inorganic fertilizers are relatively cheap and easy to handle, farmers perceive slurry as a problem of waste disposal rather than as a valuable source of nutrients.

Attempts at utilizing the ammonium (NH_4^+) or plant-available N in cattle slurry have frequently given poor and variable crop yields and N recoveries (van Dijk and Sturm, 1983; Pain et al., 1986; Unwin et al., 1986). Gaseous losses of N by volatilization and/or denitrification are largely responsible (Pain et al., 1990c). In Western Europe, 74% of the total estimated NH_3-N emission of 4016 kt NH_3-N year^{-1} comes from animals and their waste and 32% of the total NH_3-N emission is due to land spreading of slurry (ECETOC, 1994). In the UK, 90% of the total NH_3-N emission is derived from agriculture (Sutton et al., 1995). Jarvis and Pain (1990) estimated that an average of 36% of agricultural NH_3 emissions comes from surface spreading of slurry on the land.

About 70% of the N_2O emitted from the biosphere into the atmosphere is derived from soil (Bouwman, 1990). Nitrous oxide can be produced by either nitrification (Bremner and Blackmer, 1980) or by denitrification (Firestone and Davidson, 1989). When slurry is spread on the land, the NH_4^+-N supplied increases the potential for nitrification. Provided nitrate (NO_3^-) is present, the addition of readily available carbon (C) in the slurry creates conditions conducive to the production of N_2O by denitrification (Beauchamp et al., 1989). It is possible that nitrification and denitrification occur simultaneously in adjacent soil pores of different aerobicity and have an impact on the release of N_2O (Hutchinson and Davidson, 1993; Jarvis et al., 1994; Stevens et al., 1997a).

This review focuses on research work which evaluated management strategies developed to lower NH_3 and N_2O emissions. The greatest research effort has been directed towards controlling NH_3 volatilization. As different methods are developed to lower NH_3 volatilization, N_2O emission by denitrification could become more important. Few studies have measured the effect of management on NH_3 volatilization and N_2O emissions simultaneously. This chapter summarizes the factors controlling NH_3 volatilization, reviews the management strategies directed at controlling NH_3 volatilization, reviews how the main strategies directed at the control of NH_3 volatilization impact on denitrification, reviews the strategies for lowering denitrification and, finally, highlights the gaps in our knowledge which limit the development of successful strategies.

Strategies for Lowering NH$_3$ Volatilization

Factors controlling NH$_3$ volatilization

In slurry, free NH$_3$ is in equilibrium with NH$_4^+$. In solution the proportion of NH$_4^+$ present as NH$_3$ increases as pH increases. Increasing temperature decreases the solubility of NH$_3$ and increases its rate of diffusion (Freney *et al.*, 1981). Cattle slurry has a pH of 7–8 compared with most agricultural soils whose pH is about 6. Ammonia volatilization will therefore occur until the liquid fraction containing the NH$_4^+$-N infiltrates into the soil. When slurry is spread in the field, little volatilization occurs during the spreading process (Phillips *et al.*, 1991) but rapid rates of volatilization occur in the first few hours after slurry is applied to grassland (Table 21.1). Environmental factors which influence the extent of NH$_3$ volatilization are air temperature, wind speed, rainfall and relative humidity. Ammonia volatilization from surface-applied slurry increases with increasing temperature (Sommer and Olesen, 1991; Sommer *et al.*, 1991; Moal *et al.*, 1995), increases with increasing windspeed (Sommer *et al.*, 1991) and decreases with increasing rainfall (Klarenbeek and Bruins, 1991). Relative humidity of the air was not significantly correlated with NH$_3$ loss (Moal *et al.*, 1995), which confirmed previous observations by Sommer *et al.* (1991) that the apparent correlation between these two variables was related through air temperature.

The soil properties which can influence NH$_3$ volatilization are pH, cation exchange capacity (CEC), buffering capacity, moisture content, hydraulic conductivity and the nature of the vegetative cover. In a laboratory study with cattle urine, CEC was the soil property most closely correlated with NH$_3$ volatilization (Whitehead and Raistrick, 1993). In the field, conditions such as applying dilute slurry to a freely draining soil below field capacity will favour rapid infiltration of slurry into the soil and lower NH$_3$ volatilization (Pain *et al.*, 1989; Sommer and Christensen, 1991). The nature of the vegetative cover will affect the absorptive capacity of the soil. Volatilization will be greater from a grass sward than from bare soil (Thompson *et al.*, 1990a; Amberger, 1991). The proportion of NH$_4^+$-N which volatilizes will decrease as application rate increases (Thompson *et al.*, 1990b; Frost, 1994).

Ammonia volatilization from surface-applied slurry is therefore affected by a multitude of factors, hence the proportion of NH$_4^+$-N lost as NH$_3$ is highly variable. The factors which could be under management control are those associated with removal of solids, lowering the contact time between slurry and the atmosphere and shifting the NH$_4^+$/NH$_3$ equilibrium. The total solids (TS) content can be lowered by mechanical separation, dilution with water

Table 21.1.   The magnitude of NH_3-N losses after the surface application of cattle slurry to grassland.

Study	Experiment date	Days	Soil type	NH_3-N loss % NH_4^+-N applied	Pattern of loss % of total NH_3-N
Bussink *et al.*, 1994	May–July	10	Heavy clay	93	89% in 1 day
	May–Sept	4	Peat	52	85% in 1 day
	June	4	Sand	61	92% in 1 day
Klarenbeek *et al.*, 1993	Sept	4	Peat	55	80% in 1 day
Lockyer *et al.*, 1989	May–June	3–5	Sand	39	> 80% in 2 days
Moal *et al.*, 1995	March	4	Sand	75	75% in 5 h
			Sand	54	75% in 12 h
Pain and Thompson, 1989	April–June	5	Sand	35	40% in 6 h
					70% in 1 day
Pain *et al.*, 1989	May–June	3	Sand	39	24–39% in 1 h
					85% in 12 h
Thompson *et al.*, 1987	Winter	17	Sand	74	50% in 2 days
					67% in 2 days
					80% in 4 days
	Spring	17	Sand	48	33% in 6 h
					50% in 1 day
					80% in 5 days
Thompson *et al.*, 1990a,b	May	6	Sand	38	54% in 4 h
					67% in 7 h
					87% in 1 day
Stevens and Logan, 1987	Sept	7	Clay	51	76% in 1 day
Vertregt and Selis, 1990	March–Oct	3	Sand	31–78	50–85% in 1 day

or aerobic and anaerobic digestion. The contact time between slurry and the atmosphere can be shortened by application methods such as band spreading/injection, spiking and washing. The NH_4^+/NH_3 equilibrium can be shifted towards NH_4^+ by adding acids.

Lowering total solids content – mechanical separation

The TS content of slurries can be lowered by removing a proportion using mechanical separation. There are a number of different types of slurry separators available for use on farms. The prime function of these separators is to improve the flow characteristics of the slurry. The range in mesh size used for cattle slurry commonly varies from 1 to 3 mm. The efficiency of separation can be measured as the quantity of solids removed from the slurry. Solids removal efficiency for different types of separator varies from 35 to 60% (Pain *et al.*, 1978). At the Agricultural Research Institute of Northern Ireland, Frost and Stevens (1991) used a flat belt separator and demonstrated that, in the

Table 21.2. Relationship between slurry total solids (TS) content, mesh size used for separation and solids removal efficiency for a flat belt separator.

Mesh size (mm)	TS content of whole slurry (g l^{-1})		
	60	80	100
2.0	29	31	36
1.7	30	33	38
1.1	35	37	40
0.8	36	38	42
0.4	39	40	44

Table 21.3. Relationship between mesh size, slurry total solids (TS) content and separated liquid output for a flat belt separator.

Mesh size (mm)	Slurry TS (g l^{-1})	Separated liquid output	
		Fresh matter (kg h^{-1})	Dry matter (kg h^{-1})
2	60	1499	74
	80	918	58
	100	705	55
1.1	60	942	43
	80	718	43
	100	518	39
0.4	60	645	28
	80	528	30
	100	360	26

normal range of slurry TS, solids removal efficiency of this type of machine depends on mesh size and slurry TS content (Table 21.2). Solids removal efficiency increased as dilution and mesh size decreased. Frost and Stevens (1991) showed that mesh size and dilution also affect the output of liquid from separators (Table 21.3). Therefore, there is little advantage in diluting slurry as a means of improving separator throughput.

Thompson *et al.* (1990a) using a 3-mm-mesh screen found that total NH_3 losses were similar for separated (TS 4.0%) and unseparated (TS 6.5%) slurry. It is probable that the 3 mm mesh did not remove a sufficient proportion of the solids to appreciably increase the infiltration of the liquid slurry fraction into the soil. Stevens *et al.* (1992) also found that separation through a 3-mm sieve had no significant affect on NH_3 volatilization when separation lowered the TS content from 8.2 to 6.8%. Separation through a 1.1-mm sieve lowered the TS content to 5.5% and lowered NH_3 volatilization by 51% compared to unseparated slurry. Vertregt and Selis (1990) lowered the TS content of slurry from 9.6 to 8.1% which on average in four experiments lowered NH_3 volatilization by 18% compared to unseparated slurry. Separation using a practically feasible mesh size of 1–3 mm can potentially lower NH_3 loss by no more than 50%.

Lowering total solids content – dilution prior to application

An alternative to lowering TS content by separation is dilution with water. Dilution of thick slurry (10–12% TS) with as little as 30% by volume of water can dramatically improve its flow characteristics (Vetter *et al.*, 1987). Work by Vetter *et al.* (1987), Stewart (1968) and Frost and Stevens (1991) demonstrated a positive effect of dilution on crop yield. Other workers have shown a decrease in NH_3 volatilization after dilution (Vertregt and Selis, 1990; Stevens *et al.*, 1992; Frost, 1994). Stevens *et al.* (1992) showed that for equal amounts of NH_4^+ applied, diluting slurry with 87% by volume of water lowered TS from 10.5 to 5.5% and lowered NH_3 volatilization by 50%. Frost (1994) demonstrated that diluting slurry by 90–120% by volume with water lowered TS from 10.3 to 5.1% and lowered the NH_3 loss to 50% of that from undiluted slurry. Vertregt and Selis (1990) diluted slurry by three times its volume of water and lowered NH_3 volatilization by 43% compared to undiluted slurry. They also found that when diluted slurry was applied to soil which had been irrigated, NH_3 volatilization increased by 100% compared to diluted slurry being applied on dry soil. The affect of diluting slurry with water on NH_3 volatilization is therefore dependent on the original TS content of the slurry and on the soil moisture content.

Provided there is a readily available source of water for dilution and fields are close to slurry stores, dilution offers a partial means for lowering NH_3 volatilization. However, slurry collected on many farms is already diluted with yard washings and rainwater. In a survey of Irish farms the average TS content of cattle slurries was 6.7% (Stevens *et al.*, 1995b). Hence on many farms the benefit of dilution is probably already being realized. Combinations of separation plus dilution have been used to lower NH_3 volatilization by more than 50%. A 75% decrease in NH_3 volatilization was achieved by separation through a 0.4-mm mesh plus 50% dilution, or separation through a 5-mm mesh plus 100% dilution (Stevens *et al.*, 1992).

Lowering total solids content – digestion

Aerobic treatment systems and anaerobic digestion plants have been used to tackle the odour problem from slurry spreading. The effluents from these treatments may differ in their susceptibility to NH_3 loss from undigested slurry. Aerobic or anaerobic treatment increases the pH of the slurry (Pain *et al.*, 1990a,b; Paul and Beauchamp, 1989), which may contribute to increased NH₃ volatilization. During aerobic and anaerobic treatments, N transformations occur which are likely to affect NH₃ emissions when the effluent is spread on the land. The aerobic treatment of slurry slightly decreased the NH₄⁺ concentration (Paul and Beauchamp, 1989), while the anaerobic treatment increased the NH₄⁺ concentration by mineralization of organic N (Smith and Unwin, 1985). Pain *et al.* (1990b) found that aerobic treatments led to an increase in the total loss of N through NH₃ volatilization after spreading on land. Contrary to what would have been expected, Pain *et al.* (1990a) showed that anaerobic digestion had no significant effect on NH₃ volatilization from slurry following application to grassland. The digested slurry had a lower TS content than the undigested slurry, which would have increased its rate of infiltration into the soil (Pain *et al.*, 1989). However, the lowering of NH₃ volatilization by faster infiltration into the soil was negated by the higher NH₃ volatilization from the digested slurry because of its higher pH.

Decreasing the contact between slurry and atmosphere – injection

During application with a spreader fitted with a discharge nozzle and splash-plate, the contact time between slurry droplets and air is short, thus only a small part (1–4%) of the NH₄⁺ present in slurry is lost during spreading (Pain *et al.*, 1989; Jarvis and Pain, 1990). Following application to the soil surface, 24–39% of the loss can occur within 1 h, 33–75% within 6 h and 50–>90% within 24 h (Table 21.1). Consequently, rapid incorporation of slurry into the soil is essential but only feasible on arable land. The only method of incorporation suitable for grassland is injection.

Direct injection of slurry into soil to a depth of 150 mm lowers NH₃ volatilization to negligible levels (Thompson *et al.*, 1987). Pulling tines through grassland has been shown to damage sward and decrease yields (Thompson *et al.*, 1987; Frost, 1988). The use of deep injection is also limited by soil texture, soil moisture, stone content and slope. There is increasing interest in shallow injectors for grassland because they cause less mechanical damage to the sward and can be used on a wide range of soils. Phillips *et al.* (1990) showed that shallow injection gave very low rates of NH₃ volatilization. Frost (1994) used the relatively new technique of shallow channel application and the novel technique of spike injection to lower NH₃ volatilization from cattle slurry. In shallow channel application, slurry is applied into grooves 50–70 mm deep, but the grooves are not closed by press wheels or discs as they are in shallow injection (Ministry of Agriculture Fisheries and Food, 1992). Spike injection involves the placement of slurry into preformed holes in the soil. Both shallow channel application and spike injection promoted good contact between soil and slurry, and lowered NH₃ volatilization to 9% of that from surface-applied slurry. Lorenz and Steffens (1994) and Huijsmans *et al.* (1994) used an open slot method and lowered NH_3 emissions by 90% compared with surface-applied slurry.

Work in Ireland (Kiely, 1987; Carton *et al.*, 1994) showed that a utilization efficiency of 70–90% for the slurry NH_4^+-N relative to inorganic N was obtained following shallow injection of slurry to grass after the first silage harvest. The efficiency of utilization declined to 40% for the second cut and 20% for the third cut. Recent work at IGER has showed that shallow injection is not always as effective as deep injection at lowering NH_3 emissions.Misselbrook *et al.* (1996) showed that for a slurry shallow injected into wet soil during the spring, 35% of the available NH_4^+-N was volatilized compared with 58% for surface-applied slurry. Hence the efficiency of utilization of N in injected slurry may be lower than for N in inorganic fertilizer.

Decreasing the contact between slurry and atmosphere – band spreading

This application method lowers the contact area between slurry and air, which delays the rate of drying of the manure and somewhat decreases evaporization losses (Morken, 1991). Huijsmans *et al.* (1994), using a micro-meteorological technique, showed that band spreading of slurry resulted in a decrease in NH_3 volatilization of 25–46% compared to surface-applied slurry. Frost (1994) using ventilated chambers showed that, on average, band spreading of slurry lowered NH_3 loss to 40% of that from surface-applied slurry. Lorenz and Steffens (1994) using wind tunnels obtained on average a 25% reduction in NH_3 loss using band spreading. Lorenz and Steffens (1994) and Huijsmans *et al.* (1994) also used a sliding or trailing shoe technique, developed in the Netherlands, to apply slurry in bands directly to the soil beneath the herbage canopy. The technique is based on a shoe-shaped steel blade moving along the soil surface, pushing the leaves aside and letting the slurry from each outlet of a band spreader drop directly onto the soil and not onto the grass. Using the sliding shoe system they achieved a 70% decrease in NH_3 emission compared with surface-applied slurry.

Variation in the effectiveness of band spreading for controlling NH_3 loss could be due to varying environmental conditions, slurry and soil properties, width of bands or method of measurement. Another possibility is that the pattern of NH_3 loss from band-spread slurry is different from surface-applied slurry. Thompson *et al.* (1990b) found that NH_3 volatilization from the band-spread slurry was 83% of that from surface-applied slurry after 5 days. During the first 2 days, NH_3 volatilization from the band-spread slurry was lower than from the surface-spread slurry, but for the next 3 days the converse was true. It was suggested that the mechanism responsible for different patterns may be the different drying characteristics determined by the thickness of the slurry on the ground. The thinner layer of the surface-applied slurry dried more

Table 21.4.   The % efficiency of ammonium-N in band- and surface-spread cattle slurry in relation to fertilizer N (CAN).

Treatment	Efficiency (%)		
	Mean	Range	Standard error
Band-spread	58	14–161	36 (n = 39)
Surface-spread	38	0–80	23 (n = 39)

quickly than the band-spread slurry. As a slurry dries, there is an increase in the resistance to the diffusion of NH_4^+ ions within the thickening matrix of organic material. In a thinner layer, faster drying may have resulted in a comparatively greater resistance to diffusion within the slurry.

Carton *et al.* (1994) found that band spreading of slurry improved the efficiency of NH_4^+-N utilization to 30% compared with 10% for surface-spread slurry. The benefit of band spreading was greatest at the first cut and was small compared with the benefit of shallow injection. Field trials at six sites in Ireland (Stevens *et al.*, 1997b) in 1992 (cut 1), 1993 (cuts 1, 2 and 3) and 1994 (cuts 1, 2 and 3) evaluated the efficiency of N in band- and surface-spread slurry in relation to inorganic N. A summary of the results for all sites and cuts (Table 21.4) showed that efficiencies were variable but that band spreading improved NH_4^+-N utilization compared to surface spreading. As well as lowering NH_3 volatilization, band spreading will result in less sward contamination than surface spreading.

Decreasing the contact between slurry and atmosphere – washing

When slurry is surface applied, the solids stick to the grass leaves and increase the potential for NH_3 volatilization. If the slurry is washed off the leaves with high-pressure water and the liquid fraction infiltrates into the soil, NH_3 volatilization should be lowered. Stevens *et al.* (1992) used a wash-water pressure of 3 bar and found no significant decrease in NH_3 volatilization for slurries with a TS content > 8.6%. For slurries with a TS content < 8.6%, washing with 100% by volume of water lowered NH_3 loss by 25% compared to unwashed slurry. Lorenz and Steffens (1994) used a pressure of 100 bar to wash slurry off the grass leaves with a rate of water about 14% by volume and obtained an 18% reduction in NH_3 volatilization. Clearly, the effectiveness of a water wash to clean the slurry off the grass leaves and transport the slurry liquid into the soil to prevent NH_3 volatilization is dependent on the pressure and volume of the water wash.

Shifting the NH_4^+/NH_3 equilibrium – acidification

The equilibrium between NH_4^+ and NH_3 in solution is pH dependent (Husted et al., 1991). Studies in the UK have shown that acidification of animal slurry to pH 5.5–6.0 with sulphuric acid lowered NH_3 losses by 30–95% after surface application (Stevens et al., 1989; Frost et al., 1990; Pain et al., 1990c). Nitric acid was as effective as sulphuric acid at lowering NH_3 emissions (Stevens et al., 1992; Bussink et al., 1994; Huijsmans et al., 1994). Acidification of slurry with nitric acid increases the N content of the slurry, resulting in a more balanced NPK fertilizer for grass production. Acidifying with nitric acid is therefore more economically attractive than acidifying with sulphuric acid, as the costs of nitric acid could be compensated for by savings on fertilizer N. However, there are practical problems with using nitric acid to acidify slurry in the store or in the field. The addition of nitric acid to an anaerobic and carbon-rich slurry may promote denitrification, leading to an increase in pH and thus an increase in NH_3 volatilization. In addition, the high N content of the acidified slurry demands even spreading on the land at application rates of < 20 m^3 ha^{-1}. Precision spreading cannot be easily achieved by conventional surface spreaders equipped with a splash plate.

Two strategies exist for acidification of slurry with nitric acid. Either the slurry can be acidified in a storage tank prior to spreading or it can be acidified during the spreading process. Acidification of slurry stored under slatted floors required the pH to be lowered to between 4 and 4.5 to prevent denitrification (Jongebreur and Voorburg, 1992). The apparent denitrification at this pH was due to inefficient mixing of acid with slurry. Oenema and Velthof (1993) obtained a similar result for slurry at pH 4.5 and found that denitrification was rapid at pH 6. To overcome the problems of inefficient mixing in large slurry stores, Huijsmans et al. (1994) developed a system for acidifying slurry in batches of 6 m^3 in a special mixing tank. After reaching the required pH, the slurry is pumped to a tanker and applied on the field. The acidification process took about 10 min, and N analyses of slurry before and after acidification confirmed that denitrification was not taking place. Lenehan et al. (1993) developed a system for acidifying slurry during field spreading. Nitric acid and slurry were mixed at known flow rates to achieve a target pH before application either in bands or by splash-plate.

The amount of nitric acid required to lower the pH of the slurry sufficiently to prevent denitrification can increase the N content of slurry by two or three times. Hence less slurry is needed to achieve a target N application rate. The acidified slurry can potentially meet the full N requirements of the crop, so it is important that the slurry is spread evenly. Band-spreaders have been shown to give a more even distribution than surface-spreaders equipped with a splash plate (Huijsmans and Hendriks, 1992). Huijsmans et al. (1994) showed that band spreading of slurry acidified with nitric acid to pH 6.7, 6.3 and 4.9 lowered NH_3 volatilization compared with surface-applied slurry by 37, 55 and 90%, respectively. They suggested that the effects of band spreading of slurry acidified to about pH 6 on NH_3 volatilization were due to cumulative effects of spreading in bands and of acidification, whereas at pH values of about 5 the decrease in NH_3 volatilization was mainly due to acidification. Surface spreading of slurry acidified with nitric acid to pH 6.0, 5.0 and 4.5 lowered NH_3 volatilization to 55, 72 and 85%, respectively, compared with untreated slurry (Bussink *et al.*, 1994).

Field trials at six sites in Ireland assessed the efficiency of N utilization in cattle slurry treated with nitric acid (Stevens *et al.*, 1997b). The slurry was acidified to pH 5.5 in 1992 (cut 1), 1993 (cuts 1, 2 and 3) and 1994 (cuts 1, 2 and 3), and also to pH 4.5 in 1994 (cuts 1, 2 and 3). The slurry was broadcast on the surface or band spread using the system described by Lenehan *et al.* (1993). A summary of the results for all sites and cuts (Table 21.5) showed that the efficiency of the N (NH_4^+-N+NO_3^--N) in the acidified slurry was variable compared to ammonium nitrate/calcium carbonate. The large variability in the efficiency of N in acidified slurry was thought to be caused by denitrification. There was no significant difference in the efficiency of N utilization between spreading methods over all sites and cuts for slurry acidified to pH 5.5. For slurry acidified to pH 4.5, N efficiency over all sites and cuts was better for band spreading than for surface spreading. There was little difference in efficiency between acidification to pH 5.5 or pH 4.5. Nitrogen efficiencies of greater than 100% imply that some of the organic N in slurry was available to the crop, as well as the NH_4^+-N fraction.

Shifting the NH_4^+/NH_3 equilibrium – other additives

The use of base-precipitating salts of Ca and Mg (Witter and Kirchmann, 1989b) and acidic compounds such as $FeCl_3$, $Ca(NO_3)_2$ and superphosphate have been used to lower NH_3 volatilization (Husted *et al.*, 1991). Slurry is well buffered with ammonium bicarbonate (Sommer and Husted, 1995), so acidic salts are needed in too large amounts to be practically feasible. Other

Table 21.5. The % efficiency of ammonium- plus nitrate-N in band- and surface-spread slurry acidified with nitric acid in relation to fertilizer N (CAN).

Treatment	Efficiency (%)		
	Mean	Range	Standard error
Band-spread, pH 5.5	86	43–141	24 ($n = 39$)
Surface-spread, pH 5.5	82	36–136	26 ($n = 39$)
Band-spread, pH 4.5	98	60–120	16 ($n = 18$)
Surface-spread, pH 4.5	86	43–112	19 ($n = 18$)

additives such as peat or zeolite depend on adsorbing NH_4^+-N to lower NH_3 volatilization (Witter and Kirchmann, 1989a). The binding of free NH_3 by bacteria has also been reported (Pain *et al.*, 1987).The most recent novel additive is based on extracts of *Yucca schidigera* which inhibit the enzyme urease (Jarvis and Pain, 1990). It is claimed that ingestion of these products by cattle lowers the amount of N in urea and hence in manure.

Strategies for Lowering Denitrification after Slurry Application

Magnitude and variability of denitrification loss

The products of denitrification are N_2O and N_2. Nitrous oxide can also be produced by nitrification. Some studies have measured N_2O flux after slurry application, while other studies have measured N_2 plus N_2O flux. In only a few studies have the fluxes of N_2 and N_2O been quantified separately. When slurry is surface applied to grassland, denitrification losses can range from 0.01 to 29.5% of NH_4^+-N applied and are dependent on the soil type and the time of year the slurry is applied (Table 21.6). Nitrous oxide emissions after slurry has been surface applied ranged from 0.1 to 4% of NH_4^+-N applied. After inorganic fertilizer has been applied to grassland, N_2O losses ranged from 0.1 to 0.9% of the applied N (Eichner, 1990). Therefore, the conventional surface spreading of slurry on grassland tends to increase N_2O emissions per unit area.

The effect of timing of application on denitrification

When slurry was applied in the autumn or winter, denitrification losses were appreciable and ranged from 23 to 29% of the NH_4^+-N applied, whereas when slurry was applied to the same soil in the spring losses ranged from 4 to 5.5% (Thompson *et al.*, 1987; Pain *et al.*, 1990c). In the autumn and winter nitrification converts the NH_4^+-N supplied in the slurry to NO_3^--N which is not taken up by the crop and is available for denitrification. However, in the spring any NO_3^- made available by nitrification is depleted by an actively growing sward and less is available for denitrification.

The effect of soil type on denitrification

Soil type can greatly influence the magnitude of denitrification losses. When slurry was applied in the autumn, denitrification losses were 23% of applied

Table 21.6. The magnitude of N gases (kg N ha^{-1}) lost by denitrification when untreated slurry is surface applied to grassland.

Study	N_2O	$N_2 + N_2O$	Comments
Thompson *et al.*, 1987		29	Winter application 11 weeks measurement. Sandy soil pH 6.5
		4	Spring application 8 weeks measurement. Sandy soil pH 6.5
Pain *et al.*, 1990c		23	Autumn application 5 months measurement. Sandy soil pH 6.5
		5.5	Spring application 3 months measurement. Sandy soil pH 6.5
		1.3	Autumn application. 5 months measurement. Clay soil pH 5.5
		0.01	Spring application 3 months measurement. Clay soil pH 5.5
Velthof and Oenema, 1993	< 0.1		June application 7 days measurement. Sandy soil pH 4.7
	< 0.2		August application 9 days measurement. Sandy soil pH 4.7
	0.1		October application 18 days measurement. Sandy soil pH 4.7
Jarvis *et al.*, 1994	4	7	Autumn application 14 days measurement. Peat soil pH 5
Egginton and Smith., 1986b	0.5	1.3	Spring, summer and autumn application. Average for continuous measurement for 1 year. Sandy loam pH 6.3.
Christensen, 1983	1.6		May application. 10 weeks measurement. Sandy loam pH 5.3
	4.8		July application. 4 weeks measurement. Sandy loam pH 5.3

NH_4^+-N in a sandy soil and only 1.3% for a clay soil. For a spring application of slurry, denitrification losses for a sandy soil were 5.5% of NH_4^+-N applied and 0.01% for a clay soil (Pain *et al.*, 1990c). The clay soil would have been wetter and less well aerated than the sandy soil. Nitrification which is an aerobic process would therefore have proceeded faster in the sandy soil to provide NO_3^- for denitrification. No information exists on the effect of soil type on the proportioning of N-gas loss between N_2 and N_2O. Recent laboratory experiments (see Chapter 22) show that clay soils tend to emit more N_2O than sandy soils due to the more favourable moisture conditions for N_2O production in clay soils, even when both soil types were maintained at field capacity.

The effect of applying inorganic fertilizer with slurry on denitrification

Whenever inorganic N as ammonium nitrate and slurry are applied together or within a few days of each other, the potential exists for enhanced denitrification. The timings of the applications of slurry and fertilizer may be important. A period of several days may be required between slurry and fertilizer application to allow the readily metabolizable substrates in the slurry to be oxidized before NO_3^--containing fertilizers are applied. Current advice for fertilizing grass for silage is that fertilizer-N for second cut should be spread immediately after the first cut, followed in a few days by slurry (Carton and Tunney, 1989).

Studies which have determined the efficiency of utilization of N by herbage have shown that the effects of inorganic fertilizer-N and available N in slurry are cumulative (Smith and Unwin, 1985; Stevens et al., 1987). The magnitude of N-gas losses by denitrification whenever slurry or inorganic fertilizer N are applied to grassland during the growing season may be small. The loss of N due to denitrification whenever slurry and fertilizer-N interact may not be significant in terms of a loss of plant-available N and may not be detected in agronomic experiments. However, the contribution of N_2O emissions from the interaction of slurry and NO_3^- to the total N_2O flux may be important. The only study which measured N_2O emission whenever slurry and inorganic N fertilizer were applied to grassland was that conducted by McTaggart et al. (see Chapter 20). They found that in wet soil conditions N_2O emissions increased dramatically to > 1 kg N ha^{-1} day^{-1} immediately following application of ammonium nitrate and slurry, but decreased rapidly after 2–3 days. Cumulative N_2O emissions were up to four times greater from ammonium nitrate followed by slurry, than from ammonium nitrate alone. Applying the slurry prior to the ammonium nitrate significantly decreased emissions compared with the reverse treatment. Further experiments showed the increased N_2O emission was not due to the additional moisture from the slurry. Rather the volatile fatty acids in the slurry supplied readily available C to increase soil respiration rate and hence N_2O emission. More measurements of N-gas losses are needed to supplement agronomic studies, so that the fate of the NH_4^+-N and organic N in slurry and the fate of soil NO_3^- and fertilizer NO_3^- are ascertained.

The effect of rate of application of slurry on denitrification

Many studies have shown that N_2O fluxes generally increase with increasing inorganic N application (Bouwman, 1990; Eichner, 1990). In the few studies when N_2O emissions have been measured from slurry, a similar trend with increasing slurry application rate has been noted (Egginton and Smith, 1986a; Velthof and Oenema, 1993). In a laboratory experiment with manure-amended soil, Paul *et al.* (1993) found that for liquid cattle manure the N_2O emission was negligible at a manure rate of 20 g kg^{-1} soil but increased to a maximum of 15 µg N kg^{-1} soil h^{-1} when the manure rate was increased to 100 g kg^{-1} soil.

The effect of nitrification inhibitors on denitrification

Nitrification inhibitors have been added to slurry to prevent the conversion of NH_4^+ to NO_3^- hence limiting denitrification. Nitrapyrin lowered denitrification in injected slurry from 21.3 to 9.2% of the NH_4^+-N applied (Thompson *et al.*, 1987). Nitrapyrin and dicyandiamide lowered denitrification when applied to injected slurry but only dicyandiamide was effective when slurry was surface applied (Thompson and Pain, 1989). For a surface application of slurry in the autumn, Pain *et al.* (1990c) also found that nitrapyrin was not effective but that dicyandiamide at a rate of 20 kg ha^{-1} lowered denitrification by 55% of N applied. Recent work at IGER by Jarvis (see Chapter 1) has shown that nitrification inhibitors can decrease N_2O emissions.

The effect of digestion on denitrification

The aerobic and anaerobic treatment of slurry lowers the concentration of volatile fatty acids in slurry (Paul and Beauchamp, 1989; Pain *et al.*, 1990a,b). Volatile fatty acids are a labile carbon supply for denitrification, hence denitrification should be less in digested slurry than in undigested slurry. Stevens *et al.* (1995a) added quantities of nitric acid (0.7, 1.4, 2.1, 2.8 and 3.5 mg N g^{-1} slurry) to different amounts of anaerobically digested and undigested cattle slurry in incubation jars. Over all treatments they found that after 7 days the average recovery of NO_3^- from the digested slurry was 98% and from the undigested slurry 50%. In a laboratory experiment with a gel-stabilized mixture of soil and manure, Petersen *et al.* (1996) measured denitrification from fresh and anaerobically digested liquid cattle manure and found that the accumulated N loss through denitrification accounted for 17% of the NH_4^+-N applied in the fresh manure and 1.7% of the NH_4^+-N in the digested manure. The removal of volatile fatty acids by digestion prior to application of slurry to land could be an effective strategy for lowering denitrification.

Impact of the main strategies for controlling NH₃ volatilization on N₂O emission

Lowering losses of N by NH_3 volatilization from slurries applied to land may increase losses of N as N_2O and N_2. Denitrification studies have been carried out after slurry has been injected or acidified. Slurry injection and acidification have been shown to be effective strategies for lowering NH_3 volatilization but they increased denitrification and thus N_2O emissions. Denitrification losses have been found to be 12.1% of the available N in slurry surface spread on grassland but 21.3% of the N in slurry injected into grassland (Thompson *et al.*, 1987). Similarly, for slurry applied to grassland in autumn denitrification losses were 29% of the NH_4^+-N applied for unamended slurry but 41% for the slurry acidified with sulphuric acid (Pain *et al.*, 1990c).

Acidification of slurry with nitric acid increases the potential for N_2O emissions, as NO_3^- is introduced into an anaerobic and carbon-rich slurry. The extent of N-gas loss is dependent on pH. In laboratory experiments, Stevens *et al.* (1992) found that slurry had to be acidified to pH 5.5 to stop biological denitrification, although they did detect N_2 fluxes at pH values ranging from 4 to 6. These fluxes may have been due to chemodenitrification, a process which can produce NO_2, NO, N_2O, and/or N_2 by the reaction of nitrous and nitric acid with various N compounds (Chalk and Smith, 1983).

Field studies with slurries acidified with nitric acid have only measured N_2O emission. Jarvis *et al.* (1994) found that N_2O emission from untreated slurry and slurry acidified to pH 4.5 with nitric acid were < 1.5% of applied N and were similar for the first week after application to grassland. Velthof and Oenema (1993) found that N_2O emissions ranged from < 0.2% of N applied in untreated slurry to 3.4% of N applied in nitric acid-amended slurry. Nitrous oxide fluxes from slurry at pH 6 and those from slurry at pH 4.5 were similar if the slurry acidified to pH 6 was prepared 1 h before application. However, if the slurry acidified to pH 6 was stored for 1 week, N_2O losses were greater, suggesting that the denitrifiers take time to adjust to changing acidity. There was little difference in N_2O fluxes between calcium ammonium nitrate and nitric acid-amended slurry, and Oenema and Velthof concluded that N_2O emission was due to denitrification from NO_3^- irrespective of its source. In the study by Oenema and Velthof (1993), denitrification increased the pH of slurry. The denitrification process produces OH^- in equimolar amounts to NO_3^- denitrified (Delwiche, 1981), so if appreciable denitrification occurs the pH of the slurry will increase and the potential for NH_3 emission will increase.

Gaps in Knowledge

Coupling of nitrification and denitrification

Nitrous oxide can be produced by nitrification or denitrification (Firestone and Davidson, 1989). When slurry with a high NH_4^+-N content is applied to soil, nitrification will form a NO_3^- pool which can be denitrified to produce N_2O. Nitrous oxide can also be produced during the nitrification process (Bremner and Blackmer, 1980). In a laboratory study, Bergstrom *et al.* (1994)

showed that the addition of NH_4^+ and C to soil increased N_2O production under aerobic conditions but not in anaerobic conditions. Two possible explanations were given to explain this effect. First, NH_4^+ may indirectly influence N_2O production by denitrification via nitrification. Second, NH_4^+ may increase N_2O production from mixotrophic or heterotrophic growth of nitrifiers in the presence of high C concentrations. A supply of NO_3^- to denitrifiers from nitrification may be important when the denitrification rate is fast due to increased C concentrations. Petersen *et al.* (1991) observed coupled nitrification-denitrification in a gel-stabilized microcosm amended with liquid manure. Evidence from field studies (Velthof and Oenema, 1993; Jarvis *et al.*, 1994) and laboratory experiments (Paul *et al.*, 1993; Petersen *et al.*, 1996) suggests that denitrification is the dominant process for producing N_2O when slurry is applied to soil. In the carbon-rich environment created by slurry application to soil, denitrification rate and thus N_2O emission could depend on nitrification as the source of NO_3^-. Nitrification can proceed in oxygen-depleted conditions (Cole, 1994) and aerobic denitrification can occur in conditions which favour nitrification (Lloyd, 1993) or in anaerobic microsites within soil aggregates (Smith, 1980; Renault and Stengel, 1994). In a laboratory study, Paul *et al.* (1993) amended soil with cattle slurry at rates of 20, 50 and 100 g kg^{-1} soil and found that nitrification of NH_4^+ was complete after 2 days with the lowest concentration of manure and after 6 days with the highest concentration. Maximum NO_3^- production rates occurred on days 1, 2 and 3 for the low, intermediate and high concentrations of manure, respectively. The fastest rates of NO_3^- production were measured when N_2O fluxes and denitrification rates were also fastest. Emissions decreased to background levels after day 6 which was also when nitrification was complete. More information is required on quantification of the rates of nitrification and denitrification in slurry-amended soils, and on the relative contributions of each process to the N_2O flux.

Factors affecting the N_2 : N_2O ratio

The products of denitrification are N_2O and N_2 (Firestone and Davidson, 1989). The proportion of N_2O in the total N loss through denitrification may range from less than 5 to more than 50% (Ryden and Rolston, 1983). There is little information about the factors which affect the ratio of N_2 to N_2O production from soils amended with slurry. The ratio is dependent on environmental conditions and on the microbial populations present (Firestone *et al.*, 1980). In soils, a high NO_3^- concentration, a low pH and moderately reducing conditions are generally considered as beneficial for the formation of N_2O in relation to N_2 (Firestone, 1982). As slurry contains readily available C but no NO_3^-, its application to soil would tend to increase the proportion of N_2 in the N gases emitted. In a laboratory study, Paul *et al.* (1993) amended soil with liquid dairy cattle manure at a rate of 100 g kg^{-1} soil and found that total denitrification loss was 2.1% of the total N applied, compared with an emission of N as N_2O of 0.3% of the total N applied. In a field experiment Egginton and Smith (1986b) showed that most denitrification loss was as N_2O when fluxes were low but the estimated ratio of loss as N_2 to loss as N_2O increased up to 24 when fluxes were high. In a field experiment conducted on a soil of pH 4.5, Jarvis *et al.* (1994) measured N_2O losses between 40 and 74% of the total N-gas losses. In recent budget studies of N_2O losses after slurry application, Jarvis and Pain (1994) assumed that the ratio of $N_2 : N_2O$ was 3 : 1.

It is clear that the factors which affect the $N_2 : N_2O$ ratio in the N-gas fluxes subsequent to slurry being added to the soil are complex. The complexity of interacting factors and the large temporal and spatial variability associated with denitrification (Scholefield *et al.*, 1990; Jarvis *et al.*, 1994) will make the management of the $N_2 : N_2O$ ratio very difficult. All the studies to date on measuring N_2O and N_2 fluxes have used the acetylene block technique. At the concentration of acetylene necessary to block N_2O reductase, nitrification is also blocked. Since nitrification may be supplying the NO_3^- for denitrification, measuring N_2O and N_2 fluxes should allow nitrification to proceed. The ^{15}N gas flux method can measure N_2O and N_2 fluxes as well as quantifying the contributions of nitrification and denitrification to the N_2O flux (Stevens *et al.*, 1997a).

Measuring all N-gas fluxes in the same experiment

Simultaneous quantification of the fluxes of NH_3, NO, N_2 and N_2O are required to check that strategies for controlling one gaseous loss process do not result in increased losses by another process. Such studies are also necessary to help understand the interactions between N in slurry and N in soil and fertilizer.

References

Amberger, A. (1991) Ammonia emissions during and after land spreading of slurry. In: Nielsen, V.C., Voorburg, J.H. and L'Hermite, P. (eds) *Odour and Ammonia Emissions from Livestock Farming*. Elsevier, London, pp. 126–131.

Beauchamp, E.G., Trevors, J.T. and Paul, J.W. (1989) Carbon sources for bacterial denitrification. *Advances in Soil Science* 10, 113–142.

Bergstrom, D.W., Tenuta, M. and Beauchamp, E.G. (1994) Increase in nitrous oxide production in soil induced by ammonium and organic carbon. *Biology and Fertility of Soils* 18, 1–6.

Blaxter, K.L. (1977) The production of protein. In: *Proceedings of 2nd International Symposium on Protein Metabolism and Nutrition*. Centre for Agricultural Publishing and Documentation, Wageningen, pp. 4–10.

Bouwman, A.F. (1990) Exchange of greenhouse gases between terrestial ecosystems and the atmosphere. In: Bouwman, A.F. (ed.) *Soils and the Greenhouse Effect.* Wiley and Sons, Chichester, pp. 61–127.

Bremner, J.M. and Blackmer, A.M. (1980) Mechanisms of nitrous oxide production in soils. In: Trudinger, P.A., Walter, M.R. and Ralph, B.J. (eds) *Biogeochemistry of Ancient and Modern Environments.* Australian Academy of Science, Canberra, pp. 279–291.

Bussink, D.W., Huijsmans, J.F.M. and Ketelaars J.J.M.H. (1994) Ammonia volatilization from nitric-acid-treated cattle slurry surface applied to grassland. *Netherlands Journal of Agricultural Science* 42, 293–309.

Carton, O.T. and Tunney, H. (1989) Using slurry on the farm. *Farm and Food Research* 20, 8–9.

Carton, O.T., Cuddihy, A. and Lenehan, J.J. (1994) The effect of slurry spreading technique on silage dry matter yields. In: Hall, S.E. (ed.) *Proceedings of the Seventh Technical Consultation on Animal Waste Management Bad Zwischenahn, Germany.* Food and Agriculture Organisation of the United Nations, Rome, pp. 231–236.

Chalk, P.M. and Smith, C.J. (1983) Chemodenitrification. In: Freney, J.R. and Simpson, J.R. (eds) *Gaseous loss of Nitrogen from Plant-Soil Systems, Developments in Plant and Soil Sciences*, 9. Martinus Nijhoff, The Hague, pp. 65–89.

Christensen, S. (1983) Nitrous oxide emission from a soil under permanent grass: seasonal and diurnal fluctuations as influenced by manuring and fertilization. *Soil Biology and Biochemistry* 15, 531–536.

Cole, J.A. (1994) Biodegradation of inorganic N compounds. In: Ratledge, C. (ed.) *Biochemistry of Microbial Degradation.* Kluwer Academic, London, pp. 487–512.

Delwiche, C.C. (1981) The nitrogen cycle and nitrous oxide. In: Delwiche, C.C. (ed.) *Denitrification, Nitrification and Atmospheric Nitrous Oxide.* Wiley and Sons, New York, pp. 1–15.

van Dijk, T.A. and Sturm, H. (1983) Fertiliser value of animal manures on the continent. In: *Proceedings of the Fertiliser Society No. 220.* The Fertiliser Society, London.

ECETOC (1994) Ammonia emissions to air in Western Europe. In: *Technical Report No. 62.* European Centre for Ecotoxicology and Toxicology of Chemicals, Brussels.

Egginton, G.M. and Smith, K.A. (1986a) Nitrous oxide emission from a grassland soil fertilized with slurry and calcium nitrate. *Journal of Soil Science* 7, 59–67.

Egginton, G.M. and Smith, K.A. (1986b) Losses of nitrogen by denitrification from a grassland soil fertilized with cattle slurry and calcium nitrate. *Journal of Soil Science* 37, 69–80.

Eichner, M.J. (1990) Nitrous oxide emissions from fertilised soils: summary of available data. *Journal of Environmental Quality* 19, 272–280.

Firestone, M.K. (1982) Biological denitrification. In: Stevenson, F.J. (ed.) *Nitrogen in Agricultural Soils.* American Society of Agronomy, Madison, Wisconsin, pp. 289–336.

Firestone, M.K. and Davidson, E.A. (1989) Microbiological basis of NO and N_2O production and consumption in soil. In: Andreae, M.O. and Schimel, D.S. (eds) *Exchange of Trace Gases between Terrestrial Ecosystems and the Atmosphere, Life Sciences Research Report No. 47*, Wiley and Sons, Chichester, pp. 7–21.

Firestone, M.K., Firestone, R.B. and Tiedje, J.M. (1980) Nitrous oxide from soil denitrification: factors controlling its biological production. *Science* 208, 749–751.

Freney, J.R., Simpson, J.R. and Denmead, O.T. (1981) Ammonia volatilization. In: Clark, F.E. and Rosswall, T. (eds) *Terrestial Nitrogen Cycles, Ecological Bulletin No. 33*. Swedish Natural Science Research Council, Stockholm, pp. 291–302.

Frost, J.P. (1988) Effects on crop yields of machinery traffic and soil loosening: part 2, effects on grass yield of soil compaction, low ground pressure tyres and date of loosening. *Journal of Agricultural Engineering Research* 40, 57–69.

Frost, J.P. (1994) Effect of spreading method, application rate and dilution on ammonia volatilization from cattle slurry. *Grass and Forage Science* 49, 391–400.

Frost, J.P. and Stevens, R.J. (1991) Slurry utilisation. *Agricultural Research Institute of Northern Ireland, 64th Annual Report 1990–91*, pp. 20–30.

Frost, J.P., Stevens, R.J. and Laughlin, R.J. (1990) Effect of separation and acidification of cattle slurry on ammonia volatilization and on the efficiency of slurry nitrogen for herbage production. *Journal of Agricultural Science, Cambridge* 115, 49–56.

Huijsmans, J.F.M. and Hendriks, J.G.L. (1992) Slurry distribution of spreaders and injectors. *Agricultural Engineering 92 Congress*. Swedish Institute of Agricultural Engineering, Uppsala, pp. 622–623.

Huijsmans, J.F.M., Mulder, E.M. and Bussink, D.W. (1994) Acidification of slurry just before field application to reduce ammonia volatilization. *Agricultural Engineering Milano '94, Report No. 94-A-002*, p. 10.

Husted, S., Jensen, L.S. and Jørgensen, S.S. (1991) Reducing ammonia loss from cattle slurry by the use of acidifying additives: the role of the buffer system. *Journal of the Science of Food and Agriculture* 57, 335–349.

Hutchinson, G.L. and Davidson, E.A. (1993) Processes for production and consumption of gaseous nitrogen oxides in soil. In: Harper, L.A., Mosier, A.R. and Duxbury, J.M. (eds) *Agricultural Ecosystem Effects on Trace Gases and Global Climate Change*. ASA Special Publication No. 55, American Society of Agronomy, Madison, Wisconsin, pp. 79–93.

Jarvis, S.C. and Pain, B.F. (1990) Ammonia volatilization from agricultural land. In: *Proceedings of the Fertiliser Society, No. 298*. The Fertiliser Society, Peterborough, UK.

Jarvis, S.C. and Pain, B.F. (1994) Greenhouse gas emissions from intensive livestock systems: their estimation and technologies for reduction. *Climatic Change* 27, 27–38.

Jarvis, S.C., Hatch, D.J., Pain, B.F. and Klarenbeek, J.V. (1994) Denitrification and the evolution of nitrous oxide after the application of cattle slurry to a peat soil. *Plant and Soil* 166, 231–241.

Jongebreur, A.A. and Voorburg, J.H.(1992) The role of ammonia in acidification. Perspectives for the prevention and reduction of emissions from livestock operations. In: Schneider, T. (ed.) *Acidification Research Evaluation and Policy Applications*. Elsevier Science Publishers, Amsterdam, pp. 55–64.

Kiely, P.V. (1987) The effect of spreading method on slurry N utilisation by grassland. In: *Proceedings of the 12th General Meeting of the European Grassland Federation, Dublin*, pp. 353–357.

Klarenbeek, J.V. and Bruins, M.A. (1991) Ammonia emissions after land spreading of animal slurries. In: Nielsen, V.C., Voorburg, J.H. and L'Hermite, P. (eds) *Odour and Ammonia Emissions from Livestock Farming.* Elsevier Applied Science, London, pp. 107–115.

Klarenbeek, J.V., Pain, B.F., Phillips, V.R. and Lockyer, D.R. (1993) A comparison of methods for use in the measurement of ammonia emissions following the application of livestock wastes to land. *International Journal of Environmental Analytical Chemistry* 53, 205–218.

Lenehan, J.J., Carton, O.T. and Stevens, R.J. (1993) On tanker acidification system for slurry treatment. *Irish Journal of Agriculture and Food Research* 32, 103 (abstract).

Lloyd, D. (1993) Aerobic denitrification in soils and sediments: from fallacies to facts. *Trends in Ecology and Evolution* 8, 352–356.

Lockyer, D.R., Pain, B.F. and Klarenbeek, J.V. (1989) Ammonia emissions from cattle, pig and poultry wastes applied to pasture. *Environmental Pollution* 56, 19–30.

Lorenz, F. and Steffens, G. (1994) The effect of different slurry applications techniques on ammonia losses and herbage yield. In: Hall, J.E. (ed.) *Proceedings of the Seventh Technical Consultation on Animal Waste Management,* Badz-wischenahn Germany. Food and Agriculture Organisation of the United Nations, Rome, pp. 209–215.

Ministry of Agriculture Fisheries and Food (1992) *Code of Good Agricultural Practice for the Protection of Air.* MAFF Publications, London.

Misselbrook, T.H., Laws J.A. and Pain, B.F. (1996) Surface application and shallow injection of cattle slurry on grassland: nitrogen losses, herbage yields and nitrogen recoveries. Grass and Forage Science 51, 270–277.

Moal, J.F., Martinez, J., Guiziou, F. and Coste, C.M. (1995) Ammonia volatilization following surface-applied pig and cattle slurry in France. *Journal of Agricultural Science, Cambridge* 125, 245–252.

Morken, J. (1991) Slurry application techniques for grassland: effects on herbage yield, nutrient utilization and ammonia volatilization. *Norwegian Journal of Agricultural Science* 5, 153–162.

Oenema, O. and Velthof, G.L. (1993) Denitrification in nitric-acid-treated cattle slurry during storage. *Netherlands Journal of Agricultural Science* 41, 63–80.

Pain, B.F. and Thompson, R.B. (1989) Ammonia volatilization from livestock slurries applied to land. In: Hansen, J.A. and Henriksen, K. (eds) *Nitrogen in Organic Wastes Applied to Soils.* Academic Press, London, pp. 202–212.

Pain, B.F., Hepherd, R.Q. and Pittman, R.J. (1978) Factors affecting the performances of four slurry separating machines. *Journal of Agricultural Engineering Research* 23, 231–242.

Pain, B.F., Smith, K.A. and Dyer, C.J. (1986) Factors affecting the response of cut grass to the nitrogen content of dairy cow slurry. *Agricultural Wastes* 17, 189–202.

Pain, B.F., Thompson, R.B., De la Lande Cremer, L.C.N. and Ten Holte, L. (1987) The use of additives in livestock slurries to improve their flow properties, conserve nitrogen and reduce odours. In: van der Meer, H.G., Unwin, R.J., van Dijk, T.A. and Ennik, G.L. (eds) *Animal Manure on Grassland and Fodder Crops. Fertilizer or Waste.* Martinus Nijhoff, Dordrecht, pp. 229–246.

Pain, B.F., Phillips, V.R., Clarkson, C.R. and Klarenbeek, J.V. (1989) Loss of nitrogen through ammonia volatilisation during and following the application of pig or cattle slurry to grassland. *Journal of the Science of Food and Agriculture* 47, 1–12.

Pain, B.F., Misselbrook, T.H., Clarkson, C.R. and Rees, Y.J. (1990a) Odour and ammonia emissions following the spreading of anaerobically-digested pig slurry on grassland. *Biological Wastes* 34, 259–267.

Pain, B.F., Phillips, V.R., Clarkson, C.R., Misselbrook, T.H., Rees, Y.J. and Farrent, J.W. (1990b) Odour and ammonia emissions following the spreading of aerobically-treated pig slurry on grassland. *Biological Wastes* 34, 149–160.

Pain, B.F., Thompson, R.B., Rees, Y.J. and Skinner, J.H. (1990c) Reducing gaseous losses of nitrogen from cattle slurry applied to grassland by the use of additives. *Journal of the Science of Food and Agriculture* 50, 141–153.

Paul, J.W. and Beauchamp, E.G. (1989) Relationship between volatile fatty acids, total ammonia, and pH in manure slurries. *Biological Wastes* 29, 313–318.

Paul, J.W., Beauchamp, E.G. and Zhang, X. (1993) Nitrous and nitric oxide emissions during nitrification and denitrification from manure-amended soil in the laboratory. *Canadian Journal of Soil Science* 73, 539–553.

Petersen, S.O., Henriksen, K. and Blackburn, T.H. (1991) Coupled nitrification-denitrification associated with liquid manure in a gel-stabilized model system. *Biology and Fertility of Soils* 12, 19–27.

Petersen, S.O., Nielsen, T.H., Frostegård, Å. and Olesen, T. (1996) O_2 uptake, C metabolism and denitrification associated with manure hot-spots. *Soil Biology and Biochemistry* 28, 341–349.

Phillips, V.R., Pain, B.F., Clarkson, C.R. and Klarenbeek, J.V. (1990) Studies on reducing the odour and ammonia emissions during and after the land spreading of animal slurries. *Farm Buildings and Engineering* 2, 17–23.

Phillips, V.R., Pain, B.F. and Klarenbeek, J.V. (1991) Factors influencing the odour and ammonia emissions during and after the land spreading of animal slurries. In: Nielsen, V.C., Voorburg, J.H. and L'Hermite, P. (eds) *Odour and Ammonia Emissions from Livestock Farming*. Elsevier, London, pp. 98–106.

Renault, P. and Stengel, P. (1994) Modelling oxygen diffusion in aggregated soils: 1. Anaerobiosis inside the aggregates. *Soil Science Society of America Journal* 58, 1017–1023.

Royal Society (1983) *The Nitrogen Cycle of the United Kingdom: Report of a Royal Society Study Group*. Royal Society, London.

Ryden, J.C. and Rolston, D.E. (1983) The measurement of denitrification. In: Freney, J.R. and Simpson, J.R. (eds) *Gaseous Loss of Nitrogen from Plant-soil Systems, Developments in Plant and Soil Sciences 9*. Martinus Nijhoff, The Hague, pp. 91–132.

Scholefield, D., Corré, W.J., Colbourne, P., Jarvis, S.C., Hawkins, J. and de Klein, C.A.M. (1990) Denitrification in grazed grassland soils assessed using the core incubation technique with acetylene inhibition. In: *Proceedings 13th General Meeting, European Grassland Federation,Banska Bystrica, Czechoslovakia*, Vol. 2, pp. 8–12.

Smith, K.A. (1980) A model of the extent of anaerobic zones in aggregated soils, and its potential application to estimates of denitrification. *Journal of Soil Science* 31, 263–277.

Smith, K.A. and Unwin, R.J. (1985) Fertilizer value of animal waste slurries. In: *Crop Nutrition and Soil Science*. Ministry of Agriculture, Fisheries and Food RB 253(83), London, pp. 11–41.

Sommer, S.G. and Christensen, B.T. (1991) Effect of dry matter content on ammonia loss from surface applied cattle slurry. In: Nielsen, V.C., Voorburg, J.H. and L'Hermite, P. (eds) *Odour and Ammonia Emissions from Livestock Farming.* Elsevier, London, pp. 141–147.

Sommer, S.G. and Husted, S. (1995) The chemical buffer system in raw and digested animal slurry. *Journal of Agricultural Science, Cambridge* 124, 45–53.

Sommer, S.G. and Olesen, J.E. (1991) Effects of dry matter content and temperature on ammonia loss from surface-applied cattle slurry. *Journal of Environmental Quality* 20, 679–683.

Sommer, S.G., Olesen, J.E. and Christensen, B.T. (1991) Effects of temperature, wind speed and air humidity on ammonia volatilization from surface applied cattle slurry. *Journal of Agricultural Science, Cambridge* 117, 91–100.

Stevens, R.J. and Logan, H.J. (1987) Determination of the volatilization of ammonia from surface-applied cattle slurry by the micrometeorological mass balance method. *Journal of Agricultural Science, Cambridge* 109, 205–207.

Stevens, R.J., Laughlin, R.J. and Logan, H.J. (1987) Interaction between cow slurry and ^{15}N-labelled calcium nitrate as fertilizers for ryegrass production. *Journal of the Science of Food and Agriculture* 41, 309–314.

Stevens, R.J., Laughlin, R.J. and Frost, J.P. (1989) Effect of acidification with sulphuric acid on the volatilization of ammonia from cow and pig slurries. *Journal of Agricultural Science, Cambridge* 113, 389–395.

Stevens, R.J., Laughlin, R.J. and Frost, J.P. (1992) Effects of separation, dilution, washing and acidification on ammonia volatilization from surface-applied cattle slurry. *Journal of Agricultural Science, Cambridge* 119, 383–389.

Stevens, R.J., Laughlin, R.J. and O'Bric, C.J. (1995a) The fate of nitrate in cattle slurries acidified with nitric acid. *Journal of Agricultural Science, Cambridge* 125, 239–244.

Stevens, R.J., O'Bric, C.J. and Carton, O.T. (1995b) Estimating nutrient content of animal slurries using electrical conductivity. *Journal of Agricultural Science, Cambridge* 125, 233–238.

Stevens, R.J., Laughlin, R.J., Burns, L.C., Arah, J.R.M. and Hood, R.C. (1997a) Measuring the contributions of nitrification and denitrification to the flux of nitrous oxide from soil. *Soil Biology and Biochemistry* 29, 139–151.

Stevens, R.J. Laughlin, R.J., O'Bric, C.J., Carton, O.T. and Lenehan, J.J. (1997b) The efficiency of the nibrogen in cattle slurry acidified with nitric acid for grass production. Journal of Agricultural Science, Cambridge (in press).

Stewart, T.A. (1968) The effect of age, dilution and rate of application of cow and pig slurry on grass production. *Research and Experimental Record, Ministry of Agriculture, Northern Ireland,* pp. 67–90.

Sutton, M.A., Place, C.J., Eager, M., Fowler, D. and Smith, R.I. (1995) Assessment of the magnitude of ammonia emissions in the United Kingdom. *Atmospheric Environment* 29, 1393–1411.

Thompson, R.B. and Pain, B.F. (1989) Denitrification from cattle slurry applied to grassland. In: Hansen, J.A. and Henriksen, K. (eds) *Nitrogen in Organic Wastes Applied to Soils.* Academic Press, London, pp. 247–260.

Thompson, R.B., Ryden, J.C. and Lockyer, D.R. (1987) Fate of nitrogen in cattle slurry following surface application or injection to grassland. *Journal of Soil Science* 38, 689–700.

Thompson, R.B., Pain, B.F. and Lockyer, D.R. (1990a) Ammonia volatilization from cattle slurry following surface application to grassland. I. Influence of mechanical separation, changes in chemical composition during volatilization and the presence of the grass sward. *Plant and Soil* 125, 109–117.

Thompson, R.B., Pain, B.F. and Rees, Y.J. (1990b) Ammonia volatilization from cattle slurry following surface application to grassland. II. Influence of application rate, wind speed and applying slurry in narrow bands. *Plant and Soil* 125, 119–128.

Unwin, R.J., Pain, B.F. and Whinham, W.N. (1986) The effect of rate and time of application of nitrogen in cow slurry on grass cut for silage. *Agricultural Wastes* 15, 253–268.

Velthof, G.L. and Oenema, O. (1993) Nitrous oxide flux from nitric-acid-treated cattle slurry applied to grassland under semi-controlled conditions. *Netherlands Journal of Agricultural Science* 41, 81–93.

Vertregt, N. and Selis, H.E. (1990) Ammonia volatilization from cattle and pig slurry applied to grassland. In: *Dutch Priority Programme on Acidification Report 131-1.* Centre for Agrobiological Research, Wageningen.

Vetter, H., Steffens, G. and Schropel, R. (1987) The influence of different processing methods for slurry upon its fertilizer value in grassland. In: van der Meer, H.G., Unwin, R.J., van Dijk, T.A. and Ennik, G.C. (eds) *Animal Manure on Grassland and Fodder Crops. Fertilizer or Waste?* Martinus Nijhoff, Dordrecht, pp. 73–86.

Whitehead, D.C. and Raistrick, N. (1993) The volatilization of ammonia from cattle urine applied to soils as influenced by soil properties. *Plant and Soil* 148, 43–51.

Witter, E. and Kirchmann, H. (1989a) Peat, zeolite and basalt as adsorbents of ammoniacal nitrogen during manure decomposition. *Plant and Soil* 115, 43–52.

Witter, E. and Kirchmann, H. (1989b) Effects of addition of calcium and magnesium salts on ammonia volatilization during manure decomposition. *Plant and Soil* 115, 53–58.

Nitrous Oxide and Ammonia Emissions from Grassland Following Applications of Slurry: Potential Abatement Practices

22

D. Chadwick

Institute of Grassland and Environmental Research, North Wyke, Okehampton, Devon EX20 2SB, UK

Summary. The application of animal manures to grassland returns plant nutrients to the soil but can also result in significant emissions of nitrous oxide (N_2O) and ammonia (NH_3). Specific management practices are required to reduce N_2O losses following slurry applications to grasslands and several are discussed below. Factors which may influence N_2O emissions include the time of year at which slurry is applied and the rate and the method of application. Results showed lower N_2O losses in the spring, when the crop was actively growing, than in summer or autumn. Shallow injection of slurry in March resulted in significantly greater emissions of N_2O than surface broadcasting. In contrast, shallow injection appeared to reduce N_2O emissions following an application in June. Doubling the application rate significantly increased N_2O loss 3.7-fold following surface applications in November.

An 'upstream' management practice, reducing the crude protein content of pig feed and adding specific amino acids, reduced N excretion by pigs. When the resulting slurry was spread on grassland the % N added in the slurry that was lost as NH_3 and through denitrification was significantly lower than that from grassland treated with slurry from conventionally fed pigs. Nitrous oxide emissions were not significantly different.

Introduction

Approximately 85 million tonnes of slurry and solid manure are applied to UK grasslands each year (Pain *et al.*, 1995), returning essential nutrients to the soil for plant growth. It is well established that applications of animal manures to land result in significant emissions of NH_3 through volatilization (Pain *et al.*,

1995). However, these manure applications may also enhance soil conditions necessary for denitrification and the emission of N_2O through supplying readily biodegradable carbon (C) and nitrogen (N) and by creating anaerobic conditions (Eggington and Smith, 1986; Paul *et al.*, 1993; Estavillo *et al.*, 1994).

Much is known about methods for reducing NH_3 emissions from manure applications to land, e.g. shallow injection of slurry (Thompson *et al.*, 1987; Phillips *et al.*, 1990), but little attention has been paid to decreasing N_2O emissions following manure applications to land. To decrease N_2O losses from actively managed soils, it is essential to understand the impact of different agricultural practices, such as slurry applications, on emissions and to identify specific measures to reduce these emissions. One of the most obvious abatement strategies is to apply slurry only at times and rates which favour plant uptake of N. However, the 'window of opportunity' for emptying slurry stores and spreading to the land does not always coincide with crop demand and any excess N left in the soil is prone to loss. Shallow injection of slurry can be used to reduce NH_3 volatilization (Thompson *et al.*, 1987; Phillips *et al.*, 1990), but its influence on N_2O emissions has not been fully determined. A third option for reducing N losses is to reduce the amount of N excreted by animals by modifying their diet. Hobbs *et al.* (1996), for example, showed dietary control of crude protein to be successful in reducing odorous compounds in pig slurry.

This chapter presents results from preliminary experiments on some of the factors controlling emissions of N_2O following applications of slurry to grassland. Some measurements of NH_3 and denitrification are also included.

Influence of Time of Year of Slurry Application on N_2O Emissions and the Influence of Method of Application

Materials and methods

Dairy slurry was applied to small plots established on a freely draining sandy loam (Crediton series) under rye-grass. Slurry applications were made three times throughout the year (March, June and November) using two methods (surface broadcast and shallow injection with an open slot 5 cm deep). Untreated plots served as controls. Three replicates of each treatment were arranged in a randomized block design, each plot measuring $3 \text{ m} \times 5 \text{ m}$. Slurry was applied at a target rate of 25 $\text{m}^3 \text{ ha}^{-1}$ on each occasion and for each treatment; however, in November an additional surface broadcast treatment of 50 $\text{m}^3 \text{ ha}^{-1}$ was included to assess the influence of rate of application on N_2O emission.

Nitrous oxide emissions were measured using the static cover box technique (Mosier, 1989) with three cover boxes per plot, each box measuring 50 cm × 15 cm × 20 cm. At the time of sampling, clear perspex lids were clamped tightly to the cover boxes and samples of gas were removed from the cover boxes over a period of 40 min and analysed by gas chromatography. Measurements were taken once a day for the first week and thereafter once per week. On the plots which were shallow injected, two cover boxes were positioned over the injection slots and the other box was placed in the space between the slots. A weighted average rate of N_2O emission was then calculated for the injected plots based on the area of slots and spaces. Measurements of denitrification were also made on these plots using the acetylene inhibition technique (Ryden et al., 1987); NH_3 volatilization was measured following the March and June applications using wind tunnels (Lockyer, 1984).

Results and discussion

The influence of applying slurry at different times of the year on N_2O emission rates is illustrated in Fig. 22.1. The immediate emission of N_2O observed on all occasions was probably due to denitrification of the small amount of nitrate (NO_3^-) in the slurry or as a result of the extra C and moisture in the slurry providing suitable conditions for denitrification of some soil NO_3^-. Following this initial peak in N_2O emission after the March slurry application, emission rates of N_2O were very small with net emissions totalling 80 g and 33 g N ha^{-1} from the injected and surface broadcast treatments, respectively. Nitrous oxide fluxes following the June slurry application, showed a similar trend to those from the March application with the peak in emissions occurring approximately 5 days after application. However, in June the control plots also emitted significant amounts of N_2O. This peak in N_2O emission occurred 10 days after the sward was cut for silage and the N_2O emission observed from the control plots may have been as a result of denitrification of root-exudated NO_3^- or C exudates providing an energy source for denitrification of soil NO_3^- (Beck and Christensen, 1987).

In November, the surface application of slurry at 50 m^3 ha^{-1} resulted in significantly greater emissions of N_2O than at 25 m^3 ha^{-1} ($P < 0.05$). Net total emissions of N_2O were 263 g and 67 g N ha^{-1} for the 50 m^3 ha^{-1} and 25 m^3 ha^{-1} treatments, respectively. The N_2O peak was again approximately 5 days after slurry application, during which time ammonium in the slurry was nitrified to NO_3^-. A second peak in N_2O emission was also observed after 12 days from the 25 m^3 ha^{-1} treatment. Shallow injection appeared to minimize this second peak.

Table 22.1 summarizes the cumulative losses of N_2O, denitrification products and NH_3 volatilization from the small plots following the slurry

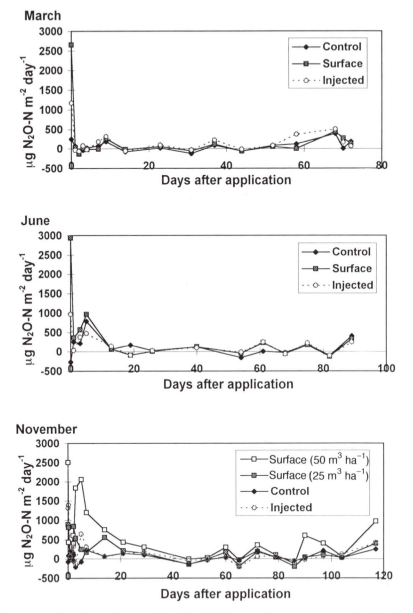

Fig. 22.1.   The effect of time and method of slurry application to a grassland soil on nitrous oxide emission.

applications in March, June and November. Cumulative net N_2O losses in March were significantly greater following injection of slurry than from the surface-broadcast treatment ($P < 0.05$). A significantly greater denitrification loss was also measured from slurry-injected plots ($P < 0.05$). This result is similar to that reported by Thompson *et al.* (1987) and is thought to be a result of larger quantities of slurry N being retained in the soil in a zone of restricted aeration. In June, the cumulative net N_2O emission and net emissions of denitrification products showed the opposite trend, with shallow injection apparently reducing emissions compared to surface broadcasting. Similar N_2O emissions and denitrification rates were observed in November from the $25 \text{ m}^3 \text{ ha}^{-1}$ surface-broadcast and injected plots. However, a doubling of the application rate to $50 \text{ m}^3 \text{ ha}^{-1}$ resulted in significantly more N_2O loss ($P < 0.05$). As a percentage of the total N applied in the slurry, 0.21% was lost as N_2O from the $50 \text{ m}^3 \text{ ha}^{-1}$ treatment and 0.11% from the $25 \text{ m}^3 \text{ ha}^{-1}$ treatment, demonstrating that once crop demand has been satisfied any extra N in the soil is at risk of being lost to the environment and that, under wet soil conditions, this can be via denitrification. Cumulative net N_2O losses expressed as a percentage of the total N applied in the slurry varied between 0.02 and 0.21%, which are similar values to those measured by Paul *et al.* (1993).

The lower three rows in Table 22.1 express the gaseous N losses as a percentage of the total ammoniacal-N (TAN) applied. Ammonia volatilization accounted for the greatest percentage lost followed by denitrification products with N_2O representing a very small loss. Shallow injection was successful at reducing NH_3 volatilization both in March and June but larger reductions (from 94 to 19%) were recorded under the warmer conditions in June. The soil was still moist enough to promote denitrification, such that nearly all the ammoniacal-N applied in the surface broadcast treatment was lost to the atmosphere as gaseous products leaving very little inorganic N for the crop.

Modifying Pig Diets to Control Gaseous N Emissions following Slurry Spreading to Land

Materials and methods

The crude protein content of pig feed was reduced by 25% after reducing the content of certain feed constituents and the addition of specific amino acids to bring the diet into line with the animal's nutritional requirements. Finishing pigs were fed either a commercial diet or a reduced crude protein diet, resulting in slurries with different NH_4^+-N contents (3.78 and 2.28 g NH_4^+-N kg^{-1}, respectively) (Hobbs *et al.*, 1996). Slurries were surface applied to

Table 22.1. The influence of time, method and rate of slurry application on gaseous N losses from grassland.

kg N ha⁻¹	March (72 days) 25S	March (72 days) 25I	June (89 days) 25S	June (89 days) 25I	November (117 days) 50S	November (117 days) 25S	November (117 days) 25I
TAN applied (kg N ha⁻¹)	44	47	32	32	56	28	28
Total N applied (kg N ha⁻¹)	72	76	44	44	124	62	62
NH₃ lost (kg N ha⁻¹)*	26	16	30	6	—	—	—
Net N₂O loss (kg N ha⁻¹)	0.03^b	0.08^a	0.05	0.01	0.26^y	0.07^z	0.05^z
Net N₂O + N₂ loss (kg N ha⁻¹)*	0.75^z	3.03^y	1.57	0.26	—	—	—
NH₃ loss (% TAN applied)	59	34	94	19	—	—	—
N₂O loss (% total N applied)	0.04^b	0.10^a	0.11	0.02	0.21	0.11	0.08
N₂O + N₂ loss (% total N applied)	1.04	3.99	3.57	0.59	—	—	—
N₂O loss (% TAN applied)	0.07^b	0.17^a	0.15	0.03	0.47	0.24	0.19
N₂O+N₂ loss (% TAN applied)	1.70	6.45	4.91	0.81	—	—	—

*Data from Misselbrook *et al.* (1996).
TAN = total ammoniacal nitrogen of the slurry.
25I, 25S and 50S represent target rates of treatments, 25 m³ ha⁻¹ injected, 25 m³ ha⁻¹ surface broadcast and 50 m³ ha⁻¹ surface broadcast, respectively.
^{a,b,y,z}Mean values with different letters are significantly different at *P* < 0.05 level.

small plots at the rate of 50 m³ ha⁻¹, resulting in an ammoniacal-N loading of 189 and 114 kg ha⁻¹ from the 'commercial slurry' and the 'reduced N' slurry, respectively. Emissions of NH₃, N₂O and denitrification were measured using the techniques described above.

Results and discussion

Ammonia volatilization and denitrification were significantly reduced in plots which received the slurry from the pigs fed the reduced crude protein diet (*P* < 0.05) (Table 22.2). Nitrous oxide emissions were also reduced but not significantly. These reductions in net emissions were to be expected, as less N was applied to those plots receiving the slurry from the pigs fed the diet with a reduced crude protein content. However, as can be seen in Table 22.2, the losses through NH₃ volatilization and denitrification expressed as a total of the TAN applied were significantly lower from plots that received the 'reduced N' slurry than from those that received standard 'commercial N' slurry, suggesting that dietary manipulation may be an effective upstream management practice to reduce gaseous N emissions from the pig industry.

Table 22.2. The influence of pig diet on gaseous N losses following slurry application to grassland.

Treatment	Ammonia		Nitrous oxide		Denitrification	
	kg N ha^{-1}	%TAN lost	kg N ha^{-1}	%TAN lost	kg N ha^{-1}	%TAN lost
Commercial N (189 kg TAN ha^{-1})	109.0**	57.7*	1.0	0.5	22.6*	12.0*
Reduced N (114 kg TAN ha^{-1})	42.8	37.5	0.7	0.6	6.0	5.3

Data from Misselbrook, Chadwick, Pain and Headon (unpublished data).
$*P < 0.05$, $**P < 0.01$.

Conclusions

The application of slurries to grassland significantly increases N_2O emissions, denitrification rates and NH_3 volatilization. The rates of these emissions depend on several factors, including rate of application, method of application, time of year of application and slurry composition. Emission factors, like those produced in the experiments described above, have been used in the construction of an NH_3 inventory (Pain et al., 1995) and are being used in the construction of an inventory for N_2O emissions from the agricultural livestock sector. Inventories such as these will allow the assessment of potential abatement practices at various stages in livestock production.

References

Beck, H. and Christensen, S. (1987) The effect of grass maturing and root decay on N_2O production in soil. *Plant and Soil* 103, 269–273.

Eggington, G.M. and Smith, K.A. (1986) Nitrous oxide emission from a grassland soil fertilised with slurry and calcium nitrate. *Journal of Soil Science* 37, 59–67.

Estavillo, J.M., Rodriguez, M., Domingo, M., Muñoz-Rueda, A. and Gonzalez-Murua, C. (1994) Denitrification losses from a natural grassland in the Basque Country under organic and inorganic fertilization. *Plant and Soil* 162, 19–29.

Hobbs, P.J., Pain, B.F., Kay, R.M. and Lee, P.A. (1996) Reduction of odorous compounds in fresh pig slurry by dietary control of crude protein. *Journal of Science, Food and Agriculture* 71, 508–514.

Lockyer, D.R. (1984) A system for the measurement in the field of losses of ammonia through volatilisation. *Journal of Science of Food and Agriculture* 35, 837–848.

Misselbrook, T.H., Laws, J.A. and Pain, B.F. (1996) Surface application and shallow injection of cattle slurry on grassland: nitrogen losses, herbage yields and nitrogen recoveries. *Grass and Forage Science* 51, 270–277.

Mosier, A.R. (1989) Chamber and isotope techniques. In: Andrae, M.O. and Schimmel, D.S. (eds) *Exchange of Trace Gases between Terrestrial Ecosystems and the Atmosphere.* John Wiley, New York, pp. 175–187.

Pain, B.F., van der Weerden, T., Jarvis, S.C., Chambers, B.J., Smith, K.A., Demmers T.G.M. and Phillips, V.R. (1995) Ammonia emission inventory for agriculture in the UK. *Final Report to the Ministry of Agriculture Fisheries and Food.* Contract OC9117, 37 pp.

Paul, J.W., Beauchamp, E.G. and Zhang, X. (1993) Nitrous oxide emissions during nitrification and denitrification from manure-amended soil in the laboratory. *Canadian Journal of Soil Science* 73, 539–553.

Phillips, V.R., Pain, B.F., Clarkson, C.R. and Klarenbeek, J.V. (1990) Studies on reducing the odour and ammonia emissions during and after the land spreading of animal slurries. *Farm Buildings and Engineering* 2, 17–23.

Ryden, J.C., Skinner, J.H. and Nixon, D.J. (1987) Soil core incubation system for the field measurement of denitrification using acetylene inhibition. *Soil Biology and Biochemistry* 19, 753–757.

Thompson, R.B., Ryden, J.C. and Lockyer, D.R. (1987) Fate of nitrogen in cattle slurry following surface application or injection to grassland. *Journal of Soil Science* 38, 689–700.

Ammonia Emissions Following the Application of Solid Manure to Grassland

23

H. Menzi[1], P. Katz[1], R. Frick[2], M. Fahrni[1] and M. Keller[1]

[1]*Swiss Federal Research Station for Agroecology and Agriculture, Institute of Environmental Protection and Agriculture (IUL), Liebefeld, CH-3003 Bern, Switzerland;* [2]*Swiss Federal Research Station for Agricultural Economics and Engineering (FAT), CH-8356 Tänikon, Switzerland*

Summary. Ammonia (NH_3) emissions following the application of solid manures were measured in field and wind-tunnel experiments. Mean emissions were about 60% of total ammoniacal nitrogen (TAN) applied. Wind-tunnel experiments with solid manures undertaken in different seasons showed a linear and highly significant relationship between the rate of ammoniacal nitrogen applied (RTAN) and emissions. This indicates that emissions were hardly influenced by climatic factors, TAN content and application rate. On farms, NH_3 emissions following spreading of solid manure can be estimated by the following equation: Emission ($g\ NH_3$-$N\ m^{-2}$) $= (0.787\ RTAN + 0.076) \times 0.75$. As solid manure is usually produced together with slurry rich in urine, total emissions of manure spreading are only about 10% lower in solid manure systems than in liquid manure systems.

Introduction

About 25% of the animal wastes produced in Switzerland are solid manures. As hill and mountain farms have no alternative, about 50% of this solid manure is applied to grassland where it cannot be incorporated into the soil. Little information exists on ammonia (NH_3) emissions from solid manure after surface spreading. In this experimental programme, treatments were included with solid manure in field experiments as well as in wind-tunnel experiments. A major aim was to identify similarities and differences in the emission behaviour of solid and liquid manures, by comparing treatments in parallel field experiments and results of field and wind-tunnel experiments.

Methods

Field experiments

Field experiments were carried out on circular plots of 20 m radius using the z_{inst}-method described by Wilson *et al.* (Wilson *et al.*, 1982; Wilson and Shum, 1992). Three plots could be measured at the same time. In 18 experiments a 'standard treatment' was included, both for comparison with other treatments in the same experiment, and for intercomparison with other experiments in the programme. In this standard treatment, liquid cattle manure from the same pit (total ammoniacal nitrogen content (TAN) 0.7–1.3 g kg^{-1}) was applied with splash-plate equipment between 9 and 12 a.m. The application rate was always around 30 t ha^{-1}. All experiments lasted for 2–5 days, depending on weather conditions.

In total, 14 solid manure treatments were included in the programme, 11 on grassland and three on arable land. Apart from one occasion when three application rates (16, 34, 56 t ha^{-1}) were compared, experiments with solid manure always included the liquid manure standard treatment. The solid manure type (origin and composition) was the main factor varied. Five of the treatments can be classified as 'normal' solid manure of Swiss farms (consisting of dung and straw removed twice daily from the cattle houses and stored on an open site with a concrete floor for 3–6 months). Five were 'special' types of cattle solid manure (fresh, strongly decomposed, loose heap, deep litter), one came from pigs and two from poultry (hens and broilers).

Wind-tunnel experiments

Wind tunnels similar to those described by Lockyer (1984) were used to study the importance of single factors such as TAN of the manure or application rate. For solid manure, three treatments with two replicates could be studied each time.

Preliminary experiments showed that the wind speed in the tunnel, varied between 1.1 and 2.7 m s^{-1}, had no great influence on emissions. All further experiments were therefore done with a constant wind speed around 1.9 m s^{-1}. These first experiments also showed that solid manure is more likely to result in NH$_3$ concentration gradients in the vent of the tunnel than liquid manure. The tunnels were therefore adjusted with a funnel to guarantee a good turbulence in the vent. Further adaptations had to be made concerning sampling and laboratory analysis of the solid manure. To achieve reproducible results, the manure had to be used directly after it was collected on the farm and the samples had to be taken at the time of spreading and to be analysed within an hour. An experiment with seven identical treatments showed a good reproducibility of the method (mean emissions 4.34 g m^{-2}, standard error ± 0.26).

Apart from experiments to improve and test the method, six wind-tunnel experiments were carried out with solid manure. In five of them, solid manures from cattle farms were used; on three occasions manures with variable TAN content were compared at a constant application rate of 5 kg m^{-2}, and twice with variable application rates (1, 3, 5, 7, 9 and 11 kg m^{-2}). These experiments lasted for 3–6 days. One experiment lasting 15 days compared poultry manure from hens and broilers with cattle manure.

Measurement technique

The NH$_3$ concentration in the air was measured with passive samplers (Blatter *et al.*, 1992) both in the field and wind-tunnel experiments. Five passive samplers per air inlet were exposed. The air to be measured was sucked through the tube at a rate of 3.5 l s^{-1}. The ammonium concentration of the acid solution (0.5 mmol HCl with 20% ethylenglycol) in the passive samplers was determined by its reaction with ortho phthaldialdehyde in the presence of sodium sulphite, yielding a fluorescent product (Genfa and Dasgupta, 1989) which was measured with a Flow Injection Analysis (FIA) system. All the field and wind-tunnel experiments measurements started within 3 min of spreading the manure. During the course of the experiment the exposure time of the passive samplers was increased from 1 to 16 h in the field experiments, and from 1 h to several days in the wind-tunnel experiments. In the field experiments, air speed was measured continuously with a cup anemometer with a threshold speed of 0.15 m s^{-1}. In the wind tunnels, air speed was kept constant and measured once a day using a small fan anemometer on 16 points distributed on perpendicular transects through the exhaust air vent.

A detailed description of the methods and results concerning liquid manure is given by Katz (1996).

Empirical model for liquid manures

For liquid manure, a simple empirical model to estimate NH$_3$-N losses was derived using the results of regression analysis on field and wind-tunnel data (Menzi *et al.*, 1995; Katz, 1996). This model uses the saturation deficit (SD; mbar) of the air, TAN content (g kg^{-1}) and application rate (AR; t ha^{-1}) as input variables:

$$\text{Emission (kg NH}_3\text{-N ha}^{-1}) \tag{1}$$
$$= (19.408 \, \text{TAN} + 1.102 \, \text{SD} - 9.506) \times (0.021 \, \text{AR} + 0.358)$$

The saturation deficit of the air depends on air temperature (T; °C) and relative humidity (RH), i.e.:

$$SD = (1 - RH) \times 6.112 \times e^{((17.67 \times T)/(243.5 + T))} \qquad (2)$$

This empirical model (equation (1)) is valid for the following conditions: liquid cattle manure applied to grassland with splash-plate; TAN content 0.7–5 g kg^{-1}; mean temperature 0–25°C; mean relative humidity of the air 0.5–0.9; no rain. The model does not yet include the time of application during the day, application technique, rainfall after manure application or water saturation of the soil.

Analysis of solid manure results

In the field experiments, emissions from the solid manure treatments were primarily compared with those from liquid manure in parallel treatments, and the emissions calculated with equation (1). Thus, differences in the emission behaviour of solid and liquid manure could be identified. Wind-tunnel results were mainly used to study the influence of TAN content and application rate using regression analysis.

Results and Discussion

Emissions in field experiments

Mean emissions during the first 2 days after application of liquid manure in the field experiments were 52% of TAN applied, with no clear difference between grassland and arable land. Slurry rich in urine (produced together with solid manure) and solid manure had higher relative emissions – 60 and 59% of TAN applied, respectively. Taking the total N applied as a basis, mean emissions were c. 25% for liquid manure and c. 10% for solid manure. In most experiments, only low-level emissions occurred over the third and fourth day irrespective of the manure type. As the experiments were often done under hot and dry conditions, when farmers would not spread manure, it is assumed that mean emissions during the first 2 days in these experiments correspond with mean total emissions to be expected after manure spreading on farms.

On average, solid cattle and pig manures had 15% higher emissions than the standard slurry treatment applied at a similar application rate in the same experiment. Per unit of TAN-applied emissions after solid manure application were 30% higher than in the standard liquid manure treatment of the same experiment.

The emissions for most solid manure treatments were underestimated by the empirical model for liquid manure. Nevertheless, agreement between measured and calculated emissions was good for the five treatments concerning 'normal' solid manure from cattle production (Fig. 23.1). The same comparison for the rest of the treatments indicated that spreading fresh solid manure (shortly after removal from stable), solid manure stacked in loose heaps or pig solid manure led to above average emissions after spreading, whilst spreading strongly decomposed, old solid manure gave below average emissions. The greater variability between solid manure than between liquid manure treatments is not surprising because solid manure is difficult to define, its composition being strongly influenced by the composition of animal excreta, the stable and manure collecting system, the amount and composition of bedding material used, the type and duration of storage, etc.

Linear regression analysis between emissions from solid manure and climatic factors indicated that the saturation deficit of the air was the most important factor. This corresponded with the model for liquid manure. In

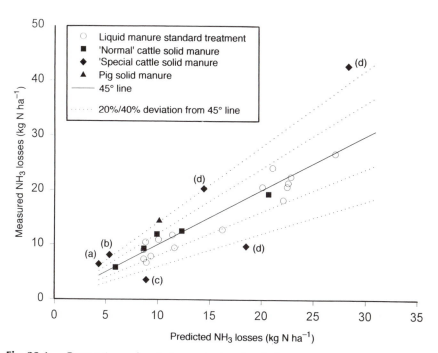

Fig. 23.1. Comparison of emissions calculated with the empirical model for liquid manure and measured values for the standard liquid manure treatments and different solid manure treatments. 'Normal' solid manure: standard on Swiss farms – consisting of dung and straw removed twice daily from the cattle houses and accumulated on an open storage site with concrete floor for 3–6 months. 'Special' solid manure: (a) fresh manure; (b) loose heap; (c) old solid manure; (d) on stubble.

general, correlations between climatic factors and emissions were lower than for liquid manure.

Emissions from solid manure in wind-tunnel experiments

In spite of different environmental conditions, three wind-tunnel experiments with solid manure of different composition used at a rate of 5 kg m^{-2} gave good agreement between the TAN content (g kg^{-1}) and NH$_3$ emissions (Fig. 23.2). The results of the three experiments (eight different solid manures) could be described by the following linear regression, i.e.:

$$\text{Emission (g NH}_3\text{-N m}^{-2}) = 3.72 \text{ TAN} + 0.15 \quad (r^2 = 0.95) \quad (3)$$

The results of the two experiments concerning the influence of application rate on emissions could be well described by linear regression (Fig. 23.3). However, due to the different composition of the manure in the two experiments, the two regressions were quite different.

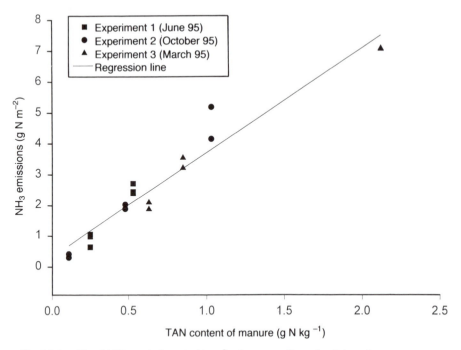

Fig. 23.2. Total NH$_3$ emissions (g N m^{-2}) after spreading of solid cattle manure on grassland in three wind-tunnel experiments with varied TAN content (g N kg^{-1}) of the solid manure. Regression line: Emission = 3.72 TAN + 0.15; $r^2 = 0.95$.

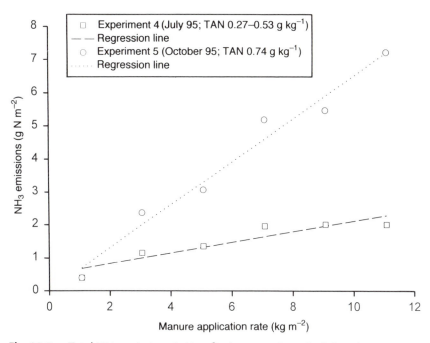

Fig. 23.3.   Total NH₃ emissions (g N m⁻²) after spreading of solid cattle manure on grassland in two wind-tunnel experiments with varied application rate (kg m⁻²). Regression lines: experiment (4) Emissions = 0.159 AR + 0.519, r^2 = 0.97; experiment (5) Emissions = 0.650 AR + 0.048, r^2 = 0.95.

Expressing emissions in relation to the rate of ammoniacal nitrogen applied (RTAN; g m⁻²) showed that the results of all five experiments done in different seasons can be summarized by one linear regression equation (Fig. 23.4), i.e.:

$$\text{Emission (g NH}_3\text{-N m}^{-2}) = 0.787 \text{ RTAN} + 0.076 \quad (r^2 = 0.95) \qquad (4)$$

This means that emissions from cattle and pig solid manures can be estimated as a function of RTAN irrespective of TAN content, application rate and climatic variables. The relatively good agreement of the results from the field experiments with equation (4) also supports this conclusion (Fig. 23.5). Nevertheless, emissions in field experiments tend to be overestimated by equation (4) because conditions in the wind-tunnel, especially the higher wind speed and turbulence at soil level, increase the emission processes. Mean relative emissions were 60% of TAN applied in the field experiments and about 80% in the wind-tunnels. To account for this equation (4) would have to multiplied by 0.75. Thus:

$$\text{Emission (g NH}_3\text{-N m}^{-2}) = (0.79 \text{ RTAN} + 0.08) \times 0.75 \qquad (5)$$

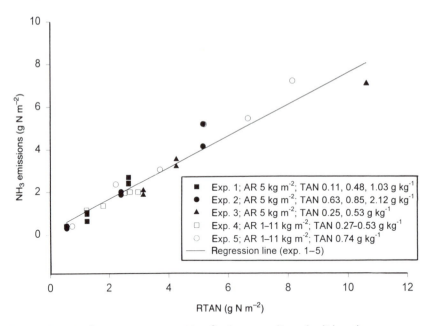

Fig. 23.4. Total NH₃ emissions (g N m⁻²) after spreading of solid cattle manure on grassland in five wind-tunnel experiments in relation to the rate of TAN applied (g N m⁻²). Regression line: Emission = 0.787 RTAN + 0.076, r^2 = 0.95.

The corrected equation (5) corresponds well with all the field experiment treatments using 'normal' solid manure (Fig. 23.5).

For liquid manure, no clear relationship could be found between RTAN and emissions. This is probably due to the infiltration of the liquid manure into the soil. Infiltration can be very variable because it is dependent on soil type and conditions, type and development of the plant cover, the viscosity and the dry matter content of the liquid manure.

Emissions from poultry manure

Two poultry manure treatments were also included in the field studies as well as in the wind-tunnel experiments. In the field experiment, emissions during 4 days after application were 19% of the total N applied for fresh laying hen manure (34% dry matter content, 22 g N kg⁻¹, application rate 22 t ha⁻¹) and about 3% for broiler manure (62% dry matter content, 32 g N kg⁻¹, application rate 17 t ha⁻¹).

In the wind-tunnel experiments, emissions from both poultry manures increased by about 50% from the fourth to the fifteenth day. Total emissions at the end of the experiment were approximately 10% of the N applied for

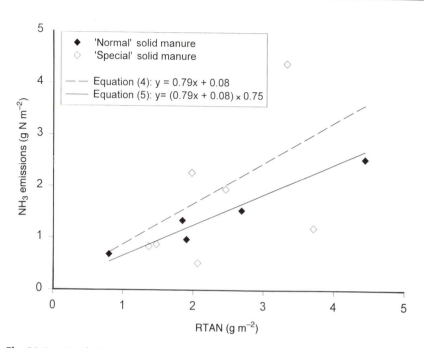

Fig. 23.5. Total NH_3 emissions (g N m^{-2}) after spreading of solid cattle manure in field experiments in relation to the rate of application TAN (g N m^{-2}). Comparison of measured values with the equations derived from wind-tunnel experiments (equation (4): regression line; equation (5): adjusted for on-farm estimates of emissions). For definitions of manures see Fig. 23.1.

stored laying hen manure (33% dry matter content, 19 g N kg^{-1}, application rate 2.5 kg m^{-2}) and 8% for broiler manure (72% dry matter content, 32 g N kg^{-1}, application rate 2.5 kg m^{-2}). As emissions had not stopped after 15 days total emissions would be somewhat higher.

Conclusions

Ammonia emissions following the application of solid manure were generally about 20% higher than from liquid manure, with a similar application rate and TAN content. Based on the field experiments and the empirical model derived, it is proposed to use an emission factor of 50% of TAN applied for liquid manure and 60% for solid cattle or pig manures in emission inventory calculations.

While emissions from liquid manure are strongly influenced by climatic factors (mainly temperature and humidity), emissions from surface-spread solid manure depend mainly on the rate of TAN applied. At the farm level,

NH$_3$ emissions from solid manure can therefore be estimated using equation (5). Seasonal differences in emissions can be expected to be small, even though emissions will tend to be more rapid under warm and dry conditions.

Due to the lower TAN content of solid manures, emissions per unit total N applied are lower than for liquid manures. Assuming that 20% of the total N in solid manure is in the form of TAN and 50% in liquid manure, losses would be 12% of the total N applied for solid manure and 25% for liquid manure. On Swiss farms solid manure is usually produced together with slurry rich in urine. This slurry has a high TAN content of 60–80% of total N and therefore high NH$_3$ emissions following its application. Model calculations show that the total NH$_3$ losses from manure spreading are therefore only about 10% lower for liquid manure systems than for solid manure/slurry systems. Furthermore, the amount of plant-available N reaching the soil per animal unit will be higher in liquid manure systems.

References

Blatter, A., Fahrni, M. and Neftel, A. (1992) A new generation of NH$_3$ passive samplers. In: *COST 611 Workshop, Rome, April 1992.*

Genfa, Z. and Dasgupta, P.K. (1989) Fluorometric measurement of aqueous ammonia ion in a flow injection system. *Analytical Chemistry* 61, 408–412.

Katz, P.E. (1996) Ammoniakemissionen nach der Gülleanwendung auf Grünland. PhD thesis 11382, ETH-Zurich, Switzerland.

Lockyer, D.R. (1984) A system for the measurement in the field of losses of ammonia through volatilization. *Journal of the Science of Food and Agriculture* 35, 837–848.

Menzi, H., Fahrni, M., Katz, P.E., Frick, R. and Keller, M. (1995) A simple empirical model based on regression analysis to estimate ammonia exmissions after manure application. In: *Proceedings of the International Conference Atmospheric Ammonia: Emission, Deposition and Environmental Impacts, Culham UK, 2–4 October 1995.*

Wilson, J.D. and Shum, W.K.N. (1992) A re-examination of the integrated horizontal flux method for estimating volatilisation from circular plots. *Agricultural and Forest Meteorology* 57, 281–295.

Wilson, J.D., Thurtell, G.W., Kidd, G.E. and Beauchamp, E.G. (1982) Estimation of the rate of gaseous mass transfer from a surface source plot to the atmosphere. *Atmospheric Environment* 16, 1861–1867.

Ammonia Emissions Following the Land Spreading of Solid Manures

24

B.J. Chambers[1], K.A. Smith[2] and T.J. van der Weerden[3]

[1]ADAS Land Research Centre, Gleadthorpe, Meden Vale, Mansfield, Nottinghamshire NG20 9PF, UK; [2]ADAS Wolverhampton, Woodthorne, Wergs Road, Wolverhampton WV6 8TQ, UK; [3]IGER, North Wyke, Okehampton, Devon EX20 2SB, UK

Summary. Ammonia (NH_3) emission measurements were made using the micro-meteorological mass balance technique following the land spreading of straw-based farmyard manures (FYM) and poultry manures. Ammonia-nitrogen (NH_3-N) losses following FYM applications were in the range 7–35 kg NH_3-N ha^{-1}, equivalent to 30–89% of the ammonium-N (NH_4^+-N) applied. Following poultry manure applications, losses were in the range 20–46 kg NH_3-N ha^{-1}, equivalent to 15–46% of the NH_4^+-N plus uric acid-N applied. The mean NH_3 emission factor for the FYMs was c. 65% of the NH_4^+-N applied, and for the poultry manures c. 35% of the NH_4^+-N plus uric acid-N. Based on these emission factors and the quantities of solid manures handled in the UK in 1993, total NH_3-N emissions following the land spreading of solid manures were estimated to be 29 kt NH_3-N, or approximately 40% of total estimated emissions from land spreading.

Introduction

Solid manures represent c. 45% of the 73 million tonnes of cattle excreta estimated to be handled in the UK in 1993, and c. 50% of the 10 million tonnes of pig excreta. Almost all poultry (c. 3.9 million tonnes) and sheep excreta (c. 1.7 million tonnes) are handled as solid manures. Overall in the UK, 42 million tonnes of excreta were estimated to be handled in a solid form and 46 million tonnes as slurry (Pain et al., 1996).

Ammonia (NH_3) emission measurements following the land spreading of manures have largely concentrated on slurries, as these have higher ammonium nitrogen (NH_4^+-N) contents (typically 50–60% of total N) than solid straw-based farmyard manures (FYM) (10–25% of total N), MAFF (1994). In addition, the most prevalent waste management systems within Northern European countries involve slurries.

The experimental work reported here measured NH_3 emissions following the application of cattle and pig FYMs, and poultry manures (broiler/turkey litters and layer manures) to arable stubbles and growing crops in the autumn–spring period. The emission measurements were then used to provide an estimate of NH_3 volatilization losses following the land spreading of solid manures in the UK.

Materials and Methods

Ammonia emission measurements were made using the micrometeorological mass balance technique to determine the vertical integration of the mean horizontal flux of NH_3. The method, which relies on differences in NH_3 concentrations between the leeward and windward extent of the experimental area, is described in more detail by Denmead (1983). In these experiments, two techniques were used to trap the NH_3: (i) an active sampling method using absorption flasks (Denmead, 1983) or 'bubblers'; and (ii) passive NH_3 samplers (Leuning et al., 1985) or 'shuttles'. Total NH_3 losses and loss patterns measured using the two techniques have been shown to be in good agreement (Moss et al., 1995).

Solid FYM and poultry manure applications were made to arable stubbles (six measurements) in the autumn and winter period (October to March during seasons 1993/94 and 1994/95), and additionally following the topdressing of a growing cereal crop with broiler litter in March 1993 (Table 24.1). The measurements were undertaken at times and in conditions considered representative of land application practices for solid manures in the UK.

The solid manures were hand applied to a circular plot area of 20 m radius, to provide a constant fetch length for measuring NH_3 losses from the applied manures. A collector mast was sited in the centre of the circular plot, with bubblers and/or shuttles located at heights of 0.2, 0.4, 0.8, 1.2, 2.2 and 3.3 m. A background mast was sited upwind of the experimental area, with the NH_3 samplers located at heights of 0.2, 0.8 and 2.2 m. Measurements commenced immediately after manure application to the windward quarter of the circular plot, and continued for between 9 and 31 days. Further details of the experimental methodology and NH_3 loss calculations are given in Moss et al. (1995).

At the time of application, representative manure samples were taken and analysed for dry matter, total N, NH_4^+-N and uric acid-N (poultry manures only), according to standard methods (MAFF, 1986), to calculate the total N and NH_4^+-N (+ uric acid-N) application rates (Table 24.1).

Table 24.1.   Details of experimental sites and manure analyses.

Manure type/ description	Site location (county)	Dry matter (%)	Application rate		
			t ha^{-1}	total N (kg ha^{-1})	NH$_4^+$-N (kg ha^{-1})*
I. Farmyard manure:					
1. Cattle	ADAS Boxworth (Cambs)	23.5	30.6	200	24
2. Pig I	Elveden (Norfolk)	26.2	35.8	277	48
3. Pig II	Elveden (Norfolk)	24.1	23.2	241	40
II. Poultry manure:					
4. Broiler litter I	Elveden (Norfolk)	63.9	6.9	215	65 (37/28)
5. Broiler litter II	ADAS Boxworth (Cambs)	58.7	9.9	302	133 (99/34)
6. Turkey litter	Freckenham (Norfolk)	50.7	8.0	257	114 (58/56)
7. Layer manure	Freckenham (Norfolk)	43.8	11.9	196	99 (71/28)

*In the case of poultry manures, ammonium and uric acid-N are combined (individual figures are in brackets).

Results

Ammonia emission measurements following the FYM applications to arable stubbles were 7.1, 35.0 and 35.4 kg NH$_3$-N ha^{-1}, which was equivalent to 30, 73 and 89% of the NH$_4^+$-N applied, respectively (Table 24.2). The mean emission factor was 64% of the NH$_4^+$-N applied or 13% of total N. Ammonia emission measurements following the cattle FYM application at ADAS Boxworth continued for 18 days, with c. 90% of losses occurring in the first 10 days. On the basis of these data, the following NH$_3$ emission measurements were continued for c. 10 days to ensure that the majority of losses were quantified. Cumulative NH$_3$ losses from the FYM applications are shown in Fig. 24.1, with c. 10% of the total measured losses occurring in the first 2 h following application, 30% in 6 h, 50% in 24 h and 80% within 5 days. Emissions appeared to increase in the immediate period following rainfall events at the Elveden sites of 13 mm and 16 mm, respectively.

Following the poultry litter dressings NH$_3$-N losses were 29.9, 19.8 and 46.4 kg NH$_3$-N ha^{-1}, respectively, and following the layer manure dressing 39.3 kg NH$_3$-N ha^{-1}. These losses were equivalent to 46, 15 and 41% of the NH$_4^+$ plus uric acid-N applied (AUN) for the poultry litters, and 40% of AUN for the layer manures. The mean emission factor was 35% of the AUN applied or

Table 24.2. Ammonia emission measurements.

Manure type/ description	Site location	Measurement period (h)	NH$_3$-N loss (kg ha^{-1})	Loss as % NH$_4^+$-N applied*	Loss as % Total N applied
I. Farmyard manure:					
1. Cattle	Boxworth	434	7.1	30	4
2. Pig I	Elveden	213	35.0	73	13
3. Pig II	Elveden	210	35.4	89	23
Mean				64	13
II. Poultry manure:					
4. Broiler litter I	Elveden	222	29.9	46	14
5. Broiler litter II	Boxworth	546	19.8	15	7
6. Turkey litter	Freckenham	713	46.4	41	18
7. Layer manure	Freckenham	739	39.3	40	20
Mean				35	15

*In the case of poultry manures includes uric acid-N.

15% of total N. Ammonia emission measurements following the first broiler litter application at the Elveden site continued for *c.* 9 days. However, it was evident that NH$_3$ losses were still occurring at the end of that period. Thereafter, emission measurements were extended to cover at least the 3-week post application period (actual measurement periods 23–31 days). Cumulative NH$_3$ losses from the poultry litters and layer manure are shown in Fig. 24.2. Approximately 10% of the total measured losses from the poultry litters occurred in the first day following application, thereafter the pattern of release differed markedly between the three litter dressings. The differences in release patterns are most likely a reflection of the rate at which uric acid-N is converted to NH$_4^+$-N. Following the layer manure application, *c.* 30% of losses occurred in the first day and 70% within 5 days, indicating a more rapid emission pattern than from the poultry litters.

Discussion and Conclusions

There have been relatively few measurements of NH$_3$ emissions following the land application of solid manures in Northern European Countries. The mean NH$_3$ emission factor for the FYM measurements reported here was *c.* 65% of NH$_4^+$-N applied. This was similar to the 60% emission factor reported by Menzi *et al.* (see Chapter 23), although there was considerable variation about this mean factor in both sets of experimental work. In the case of poultry manures, the mean emission factor was *c.* 35% of AUN applied or 15% of total N, which is higher than measurements reported by Jarvis and Pain (1990) at 7% of the total N applied or 30% of NH$_4^+$-N over a 5-day period following

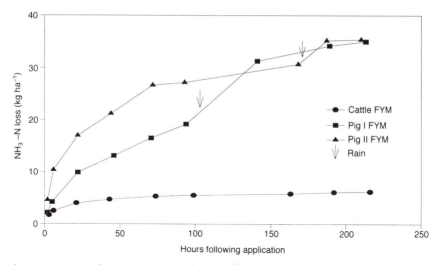

Fig. 24.1. Cumulative ammonia-N losses following farmyard manure applications.

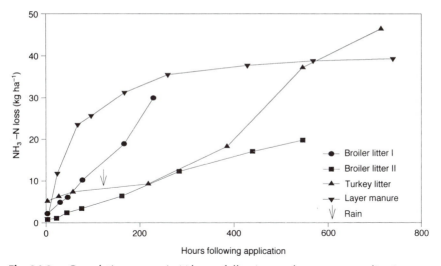

Fig. 24.2. Cumulative ammonia-N losses following poultry manure applications.

application. However, it is probable that in these latter studies NH_3 emissions continued beyond the 5-day measurement period, leading to an underestimate of losses.

Based on these emission measurements and the quantities of solid manures handled in the UK in 1993, NH_3-N losses following the land spreading of FYM (cattle, pig and sheep) were estimated at 16 kt N year^{-1} and following poultry manure applications 13 kt N year^{-1}. NH_3-N emission estimates from solid manures (29 kt N year^{-1}) were lower than estimates from slurries (cattle and pig) at 41 kt N year^{-1}. Emissions estimated from manures land spread in the UK at 70 kt N year^{-1}, were responsible for *c.* 36% of the total estimated emissions from UK agriculture in 1993 at 197 kt NH$_3$-N year^{-1} (Pain *et al.*, 1997).

Acknowledgements

Funding for this work from the Ministry of Agriculture, Fisheries and Food is gratefully acknowledged.

References

Denmead, O.T. (1983) Micrometeorological methods for measuring gaseous losses of nitrogen in the field. In: Freney, J.R. and Simpson, J.R. (eds) *Gaseous Losses of Nitrogen from Plant-soil Systems.* Martinus Nijhoff, The Hague, pp. 133–157.

Jarvis, S.C. and Pain, B.F. (1990) Ammonia volatilisation from agricultural land. *Proceedings of the Fertiliser Society No. 298, Peterborough*, 35 pp.

Leuning, R., Freney, J.R., Denmead, O.T. and Simpson, J.R. (1985) A sampler for measuring atmospheric ammonia flux. *Atmospheric Environment* 19, 1117–1124.

MAFF (1986) *The Analysis of Agricultural Materials.* Ministry of Agriculture, Fisheries and Food. Reference Book 427, HMSO, London, 248 pp.

MAFF (1994) *Fertiliser Recommendations for Agricultural and Horticultural Crops.* MAFF, Reference Book 209, 6th edn. HMSO, London, 112 pp.

Moss, D.P., Chambers, B.J. and van der Weerden, T.J. (1995) Measurements of ammonia emissions from land application of organic manures. *Aspects of Applied Biology* 43, Field Experiment Techniques, pp. 221–228.

Pain, B.F., van der Weerden, T.J., Chambers, B.J., Phillips V.R. and Jarvis, S.C. (1997) A new inventory of ammonia emissions from UK agriculture. *Atmospheric Environment* (in press).

Reduction of Ammonia Emission by New Slurry Application Techniques on Grassland

J.F.M. Huijsmans[1], J.M.G. Hol[1] and D.W. Bussink[2]

[1]Institute of Agricultural and Environmental Engineering (IMAG-DLO), PO Box 43, 6700 AA Wageningen, The Netherlands; [2]Nutrient Management Institute, Runderweg 6, 8219 PK Lelystad, The Netherlands

Summary. Surface spreading of slurry has the disadvantage of causing considerable emission of ammonia (NH_3) into the environment. In recent years, new slurry application techniques have been developed for grassland in order to reduce NH_3 emissions. In the past 5 years, injection (closed slot), shallow injection (open slot) and narrow band spreading by trailing feet (or 'shoes') have been investigated in field trials to evaluate their effect on emissions compared to surface-spread slurry. The results show that a significant reduction in emission can be obtained by the use of these techniques, with average NH_3 emissions being 31, 14 and 1% of those from surface spreading for trailing feet, shallow injection (open slot) and injection (closed slot), respectively.

Introduction

In practice, surface spreading of slurry is usually carried out with a tanker fitted with a splash-plate, the slurry being pumped through an orifice onto the splash-plate and spread onto the soil or grass surface. This surface spreading of slurry has the disadvantage of causing considerable emission of ammonia (NH_3) into the environment. Injection with a goose-foot injector reduces the NH_3 emission almost completely. The injector applies the slurry at 0.15 m depth through hollow, rigid tines at 0.5-m spacing and equipped with lateral wings. The wings lead to a better distribution of the slurry under the grassland surface. However, this type of injection is not applicable in all situations. To reduce NH_3 emissions at reasonable costs and on most soil types, new slurry application techniques were developed for grassland. These new techniques are narrow band application and shallow injection. Compared to conventional injection, these techniques cause less soil disturbance and hence the draught requirement is lower (Huijsmans and Hendriks, 1994). Narrow band application can be achieved by trailing a narrow sliding foot or shoe over the soil surface, without cutting the sward but pushing aside the grass leaves. Slurry is released at the back of the foot leaving narrow bands of slurry on the soil surface at 0.20-m spacing. Contamination of the grass with slurry is minimal. Unlike the conventional deep injector, shallow injector designs do not cut the soil underneath the sward horizontally because there are no lateral wings. Different shaped knives or disc coulters are used to cut a vertical slot in the grass sward and slurry is released into the slot, which may remain open (open slot) or is closed by a press wheel (closed slot) operating behind the tine or coulter(s). When the slot is closed the sward is cut to a depth of 0.08–0.10 m and slurry is found in the slot at a depth below 0.03 m under the soil surface; slot spacing is 0.30 m. When the slot remains open after releasing the slurry, the slots are up to 0.05 m deep and, depending on the application rate, the open slot is more or less filled with slurry.

By injection (open and closed slot) and narrow band application the emitting surface is reduced compared to surface spreading and soiling of the grass is prevented. In the past 5 years, conventional and shallow injection (closed slot), shallow injection (open slot) and narrow band application by trailing feet have been investigated to evaluate their effectiveness for reducing emissions compared to surface-spread slurry. This chapter describes the effects of application with these techniques on NH_3 emission.

Materials and Methods

Experiments

In the period 1989–93, experiments were carried out to determine the NH_3 emission from applying slurry with different techniques. Within each experiment, NH_3 measurements were made on up to five comparable plots, each differing only in application technique, grass height and/or application rate. In total 46 experiments were carried out under Dutch conditions at different times during the growing seasons (March to September) to meet different soil (i.e. soil water content) and weather conditions. Thirty-three experiments were conducted on a clayey, nine on a peaty, and four on a sandy soil. The swards were all well established and had been intensively managed in previous years. Perennial ryegrass was the dominant grass species. Cattle slurry was used in 39 experiments and pig slurry in seven. On average, the slurries contained 2.7 g kg^{-1} ammoniacal nitrogen (NH_4^+-N), 8.1% dry matter and had an average pH of 7.3.

The techniques investigated were: surface spreading; narrow band spreading; shallow injection (open slot) and injection (closed slot).

Surface spreading was carried out by a tanker with one outlet and a splash-plate and a net working width of approximately 8 m. Narrow band spreading was carried out by a tanker equipped with 20 trailing feet with a total working width of 5 m. The foot had a length of 0.37 m and width of 0.02 m; the foot was kept in the horizontal position by a parallellogram construction. The slurry bands had an actual width of approximately 0.03 m. Shallow injection (open slot) was carried out by either a knife coulter or a combination of two disc coulters. Depth of the slots was approximately 0.05 m; slot spacing was 0.20 m and the total working width was 4–5.6 m. Injection (closed slot) was carried out either with a goose-foot injector (six tines; spacing 0.5 m) or a shallow injector (14 tines; spacing 0.3 m). After releasing the slurry, the slots were closed by press wheels. The total number of plots on which the NH_3 emission was measured and the slurry application rates for each technique are given in Table 25.1.

Measurement of ammonia emission

The NH_3 emission after application of the slurry was determined per plot using the micrometeorological mass balance method as described by Denmead (1983) and Ryden and McNeill (1984). By applying the slurry in parallel passes varying in length over a pre-marked area, circular plots were achieved with a radius varying between 20 and 24 m. As soon as the the slurry was applied to the first half of a plot (usually within 5 min), a mast supporting seven to eight NH_3 traps at heights between 0.25 and 3.30 m was placed at the centre of the plot. At the windward boundary of the plot, another mast was placed with four to five NH_3 traps at heights between 0.40 and 2.30 m. Fewer traps were placed at the boundary because the background concentration was low and essentially uniform with height. Each trap contained 20 ml 0.02 M HNO_3 held in 100-ml collection tubes. Air was drawn through the acid via a sintered gas dispersion tube at rates of 2–4 l min^{-1}, measured with flow

Table 25.1. Number of plots and the slurry application rates from which the NH_3 emission was measured.

Application method	Plots (no.)	Application rate (m^3 ha^{-1})*
Surface spreading	45	13.8 (8.3–24.9)
Narrow band spreading	27	13.9 (6.6–28.1)
Shallow injection; open slot	32	22.0 (14.0–45.5)
Injection; closed slot	6	37.8 (21.9–50.0)

*Data are means with range in parentheses.

meters. Ion-chromatography was used to analyse the NH_4^+ concentration in the HNO_3 solution.

All plots in an experiment received slurry at approximately the same time before noon to avoid the influence of changing soil and weather conditions on the NH_3 emission and measurements on each plot continued for at least 4 days (96 h) after slurry application. Bussink *et al.* (1994) showed that the remaining NH_3 emission after 96 h was zero and this was confirmed in these experiments. Traps were replaced five times in the first 24 h when the highest NH_3 loss rates occurred. From day 2 until day 4 the traps were replaced in early morning and late afternoon. Wind speed was measured on a mast outside the plot, at six different heights between 0.40 and 3.30 m. Temperature, humidity and radiation were recorded on all sites during the experiments.

Statistical analysis

A non-orthogonal block design was used with the experiments being considered as the blocks. Cumulative NH_3 emissions 6 h and 96 h after slurry application were analysed to examine differences between application techniques. In order to get homogeneous variance, the logarithms of both cumulative NH_3 emissions were taken as the response variables. The response was analysed by regression analysis, with blocks and application techniques as the prediction factors. The significance of differences between application techniques was evaluated by pairwise t-tests.

Results and Discussion

In Table 25.2, the cumulative emissions 6 h after the application and the overall NH_3 losses per technique (after 96 h) are presented with the co-efficients of variation (c.v.) These figures are back transformed means of the response variate analysis. Pairwise differences between application techniques were all significant ($P < 0.05$), both for the emission after 6 and 96 h.

On the basis of these results, which are considered to be representative for the average conditions in the Netherlands, the emission reduction achieved by narrow band application was 69% compared with surface spreading. The open slot shallow injectors achieved average emission reduction scores of over 80%. The injectors, closed slot, reached an average emission reduction of more than 95%.

The actual NH_3 emissions for each application technique depend on the slurry composition (NH_4^+-N concentration, pH and dry matter content), the environmental conditions (weather, soil type and soil condition) and the farm management factors (amount of manure applied, grass height at the time of

Table 25.2. Cumulative NH_3 losses 6 h and 96 h after application following different application techniques and the coefficients of variation (c.v.).

	NH₃ losses			
	After 6 h		After 96 h	
Application method	%NH₄-N applied	c.v. (%)	%NH₄-N applied	c.v. (%)
Surface spreading	55.2	8	65.6	8
Narrow band spreading	10.7	12	20.4	11
Shallow injection; open slot	4.0	12	9.4	11
Injection; closed slot	0.8	28	0.9	24

application). Currently, research aims to model the effect of these factors for each application technique, using the data collected in the experiments. Some of these factors are expected to have a major effect on the emission. The most suitable technique, in a specific case, will depend on application rate, soil type and condition, weather conditions and time of the year. The farmer will choose the emission controlling technique most suitable in terms of costs and soil type.

References

Denmead, O.T. (1983) Micrometeorological methods for measuring gaseous losses of nitrogen in the field. In: Freney, J.R. and Simpson, J.R. (eds) *Gaseous Loss of Nitrogen from Plant-soil Systems. Developments in Plants and Soil Sciences,* Vol. 9. Martinus Nijhoff, Den Haag, pp. 1–29.

Bussink, D.J., Huijsmans, J.F.M. and Ketelaars, J.J.M.H. (1994) Ammonia volatilization from nitric-acid-treated cattle slurry, surface applied to grassland. *Netherlands Journal of Agricultural Science* 42, 293–309.

Huijsmans, J.F.M. and Hendriks, J.G.L. (1994) Draft requirement of new ammonia emission reducing slurry application techniques. *International Conference on Agricultural Engineering, Milano, 29 August to 1 September,* pp. 149–150.

Ryden, J.C. and McNeill, J.E. (1984) Application of the micrometeorological mass balance method to the determination of ammonia loss from a grazed sward. *Journal of the Science of Food and Agriculture* 35, 1297–1310.

Effect of Application Techniques on Ammonia Losses and Herbage Yield Following Slurry Application to Grassland

26

F. Lorenz and G. Steffens

Landwirtschaftskammer Weser-Ems, Landwirtschaftliche Untersuchungs- und Forschungsanstalt (LUFA), PO Box 2549, D-26015 Oldenburg, Germany

Summary.   The effects of five application techniques (conventional broadcast spreading, broadcast spreading with water wash treatment, band spreading, 'sliding shoes' and shallow injection) on ammonia (NH_3) losses, herbage yield and on slurry nitrogen (N) efficiency were investigated on three soil types (clay, sand and peat). Treatments were applied to first and third herbage cuts, with residual effects measured at the second cut. Compared to conventional broadcast application, NH_3 losses after application were decreased by 25% (water wash treatment), 30% (band spreading), 70% (sliding shoes) and 90% (shallow injection). Herbage yields were similar for each application technique, but under dry and warm weather conditions significantly higher yields were recorded for sliding shoe or shallow injection. Herbage N offtake increased with decreasing NH_3 losses. Similar results were recorded for broadcast application (with or without washing) and band spreading, but N offtake increased after 'sliding shoe' application and particularly after shallow injection. It was concluded that sliding shoes was the most suitable application technique for use on grassland under North German conditions.

Introduction

Slurry application to grassland is sometimes associated with problems such as high ammonia (NH_3) loss and scorching or soiling of the sward, which may give rise to yield depression and poor silage quality. For these reasons, farmers tend to apply slurry to arable land rather than to grassland. In recent years, a number of new slurry application techniques have been developed with potential to reduce these problems. Studies were carried out at LUFA, during 1991–94, to investigate the effects of five application techniques on NH_3 losses after application, grass herbage yield and nitrogen (N) removal.

Materials and Methods

The experiments included the following five application techniques: (i) conventional broadcast spreading; (ii) broadcast spreading with water wash treatment (washing off slurry immediately after spreading with water at a pressure of 80–100 bar and an application rate of 15% of the slurry application rate); (iii) band spreading; (iv) 'sliding shoes' (slurry placed on the soil surface and beneath the grass canopy); and (v) shallow injection (5 cm deep, open slots). The experiments were conducted on three soil types typical for dairy farming in Lower Saxony in North-West Germany. The soil types were clay, sand and peat soil; the clay site was permanent grassland, the sand and the peat site were temporary grass leys. Cattle slurry was applied at two rates (12.5 and 25.0 m^3 ha^{-1}; Table 26.1), without or with fertilizer N at 40 kg ha^{-1}. These treatments were applied to first and third herbage cuts, with residual effects measured at the second cut. To the second cut, a small amount of fertilizer N (40–60 kg ha^{-1}) was applied, uniformly, to all plots. Ammonia losses were measured for 2 days after application, using the wind-tunnel method described by Lockyer (1984) on two replicates of each treatment. Grass herbage yield was recorded, and a sample of the harvested crop was analysed for dry matter and N content. Plot size was 24 m^2; treatments were replicated four times, within a randomized block experimental design.

Results and Discussion

Ammonia losses

It was not possible to match the wind speed in the wind-tunnels to the actual outside wind speed. This resulted in a tunnel wind speed which was usually

Table 26.1. Range of slurry dry matter, total nitrogen (N) and ammonium nitrogen (NH_4^+-N) rates.

Application time	Slurry rate	DM (%)	Total N (%)	Total N (kg ha^{-1})	NH_4^+-N (%)	NH_4^+-N (kg ha^{-1})
Before first cut	12.5	8.5–10.2	0.46–0.56	55–66	0.26–0.30	31–36
Before first cut	25.0			117–142		65–77
Before third cut	12.5	5.1–10.1	0.29–0.53	36–66	0.17–0.30	22–37
Before third cut	25.0			76–136		45–77

Table 26.2.   Ammonia losses (%) on average for all experiments.

Application technique	NH_3 losses in % of losses of broadcast application
Broadcast application	100.0^a
Broadcast application + washing	75.0^b
Band spreading	70.0^b
Sliding shoe	30.4^c
Shallow injection	10.9^c

[a,b,c]Values assigned a different letter, significantly different according to SNK test.

higher than that outside. For this reason, the average NH_3 losses of the application techniques in Table 26.2, over the 3 years of the experiment, are compared to conventional broadcast application on a relative basis. With broadcast application + washing and with band spreading, significant reductions of 25 and 30%, respectively, were achieved. The best results, with significantly higher reduction of NH_3 emissions, were obtained with sliding shoe application (70% lower than broadcast) and shallow injection (90%).

These figures are in the range of losses found by other researchers (e.g. Klarenbeek and Bruins, 1991). However, there was sometimes a great variation in NH_3 losses within an individual application technique, depending on the technique itself (higher variation with broadcast application than with shallow injection), the weather conditions during and after application and slurry dry matter (DM) content.

Grass herbage yield

Figure 26.1 summarizes the treatment average for the herbage DM yield at each cut. At first and third cuts, slurry application by sliding shoe or shallow injection led to significantly higher yields than broadcast slurry application, with or without washing. Only at the third cut was a significant yield difference recorded between application by sliding shoe or shallow injection.

The results indicated a clear yield response with slurry applied onto or into the soil. The yield achieved with band spreading was lower than expected, especially in those cases where slurry with a high DM content (about 10%) had been used. It was generally observed that this high DM was not able to penetrate the sward and reach the soil, but remained on the top of the sward for some days or even weeks. The washing treatment also yielded poorly, even though it was effective in removing most of the slurry solid material from the grass. One reason could be increased NH_3 volatilization during washing. In the short term over 3 years, there was no visible residual effect (at the second cut) between application techniques.

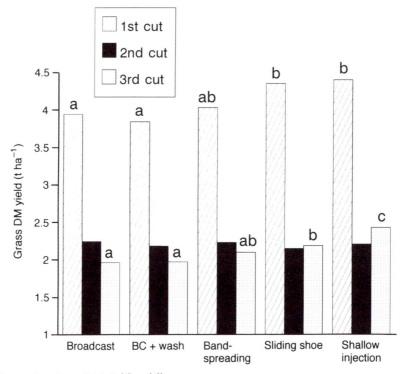

Fig. 26.1.   Grass DM yield at different cuts.

The results shown in Fig. 26.1 are averages for eight harvests for each cut. Results from single cuts indicate that there is potential for the farmer to use conventional equipment quite effectively if he takes into account weather conditions. An example is given in Fig. 26.2. Slurry was applied before the first cut 1992, when it was cool and cloudy with a high relative air humidity. Only a small but non-significant yield increase was recorded when slurry was applied by sliding shoe and shallow injection compared to broadcast application and band spreading. On the other hand, slurry application before the third cut, when it was hot and dry, with sliding shoe and shallow injection led to a significant yield increase.

Nitrogen offtakes

Table 26.3 shows an example, in this case on the clay soil, for the N removal by the harvested crop. In general, there were only small differences in N content and removal after broadcast slurry application (with and without washing) or band spreading. Nitrogen removal, however, was higher after

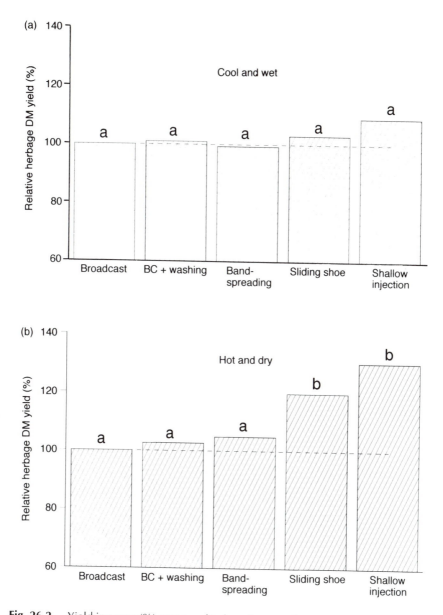

Fig. 26.2. Yield increase (%) compared to broadcast application (= 100%) after slurry application at different weather on peat soil. (a) Cool and wet: first cut 1992, 100% = 4.32 t ha^{-1}. (b) Hot and dry: third cut 1992, 100% = 2.29 t ha^{-1}.

Table 26.3.   Nitrogen removal from three herbage cuts* on heavy clay soil (average of 3 years).

Application technique	N applied (kg ha^{-1})	N removed (kg ha^{-1})
Broadcast	211	202
Broadcast + washing	211	188
Band spreading	211	196
Sliding shoe	211	215
Shallow injection	211	235

*First and third cut: 12.5 m^3 ha^{-1} cattle slurry + 40 kg ha^{-1} N; second cut: 40 kg ha^{-1} N.

sliding shoe application and particularly after shallow injection. This was also the case if slurry was applied under favourable conditions (i.e. cool and wet weather) where only small differences in yield occurred; increasing grass N offtake reflected more closely the reduction in NH$_3$ losses, with treatment, than did increasing grass DM grass yield. The yield was not only dependent on N response and, therefore, on the level of NH$_3$ losses, but also on other effects like scorching and soiling (e.g. from surface-applied slurry, Smith *et al.*, 1995) or on aeration and the damaging effects of shallow injection (van der Meer *et al.*, 1987; Long and Gracey, 1990).

References

Klarenbeek, J.V. and Bruins, M.A. (1991) Ammonia emissions after land spreading of animal slurries. In: Nielsen, V.C., Voorburg, J.H. and L'Hermite, P. (eds) *Odour and Ammonia Emissions from Livestock Farming.* Elsevier Science Publishers, Barking, Essex, pp. 107–115.

Lockyer, D.R. (1984) A system for the measurement in the field of losses of ammonia through volatilisation. *Journal of the Science of Food and Agriculture* 35, 837–848.

Long, F.N.J. and Gracey, H.I. (1990) Herbage production and nitrogen recovery from slurry injection and fertilizer nitrogen application. *Grass and Forage Science* 45, 77–82.

van der Meer, H.G., Thompson, R.B., Snijders, P.J.M. and Geurink, J.H. (1987) Utilization of nitrogen from injected and surface-spread cattle slurry applied to grassland. In: van der Meer, H.G. *et al.* (eds) *Animal Manure on Grassland and Fodder Crops.* Martinus Nijhoff Publishers, Dordrecht, pp. 47–71.

Smith, K.A., Jackson, D.R., Unwin, R.J., Bailey, G. and Hodgson, I. (1995) Negative effects of winter- and spring-applied cattle slurry on the yield of herbage at simulated early grazing and first-cut silage. *Grass and Forage Science* 50, 124–131.

Sources of Variation in Ammonia Emission Factors for Manure Applications to Grassland

27

B.F. Pain and T.H. Misselbrook

Institute of Grassland and Environmental Research, North Wyke, Okehampton, Devon EX20 2SB, UK

Summary.   Agriculture is the major contributor to ammonia emissions to the atmosphere in the UK, with spreading of livestock manures on land the single largest source. Variation in the emission factor (the loss expressed as % total ammoniacal N applied in the manure) for manures applied to grassland could have a significant effect on estimates of total emission and have implications for assessing the potential worth of methods for reducing emission. Factors influencing the emission from manure applied to land include manure composition, method of application, environmental conditions and method of measurement.

Introduction

It is generally accepted that, for many EU countries, about 90% of the total emission of ammonia (NH_3) to the atmosphere comes from agriculture. Estimates for the UK range from 186 to 405 kt NH_3-N per year (Sutton *et al.*, 1995), suggesting that there may be large differences in the emission factors used in the calculations for each source. A recent NH_3 emission inventory for agriculture in the UK estimated total annual emission to be 197 kt NH_3-N (Pain *et al.*, 1997) and confirmed that spreading manures from housed livestock onto land, particularly grassland, was the single largest source (Fig. 27.1). Inventories are often constructed using emission factors, based on experimental data, for each of the main sources of NH_3 emission. Mean emission factors often conceal the wide range of values obtained in experiments, the rate and extent of NH_3 loss following manure application to

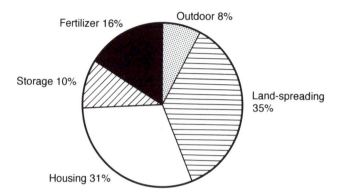

Fig. 27.1. Ammonia emissions from components of farm management, UK.

grassland being influenced by many factors. The emission factor for losses following application of cattle slurry to grassland, expressed as a percentage of the total ammoniacal nitrogen (TAN) applied, ranges from 12.7 to 111.4% (Fig. 27.2), showing a skewed distribution with the majority of values within the 30–60% categories.

Factors Influencing Ammonia Loss

Slurry dry matter content

The inventory constructed by Pain *et al.* (1997) uses emission factors for losses from slurry applications to land based on a relationship between slurry dry matter (DM) content and loss expressed as %TAN applied. This relationship was based on that derived from UK data by Smith and Chambers (1995) with some recent IGER data added (Fig. 27.3). To calculate total emission, slurries were divided into three categories, based on DM content, and the estimated volume of slurry in each category multiplied by the appropriate emission factor as derived from the fitted relationship. Total emissions from cattle slurry applied to grassland are given in Table 27.1, together with 95% confidence limits. If a mean of all values in Fig. 27.1 is used as the emission factor, the calculated loss lies within the 95% confidence limits of the value calculated from the three DM categories. Including all the data used in Fig. 27.2, a significant relationship still exists between DM content and NH_3 loss ($P < 0.01$), but with much less of the variation accounted for:

loss (% TAN applied) = 2.07 × slurry DM content + 30.3 $r^2 = 0.089$

Using the same method of calculation, this give a higher total loss and again, using the mean value, gives a total loss which lies within the 95%

confidence limits (Table 27.1). Also shown in Table 27.1 is the effect on total loss of using the extreme values from the experimental data, illustrating the variation which can exist depending on the data used to estimate the emission factor. There is clearly large variation in emission factor within each slurry DM category, and it may be more appropriate to use one emission factor based on the mean value of all data rather than three values based on a relationship which accounts for so little of the variation and also involves estimating volumes of slurry within each DM category.

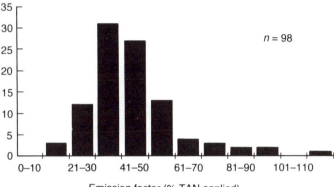

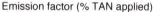

Fig. 27.2. Distribution of emission factor following land spreading of cattle slurry. (Source: Stevens and Logan, 1987; Thompson *et al.*, 1987, 1990a,b; Lockyer *et al.*, 1989; Pain *et al.*, 1989, 1990, 1994; Sommer and Olesen, 1991; Sommer *et al.*, 1991; Klarenbeek *et al.*, 1993; IGER/ADAS, unpublished data.)

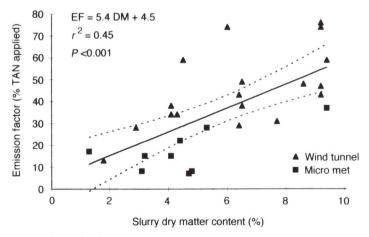

Fig. 27.3. Relationship between emission factor following slurry application and slurry dry matter content. Fitted line with 95% confidence intervals.

Table 27.1. Effect of changing the emission factor for ammonia losses following slurry application to grassland on total UK emission.

Data source	Slurry DM content			Total annual loss (kt)	% UK total emission from agriculture
	< 4%	4–8%	> 8%		
IGER inventory (Pain *et al.*, 1996)					
Fitted line	15.2	37.0	58.9	28.9	14.6
Lower 95% CL	4.3	31.2	47.0	22.7	11.5
Upper 95% CL	26.1	42.9	70.8	35.3	17.9
Overall mean		35.8		24.9	12.6
All available data (see Fig. 27.2 for source)					
Fitted line	34.4	42.7	51.0	30.9	15.7
Lower 95% CL	27.4	39.4	45.4	27.6	14.0
Upper 95% CL	41.5	46.1	56.6	34.3	17.4
Overall mean		44.0		30.6	15.5
Extremes					
Minimum	21.5	12.7	33.3	15.4	7.8
Maximum	95.9	79.6	111.4	65.9	33.5

Application rate

Experimental data from the previously quoted sources (Fig. 27.2) included slurry application rates which ranged from 8 to 113 m^3 ha^{-1}. No relationship was found between NH_3 emission and slurry application rate. However, in a series of experiments by Thompson *et al.* (1990b), where other factors were kept constant, increasing application rate was associated with a decreasing emission factor between 20 and 80 m^3 ha^{-1}, with no further decrease at application rates above 80 m^3 ha^{-1}.

Temperature

Sommer *et al.* (1991) reported an increase in the rate of NH_3 loss with increasing temperature, but that accumulated loss could still be as great at lower temperatures due to a longer emission period.

Rainfall

There is some experimental evidence (unpublished data) that rainfall soon after slurry application decreases NH_3 emission from surface-spread slurry (Table 27.2).

Table 27.2. Effect of rainfall on ammonia loss following surface application of cattle slurry to grassland.

Application rate ($m^3 ha^{-1}$)	Rainfall	Emission (% TAN applied)
26	Dry	48.4
26	18 mm	32.1
52	Dry	33.4
52	18 mm	21.9

Season

Summer applications tend to be associated with higher losses than spring, autumn or winter, probably due to a combination of temperature, rainfall and soil condition effects. Using the data presented in Fig. 27.2, mean losses from each season were 55, 39, 40 and 42% applied TAN for summer, spring, autumn and winter, respectively.

Wind speed

There is a trend for NH_3 emission to increase with increasing wind speed. Thompson *et al.* (1990b) showed emission to increase by a factor of 0.3 when increasing wind speed from 0.5 to 3 m s^{-1} and Sommer *et al.* (1991) reported an increase in emission with wind speed up to 2.5 m s^{-1}, but no consistent effect at higher wind speeds.

Method of application

Alternative slurry application methods, such as band spreading, trailing shoe and shallow injection have been developed, with the aim of reducing NH_3 emission from slurry applications to land. Deep injection of slurry (150–300 mm) can achieve large reductions in NH_3 losses (Hoff *et al.*, 1981; Thompson *et al.*, 1987), but is not popular on grassland as it can result in uneven field surface and yield reductions due to root damage (Rees *et al.*, 1993). Shallow injectors (50 mm depth) have been developed which cause minimal surface disruption and sward damage (Jarvis and Pain, 1990; Rees *et al.*, 1993), while still reducing NH_3 emission to a certain extent (Phillips *et al.*, 1991). Band spreading is less effective at reducing NH_3 emissions; Thompson *et al.* (1990b) reported losses following band spreading to be 85% of those following surface application. Experiences in other parts of northern Europe, particularly in the Netherlands, indicate that shallow injection can reduce NH_3

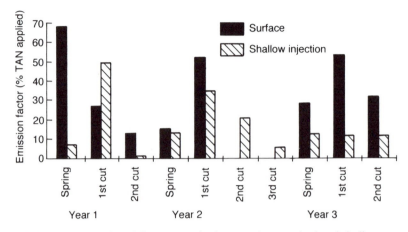

Fig. 27.4. Ammonia loss following cattle slurry surface applied and shallow injected to grassland.

loss by 70–80% compared with surface application of slurry. Recent experiments in the UK have given much lower efficiencies.

Experiences with a shallow injector at IGER over a 3-year period, with NH_3 emission being measured using a micrometeorological mass balance method, gave losses of 0.7–49.3% applied TAN, compared with 12.7–68.1% applied TAN from slurry applied to the surface over the same period, although not always on the same dates (Fig. 27.4). Respective mean values were 16.4 and 35.7% applied TAN, with shallow injection reducing NH_3 emission by 54%. There was some evidence that rainfall following shallow injection led to a reduced efficiency. Application rates ranged from 20 to 61 $m^3 ha^{-1}$ and there was some evidence that higher application rates with shallow injection led to higher losses, presumably due to more spillage of slurry out of the injection slots.

Small plot trials using a small plot slurry applicator have also shown the variation in efficiency of different application methods (Fig. 27.5). Slurry was applied at 30 $m^3 ha^{-1}$ for all experiments and NH_3 emission measured using a system of small wind-tunnels. Experiments 1 and 2 were conducted in the summer – experiment 1 with a low soil moisture content and experiment 2 with a higher soil moisture content. Experiment 3 was conducted in November and experiment 4 in March. In the summer applications, shallow injection was no more efficient than either band spreading or trailing shoe at reducing emissions, due to soil conditions. In the autumn and spring applications shallow injection was more efficient and in experiment 4, where grass sward height was greater than in other experiments, the trailing shoe was as effective as shallow injection.

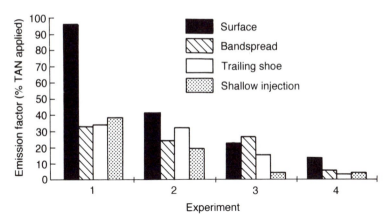

Fig. 27.5. Effect of application method on emission factor following slurry application to grassland.

Other factors

Other factors which will affect the rate and extent of NH_3 emission following manure application to land include manure pH, soil conditions, crop cover, manure TAN content and also the method of measurement. Emission values tended to be greater when measured using wind-tunnels than micrometeorological mass balance methods (Fig. 27.3).

Conclusions

The best relationship that could be fitted between any single factor and NH_3 emission following slurry application to grassland was for slurry DM content, but even this accounted for very little of the variation. It is clear that interaction between factors can result in large variation in measurements of NH_3 emission. Estimates of total emission in inventories may therefore vary greatly depending on the data used in estimating the emission factors.

Acknowledgements

This work was funded by the Ministry of Agriculture Fisheries and Food, London.

References

Hoff, J.D., Nelson, D.W. and Sutton, A.L. (1981) Ammonia volatilization from liquid swine manure applied to cropland. *Journal of Environmental Quality* 10, 90–95.

Jarvis, S.C. and Pain, B.F. (1990) *Ammonia Volatilization from Agricultural Land. The Fertiliser Society, Proceedings no. 298, 13 December 1990.*

Klarenbeek, J.V., Pain, B.F., Phillips, V.R. and Lockyer, D.R. (1993) A comparison of methods for use in the measurement of ammonia emissions following the application of livestock wastes to land. *International Journal of Environmental Analytical Chemistry* 53, 205–218.

Lockyer, D.R., Pain, B.F. and Klarenbeek, J.V. (1989) Ammonia emissions from cattle, pig and poultry wastes applied to pasture. *Environmental Pollution* 56, 19–30.

Pain, B.F., Misselbrook, T.H. and Rees, Y.J. (1994) Effects of nitrification inhibitor and acid addition to cattle slurry on nitrogen losses and herbage yields. *Grass and Forage Science* 49, 209–215.

Pain, B.F., Phillips, V.R., Clarkson, C.R. and Klarenbeek, J.V. (1989) Loss of nitrogen through ammonia volatilisation following the application of pig or cattle slurry to grassland. *Journal of the Science of Food and Agriculture* 47, 1–12.

Pain, B.F., Thompson, R.B., Rees, Y.J. and Skinner, J.H. (1990) Reducing gaseous losses of nitrogen from cattle slurry applied to grassland by the use of additives. *Journal of the Science of Food and Agriculture* 50, 141–153.

Pain, B.F., van der Weerden, T.J., Chambers, B.J., Phillips, V.R. and Jarvis, S.J. (1997) A new inventory for ammonia emissions from UK agriculture. *Atmospheric Environment* (in press).

Phillips, V.R., Pain, B.F. and Klarenbeek, J.V. (1991) Factors influencing the odour and ammonia emissions during and after the land spreading of animal slurries. In: Neilsen, V.C., Voorburg, J.H. and L'Hermite, P. (eds) *Odour and Ammonia Emissions from Livestock Farming.* Elsevier Applied Science, London, pp. 98–106.

Rees, Y.J., Pain, B.F., Phillips, V.R. and Misselbrook, T.H. (1993) The influence of surface and sub-surface application methods for pig slurry on herbage yields and nitrogen recovery. *Grass and Forage Science* 48, 38–44.

Smith, K.A. and Chambers, B.J. (1995) Muck from waste to resource-utilization: the impacts and implications. *Agricultural Engineering* Autumn, 33–38.

Sommer, S.G. and Olesen, J.E. (1991) Effects of dry matter content and temperature on ammonia loss from surface-applied cattle slurry. *Journal of Environmental Quality* 20, 679–683.

Sommer, S.G., Olesen, J.E. and Christensen, B.T. (1991) Effects of temperature, wind speed and air humidity on ammonia volatilization from surface applied cattle slurry. *Journal of Agricultural Science, Cambridge* 117, 91–100.

Stevens, R.J. and Logan, H.J. (1987) Determination of the volatilization of ammonia from surface-applied cattle slurry by the micrometeorological mass balance method. *Journal of Agricultural Science, Cambridge* 109, 205–207.

Sutton, M.A., Place, C.J., Eager, M., Fowler, D. and Smith, R.I. (1995) Assessment of the magnitude of ammonia emissions in the United Kingdom. *Atmospheric Environment* 29, 1393–1411.

Thompson, R.B., Ryden, J.C. and Lockyer, D.R. (1987) Fate of nitrogen in cattle slurry following surface application or injection to grassland. *Journal of Soil Science* 38, 689–700.

Thompson, R.B., Pain, B.F. and Lockyer, D.R. (1990a) Ammonia volatilization from cattle slurry following surface application to grassland. I. Influence of mechanical separation, changes in chemical composition during volatilization and the presence of the grass sward. *Plant and Soil* 125, 109–117.

Thompson, R.B., Pain, B.F. and Rees, Y.J. (1990b) Ammonia volatilization from cattle slurry following surface application to grassland. II. Influence of application rate, windspeed and applying slurry in narrow bands. *Plant and Soil* 125, 119–128.

Posters 14–15

IV

Contribution of ^{15}N Labelling Techniques to the Study of Nitrogen Gaseous Emissions from Animal Manures

J. Martinez[1], J.F. Moal[1], C. Marol[2] and G. Guiraud[2]

[1]Cemagref, 17 avenue de cucillé, 35044 Rennes Cedex, France; [2]CEA, Centre de Cadarache DEVM, 13108 Saint Paul lez Durance Decex, France

The stable isotope ^{15}N has been widely used in investigations of nitrogen (N) transformations in soil–plant systems, particularly from N fertilizer applications (Martinez and Guiraud, 1990). There is less information on its use for studying losses from animal manures. Direct labelling of animal manure with ^{15}N is possible by incorporating ^{15}N into animal feed, but this is very expensive, labour intensive and time consuming. It is however particularly relevant when the organic N fraction is investigated (Sorensen *et al.*, 1994). More frequently, authors have labelled the ammonium-N (NH_4^+-N) pool in slurry by adding a small amount of highly enriched NH_4^+-N. Moal *et al.* (1994) conducted a preliminary study that demonstrated that the added N behaves as the endogenous NH_4^+ initially present in the slurry. This abstract describes briefly field and laboratory ^{15}N balance studies conducted to determine the fate of labelled ammonium sulphate in a pig slurry.

A first study was conducted using cylindrical glass volatilization chambers. Each chamber contained 250 g of fresh soil and 40 g of ^{15}N-labelled pig slurry (equivalent to 50 m^3 ha^{-1} on a surface basis calculation). The fate of total ammoniacal nitrogen (TAN) was followed for 48 h from surface application to soil. During the experiment, air was drawn from the chamber to a trap containing 0.1 N H_2SO_4 to measure NH_3. At each sample time (0.25, 2, 6, 12, 24 and 48 h), all the soil/mixture in each chamber was extracted for 1 h with 1 M KCl solution and filtered through Whatman No. 41 filter paper. Finally, NH_3 and NH_3-^{15}N volatilized were determined in the acid-trapping solution. Laboratory incubations (0, 3, 7, 14, 21 days) were designed to determine the nitrifying activity of a soil receiving surplus slurries. Different N rates (100, 600, 1200 mg NH_4^+-N kg^{-1} soil) were applied as ^{15}N-labelled ammonium sulphate (AS) or pig slurry (PS) N.

A field experiment was also conducted using a wind-tunnel system based on the design described by Lockyer (1984). This method measures NH_3 emission from 1 m^2 experimental plots. Rye-grass on the plots was cut just before slurry application. Labelled slurry was applied at a rate of 100 m^3 ha^{-1}, equivalent to 290 kg TAN ha^{-1}. Ammonia volatilization was measured for 48 h and then the tunnels were removed from plots. Five weeks later, soil and plants were sampled and analysed. Isotopic analysis was performed using a mass spectrometer (Micromass VG 622).

During the 48 h preliminary study, biological N transformations were not apparent. Figure P14.1 shows that 40% of the applied TAN was volatilized after 48 h. All the labelled ammoniacal-N applied in slurry was distributed between volatilized NH_3 and ammoniacal N remaining in the soil. An average TAN-^{15}N recovery of 97.5% (c.v. 2%) was found.

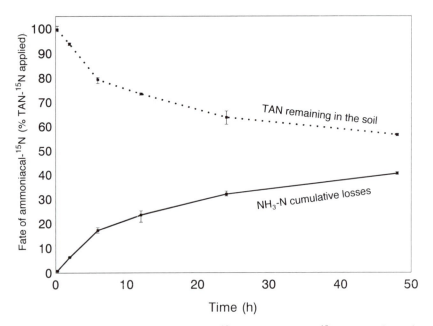

Fig. P14.1. Cumulative NH_3 losses (NH_3-^{15}N) and remaining ^{15}N TAN in the soil as a percentage of TAN-^{15}N applied in the slurry.

During the 21 days' laboratory incubation, 5% and 10.5% of N applied in AS or slurry application was measured, respectively, in the soil. Moreover, at the 600-mg rate, 72% and 96% of the ^{15}N was nitrified following the application of pig slurry or AS respectively. At the higher application rate of 1200 mg, the corresponding figures were 35% and 56%. Further, gaseous losses, up to 50% of ^{15}N applied, were observed following high rates of slurry application.

At the start of the field ^{15}N balance (Fig. P14.2) experiment, all ^{15}N was in the form of ammoniacal nitrogen ($NH_3 + NH_4^+$). Ammonia loss, measured during 48 h after slurry spreading, accounted for 24% of the ammoniacal ^{15}N applied. Five weeks later, 18% of N applied had been taken up by the grass (5 tonnes of dry matter production), 38% was found in the soil as labelled nitrate (14%) and as labelled immobilized N (24%). Less than 1% remained as labelled ammoniacal N. The recovery as a percentage of applied ^{15}N varied from 75 to 85% for each individual mass balance.

Further experiments have demonstrated that a new method developed to measure NH_3 volatilization in field studies using passive samplers (shuttles) compares well with ^{15}N mass balance recovery methods (van der Weerden, 1995, personal communication).

From the first experiment, the analysis of isotopic excess (E%) of volatilized NH_3 after labelled slurry application demonstrated that the added NH_4^+

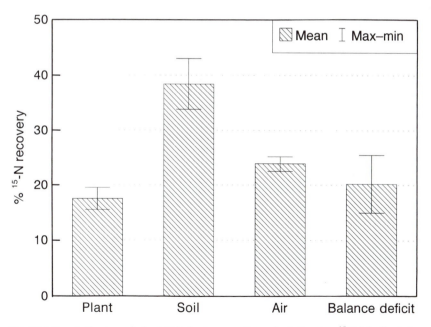

Fig. P14.2. Soil, air and plant N balance over 5 weeks following ^{15}N-labelled pig slurry application on grass.

behaved as the endogenous NH_4^+ initially present in the slurry. During this study, no immobilization or nitrification of slurry-derived N was noticed over 48 h. The laboratory incubation shows gaseous losses, up to 50% of ^{15}N applied, with high rates of slurry application. Nitrite accumulation and the presence of free amines in the pig slurry may suggest that a 'van Slyke' reaction or chemical denitrification could explain such losses. This unusual process may occur in soils in combination with biological denitrification.

In conclusion, the reliability of the direct incorporation of the tracer in pig slurry to study the fate of ammoniacal-N was satisfactorily demonstrated. Within a short period following pig slurry application (48 h), soil N transformations were not apparent: NH_3 emission was the main fate of ^{15}N applied in slurry. Over longer periods, several processes took place (nitrification, immobilization) which illustrate a much more complex situation. In this case, the use of stable isotope ^{15}N is particularly relevant in explaining the pathways of the slurry N cycle and for making quantitative measurements of N transformation processes.

References

Lockyer, D.R. (1984) A system for the measurement in the field of losses of ammonia through volatilisation. *Journal of the Science of Food and Agriculture* 35, 837–848.

Martinez, J. and Guiraud, G. (1990) A lysimeter study of the effects of a ryegrass catch crop, during a winter wheat/maize rotation, on nitrate leaching and on the following crop. *Journal of Soil Science* 41, 5–16.

Moal, J.F., Martinez, J., Marol, C. and Guiraud, G. (1994) A direct incorporation of N-15 labelled ammonium sulphate into pig slurry: a laboratory experiment on NH_3 volatilization. *Bioresource Technology* 48, 87–89.

Sorensen, P., Jensen, E.S. and Nielsen, N.E. (1994) Labelling of animal manure nitrogen with ^{15}N. *Plant and Soil* 162, 31–37.

Nitrous Oxide Emissions Following Fall Application of Whole and Separated Liquid Dairy Cattle Manure on Grassland and Arable Land in Southern British Columbia, Canada

J.W. Paul and B.J. Zebarth
Agriculture and Agri-Food Canada, Pacific Agriculture Research Centre, Box 1000, Agassiz, BC, Canada V0M 1A0

Dairy cattle manure is an important nutrient source for both maize and grass production in south coastal British Columbia. Most of the dairy cattle manure is handled as a slurry, but there is some interest in liquid/solid separation of the slurry in order to apply the separated liquid to grassland. Application of the liquids on grassland improves nitrogen (N) utilization, and reduces odour and ammonia (NH_3) emission (Bittman, 1996 personal communication). A significant portion of the dairy cattle manure is applied during September and

October to fields cropped to grass or where maize was harvested. Manure application to soil results in significant nitrous oxide (N_2O) emissions from soil from both the nitrification and denitrification processes (Paul *et al.*, 1993). Soil conditions in south coastal British Columbia in October and November are conducive to denitrification because of high rainfall and relatively warm soils.

The objectives of this research were to quantify N_2O emissions from manure-amended soil during the fall, and to determine how N_2O emissions are affected by manure application to grassland compared to arable land cropped to maize, and by manure type, specifically whole dairy slurry and separated dairy liquids (liquids from liquid–solid separation of whole slurry).

Experimental plots (3×3 m) were established on two adjacent sites at the end of September 1994 on a silt loam soil at Agassiz, BC. One site was cropped to perennial forage grass (grassland) and the second site had been cropped to forage maize (arable soil). The three treatments included four replicates of whole dairy slurry or separated dairy liquids (75 kg NH_4^+-N ha^{-1}) as well as a control. Whole dairy cattle slurry was obtained from a local dairy farm after thorough mixing of the manure storage tank. Separated liquids were obtained following separation of some of the whole slurry using a screw press separator. Manure was surface applied on 27 September. Steel collars ($0.7 \times 0.7 \times 0.15$ m deep) were inserted into each of the plots the next day following manure application. Nitrous oxide emission was measured by covering the soil with vented covers for 30 min twice per week for 4 weeks, then once weekly for an additional 5 weeks. Gas samples were taken using a two-way needle into a pre-evacuated vacutainer at 0, 15 and 30 min. Nitrous oxide was analysed on a Varian GC equipped with a ^{63}Ni electron capture detector. The N_2O emission rate for each plot at each sampling was obtained from the slope calculated by the increase in N_2O concentration. Total N_2O emission for the 9-week sampling period was calcuated from the mean N_2O rates using Simpson's rule.

Soil samples (ten replicate cores from each plot) were taken from 0–15, 15–30 and 30–60 cm depths before manure application, and 4 and 9 weeks following application. Extractable ammonium (NH_4^+) and nitrate (NO_3^-) were analysed by flow injection analyser following extraction with 2 M KCl. The experimental design was a split-plot with cropping type as the main plots and manure treatment as the subplots. The N_2O emission rates and the soil inorganic N contents were analysed using the General Linear Models program of SAS (SAS Institute Inc., 1985). For the N_2O emission rates, sampling date was the repeated measure.

Manure application caused a significant increase in N_2O emission (Fig. P15.1). The emissions occurred during the first 4 weeks following manure application. Emissions were higher from the arable soil than from the grassland soil. In the grassland soil, N_2O emissions were higher from separated liquids than from whole slurry. Total N loss as N_2O was also higher on the

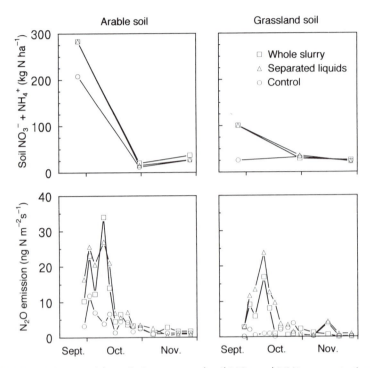

Fig. P15.1. Nitrous oxide emission rates and soil NO_3^- and NH_4^+ concentrations following whole slurry and separated liquid manure addition to grassland and arable soil.

arable soil than on the grassland soil. (Table P15.1). Nitrous oxide emission from the control plot on the arable soil was 2.5 times higher than from the control plot of the grassland soil. Soil NO_3^- and NH_4^+ concentrations to 30 cm were very high in the maize soil at the beginning of the experiment (Fig. P15.1). After 4 weeks, and at 9 weeks, there were no significant differences in soil NO_3^- and NH_4^+ concentrations between manure treatments. Rainfall following manure application exceeded 5 cm, which probably leached the NO_3^- to below 60 cm depth after nitrification of the NH_4^+ from the manure.

These results show that N_2O emissions were higher from the arable soil than from the grassland soil, probably because of higher initial NO_3^- concentration in the arable soil. The higher N_2O emissions with separated dairy liquids than with whole slurry may be due to better infiltration of the liquids into the soil, resulting in less loss of N via NH_3 volatilization. Total N_2O emissions during the 9 weeks were lower than expected. A significant amount of N_2O may have been washed down through the soil profile because of the very high rainfall that occurred following manure application,

Table P15.1. Total loss of N as N_2O from grassland and arable soils following manure application in the fall.

Treatment	Grassland (g N ha^{-1})	Arable soil (g N ha^{-1})
Whole slurry	210	418
Separated liquids	321	477
Control	73	246

as indicated by the rapid loss of inorganic N from the surface 60 cm. Nitrous oxide emission from soil following manure application during the fall represents a significant source of greenhouse gas. Supplying N to match the crop requirement is the best strategy to reduce N_2O emissions.

Acknowledgements

This project was funded by Agriculture and Agri-Food Canada and Environment Canada under the Greenhouse Gas Initiative. Tissa Kannangara, Shaobing Yu and Corey Giesbrecht provided the capable technical expertise.

References

Paul, J.W., Beauchamp, E.G. and Zhang, X. (1993) Nitrous and nitric oxide emissions during nitrification and denitrification from manure-amended soil in the laboratory. *Canadian Journal of Soil Science* 73, 539–553.

SAS Institute Inc. (1985) *SAS User's Guide: Statistics,* Version 5 edn Cary, North Carolina.

Scaling of Nitrogen Gas Fluxes from Grasslands

28

A.F. Bouwman[1] and W.A.H. Asman[2]

[1]National Institute of Public Health and the Environment, PO Box 1, 3720 BA Bilthoven, Netherlands; [2]National Environmental Research Institute, Frederiksborgvej 399, 4000 Roskilde, Denmark

Summary. Techniques used for extrapolating measurements or properties and constraining results between different temporal and spatial scales are nowadays referred to as 'scaling'. Scaling should always explicitly account for the heterogeneity of the system studied. This chapter discusses general aspects and problems of scaling the gaseous fluxes of ammonia (NH_3), nitrous oxide (N_2O) and NO_x that are associated with livestock production and grasslands. Examples are cited from different scales, ranging from individual processes on soil plots to fields, farms, landscapes and larger scales. The following aspects of scaling are discussed:

1. *Data availability.* Any scaling exercise relies on the availability of data. Quantitative data with adequate spatial and temporal resolution on the use and management of fertilizers, and animal excreta and environmental data (including soil, climate and weather data) are very sparse. In many cases measurements do not cover the full range of the environmental and management conditions met in the field.

2. *Delineation of functional types.* Disaggregation of a population, or delineation of functionally different types, should be done where distinct and easily identified differences in structure and composition of a landscape coincide with the functions or management conditions relevant for N fluxes.

3. *Measurement strategy.* Delineation is a useful basis for measurement strategies and scaling. Measurement strategies should aim to determine relations between fluxes and key processes and variables which can be described and used in models at the next higher scale level.

4. *Model development.* Correlations between fluxes and key variables play an invaluable role in suggesting candidate mechanisms for investigation and model development. For grassland, microscale process, field scale, landscape and whole-farm approaches exist. Dynamic whole-farm approaches are necessary tools for extrapolation and the assessment of strategies to reduce N gas fluxes and other adverse environmental effects.

5. *Boundary conditions.* An increasingly important tool is the top-down scaling approach, whereby boundary conditions derived from the next higher scale level are used to constrain or test the possible variation in the results of scaling from smaller scales.

It is hard to separate the concepts of scale at the local level from those at any other level. While the importance of such concepts may be manifested differently at each scale level, the basic concepts apply across all scales.

Introduction

A key problem in determining nitrogen (N) gas fluxes from terrestrial eco-systems is the high degree of spatial and temporal heterogeneity. Gas fluxes vary at both fine and coarse scales of resolution. The techniques used for extrapolating measurements or properties and constraining results between different temporal and spatial scales are referred to nowadays as 'scaling'. Scaling should always explicitly account for the heterogeneity of the system studied. Bottom-up and top-down approaches both play important roles in scaling (Fig. 28.1). Bottom-up approaches, i.e. from lower to higher scales, involve extending calculations from an easily measured and reasonably well understood unit to processes on a more encompassing scale. Top-down approaches can mean using the coarser scale measurements that set the boundary conditions for problem identification, and they stimulate the testing of general relationships for specific cases. For example, fluxes determined by bottom-up scaling can be used to drive atmospheric transport models. Comparison of the concentrations or deposition velocities simulated by transport models with observations can result in an expression of scientific confidence or a warning that crucial information is still missing. Between these two extremes, scaling can be very useful for testing hypotheses and identifying missing information, in particular when atmospheric pollutants such as ammonia (NH_3), nitrous oxide (N_2O) and nitric oxide (NO) and nitrogen dioxide (NO_2) (together denoted as NO_x) are concerned.

The atmospheric residence time of NH_3 is of the order of a few hours, and the lifetime of its reaction product NH_4^+ aerosol is of the order of a few days. A minor fraction of NH_3 is oxidized in the atmosphere (Dentener and Crutzen, 1994), but this process may be a significant source of N_2O. About half of the emitted NH_3 is re-deposited as NH_3, mainly as dry deposition. The other half reacts with aerosols and gases to NH_4^+ aerosol, which is mainly removed from the atmosphere as wet deposition. Ammonia may be important to tropospheric chemistry even at emission rates not significant to ecosystem N balances (Schimel *et al.*, 1986). Finally, oxidation of deposited NH_3 contributes to soil acidification (Van Breemen *et al.*, 1982).

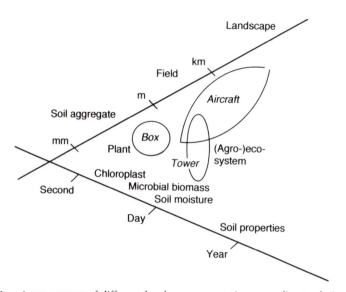

Fig. 28.1. Arrangement of different landscape properties according to their characteristic spatial and temporal scale. Bottom-up scaling approaches proceed to the right and top-down scaling approaches proceed from the right to the left. The spatial scale points are to the southeast of the figure and the temporal scale to the northeast. From the bottom to the top the diffusion coefficient increases. Measurement techniques used at different spatial and temporal scales are printed in italics. (Modified from Goudriaan, 1993.)

The residence time of N_2O in the whole atmosphere (including the stratosphere) is 120 years (Bouwman *et al.*, 1995), whereas the reactive NO_x is broken down within 1 day in the troposphere (Ramanathan *et al.*, 1985). In the troposphere N_2O is inert, whereas NO_x is a precursor of ozone (O_3) and involved in conversions of methane (CH_4), carbon monoxide (CO) and other compounds. In the stratosphere the oxidation of N_2O produces NO_x, which is involved in the catalytic destruction of stratospheric O_3 (Crutzen, 1983).

Emissions from livestock production systems contribute about 40% to the global emission of NH_3, 20% to the global budget of N_2O (Table 28.1) and 50% to the atmospheric increase of N_2O. The latter is a better way to express the contribution of animal production systems to atmospheric N_2O, which is a long-lived greenhouse gas. No reliable estimates for NO_x emission from livestock production systems exist. The above estimates do not include effects caused by the production of animal feed. Losses of NH_3, N_2O and NO_x from fertilizers used for the production of animal feed may be substantial (Bouwman, 1997).

The species NH_3, N_2O and NO_x have different mechanisms of formation and different properties. In livestock production systems most NH_3 is formed

Table 28.1. Emissions in 10^{12} g year^{-1} of NH_3, N_2O and NO_x associated with livestock production*.

	NH_3	N_2O	NO_x
Grasslands, decomposition	3.4[†]	1.4	2.6
Synthetic fertilizer[‡]	0.6	0.1	n.d.
Animal excreta[§]	18.4	0.8	n.d.
Savanna burning	1.8	0.1	3.2
Sum	24	2	6
% of global source	40	20	10

Based on Bouwman *et al.* (1997), Bouwman *et al.* (1995) and Lee *et al.* (see Chapter 31).
n.d. = no data.
*In this chapter the term grasslands represents landscapes in which animal production forms the major activity, including both confined animals and extensive and intensively used grasslands.
[†]For intensively used grasslands the NH_3 emission from decomposition is included in the estimate for animal excreta.
[‡]Based on synthetic fertilizer use on global grasslands of about 10% of total fertilizer use of 80 million t N year^{-1} (FAO/IFA/IFDC, 1994). This excludes emissions associated with synthetic fertilizer use for feed production. About 40% of global cereal production, 25% of global production of starchy foods, an important part of production of oilseeds, and unknown amounts of fodder crops and roughage, is fed to animals (Bouwman, 1997).
[§]Including emissions from stables, grazing and emissions resulting from decomposition of grass, and excluding emissions from application of manure on croplands of about 20 million t N year^{-1} worldwide.

from urea in animal urine, NH_3-based fertilizers, and from decomposition of soil organic matter and faeces whereby reduced forms of N are released. The exchange of NH_3 is a physicochemical process, whereby volatilization is determined by chemical factors such as the pH of soil and manure and the cation exchange capacity, whereas the exchange between the surface and the atmosphere is influenced by temporal and spatial variation of physical parameters such as soil porosity, temperature, wind speed and precipitation. The variability of NH_3 fluxes is as great as the heterogeneity of the terrain and the spatial and temporal variation of weather conditions.

The nitrogen oxides N_2O and NO are formed in soils during biological nitrification of NH_4^+ and denitrification of nitrate (NO_3^-) and nitrite (NO_2^-). Very high fluxes can be expected where N concentrations are high (Bouwman, 1996). For example, in intensively fertilized grazing areas high NH_3 concentrations may occur, which through nitrification may lead to accumulation of NO_2^-. Nitrite may reach levels toxic to microorganisms and nitrifiers may react by denitrifying this (Poth and Focht, 1985). Apart from biological formation, abiotic formation of nitrogen oxides may occur in soils. The formation of N_2O and NO is extremely variable in time and space, and there is additional variability caused by the factors regulating the exchange between the soil and the atmosphere. Other sources of NH_3, N_2O and NO not related to livestock production systems include biomass burning, plants, oceans, human excretion, exhalation and households, combustion and traffic. For global budgets of N gas fluxes from grasslands and livestock production, refer to Chapter 31.

The objective of a study is the first determinant of the level of organization to be studied. For example, the objective of scaling gas fluxes from a field requires information on the heterogeneity within plots, due to processes operating at 'hot spots' within soil aggregates (see Chapter 29). In this chapter it is assumed that this heterogeneity is well understood, and that it is properly recognized and described at the plot or field level. General aspects and problems of scaling gaseous fluxes of NH_3, N_2O and NO_x will be discussed, on the basis of a number of examples covering different scales, ranging from individuals or soil plots to fields, farms, landscapes and even larger scales (see also Chapters 29–31). One of the problems in scaling exercises is the sparsity of statistical data and data on management systems. Aspects of scaling at all levels will be discussed, starting with data availability, an essential aspect in any scaling exercise. Important steps involved in the scaling of nitrogenous gas fluxes include the delineation of functional types (sometimes referred to as stratification, e.g. Matson *et al.*, 1989), measurement strategy, correlation and model development, and the setting of boundary conditions.

Data Availability

The starting point of any scaling exercise must be embedded in the data, including those on environmental conditions and those on management, including the temporal and spatial distributions. For grasslands, N gas losses can be obtained only if the complete cycle of N is known (see Chapter 33), from feed to excretion and emission in the stable and during grazing. For example, to estimate NH_3 emissions from animal manure during storage and during and after application, the following data are required: animal category (age, weight); N content and relative content of the various amino acids; N use efficiency (conversion to milk or meat); housing system and period of confinement; form, and mode and period of storage; weather conditions during spreading (turbulence, air temperature, air humidity and rainfall); properties of the soil on which the manure is applied; amount of manure per unit area; mode of application; and period between application and cultivation and use of synthetic fertilization of grasslands.

To estimate the emissions during periods of grazing it is important to know the N consumption, the fraction of the excretion in grazing periods, and weather conditions. This information is required for each individual farm within an area or landscape. Usually the information is available as a country or regional average, or at best as an average with a range of uncertainty. Some of the data are simply not available, even for leading countries in NH_3 research with good national statistics.

Outside Europe and North America all these data pose difficulties. For example, data on animal populations by category, age and weight class are scarce. For many countries only the total number of animals within a category is available. Data on some animal categories, such as pets, horses, buffaloes, donkeys, camels, man, and on housing and the type and form of manure, are not available. Estimates for regions within countries may be available, but sometimes they do not correspond to the official statistics, or they are out-dated. Data on the coverage of stored manure, which may vary strongly in effectiveness, are lacking. Geographic data on the application rate of manure, soil conditions, the variability in timing of application and weather conditions during application are not available (ECETOC, 1994).

A very detailed inventory from Denmark shows the heterogeneity of the NH_3 emission for a landscape dominated by livestock production (Fig. 28.2). In addition to spatial variability, manure application rates and mode and timing of application show a strong interannual variability, which is not easy to include in scaling exercises. Storage and spreading of manure are regulated by law to reduce emissions in a number of countries. It is difficult to obtain information on the actual observance of these laws and the emission reductions achieved.

The paucity of data makes it necessary to generalize. This is done usually by treating a landscape as a composite of average farms with average waste management and average weather conditions, or by treating populations as a group of identical average members. Such generalizations may lead to errors due to averaging procedures. For example, the result from a population of average farms may differ geographically and temporally from the sum of all individual farms. Much of the detail can be lost if different scales are used to distribute the data. In general, it is better to aggregate the model results than to aggregate the data before modelling. Nevertheless, the purpose of the scaling often dictates which method to use. For example, a simple data aggregation may be enough to show in which provinces or regions certain environmental problems are most important.

Additional problems occur when the experimental data do not fully cover the range of field conditions. One example concerns the emission coefficients used for animal houses in Europe (ECETOC, 1994). In houses with mechanical ventilation the emission can easily be calculated from the NH_3 in the ventilation air and the flow rate. The emission from naturally ventilated houses can only be determined indirectly and with a larger uncertainty. In such 'open' houses the emission depends on the opening and closing of doors. For these reasons, many measurements have been made of emission from mechanically ventilated houses. Measurements of emissions from naturally ventilated houses, on the other hand, are very scarce. In large

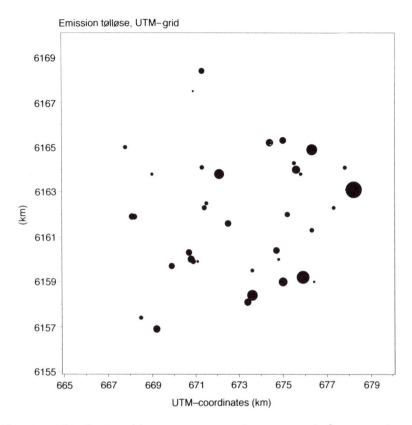

Fig. 28.2. Distribution of the NH_3 emission within a 13×13 km^2 area at Tølløse, Zealand, Denmark. The area of the circles is a relative measure for the emission from a farm, including the farmland. This figure shows that emissions are not distributed homogeneously.

parts of Europe, housing for cattle – the most important category for N excretion and associated emissions of NH_3, N_2O and NO_x – is naturally ventilated (Asman, 1992). Besides being scarce, the available data may also be unreliable. The NH_3 emission per animal may vary by a factor of 4 within the same type of housing. This difference is larger than the day-to-day variation of the emission from one single house, as indicated by measurements from Denmark (Pedersen *et al.*, 1996). The variation may be caused by differences in the ventilation over the slurry between different houses and by differences in waste management practices such as cleaning.

Another example of the range of field conditions not being completely covered by measurements concerns N_2O fluxes from fertilized fields. Despite the wealth of literature data, Bouwman (1996) concluded that the range of environmental and management conditions is too limited to estimate emission coefficients by climatic region, soil type or fertilizer type, and he proposed using a wide range of uncertainty around one single emission coefficient for all fertilizers worldwide.

Delineation of Functional Types

Where distinct and easily identified differences in structure and composition of a landscape coincide with the functions or management conditions relevant to N fluxes at the scale considered, the delineation of functionally different types or production/management systems is a useful basis both for measurement strategies and scaling (Matson *et al.*, 1989; Groffman, 1991). Appropriate selection of classes may lead to reducing the number of sites to be sampled in order to derive a reliable flux value. Maps provide a useful basis for delineation, and in recent years remote sensing of ecosystem characteristics has been used increasingly for classification and modelling.

Such approaches use the variability of the systems or landscape instead of averaging out that variability, sometimes with unexpected consequences. It is very important to select appropriate functional types for the scale of the exercise. Scaling based on delineations with the finer spatial data may be completely different from that derived from the data with lower resolution. Aggregating the functional types may not yield reliable results for some gaseous compounds with concentration patterns that are variable in space and time. Results for other compounds with more homogeneous mixing in the atmosphere are influenced much less by the resolution of the basic data (Figs 28.3, 28.4).

The temporal dimension can be influenced by factors such as management and climate. Schimel *et al.* (1986) illustrated the effect of averaging the geographic information on the resulting spatial and temporal flux estimates. In their study, Schimel *et al.* (1986) analysed the cycling and volatile loss of N derived from cattle urine at lowland and upland sites in a shortgrass steppe in Colorado, USA. The NH_3 losses observed in the microplots representing three soil types were scaled to a typical shortgrass steppe landscape. For this, seasonal rates of urine and faeces deposition were mapped by landscape position, allowing for responses of animals to microclimate and forage availability and differential use of upland and lowland pastures. This gave variation in the proportion of total deposition vulnerable to loss. Urine deposition was higher during the growing season when forage-N levels were high, and highest in lowland soils. The results of the spatial and temporal stratification indicated that 0.1 kg NH_3-N ha^{-1} year^{-1} was lost as NH_3, accounting for only 5% of the N excretion. Nevertheless, if the study had not included spatial patterns of deposition and loss, the calculated loss of NH_3 would have been a factor of seven times higher. Studies of gaseous fluxes are vulnerable to this

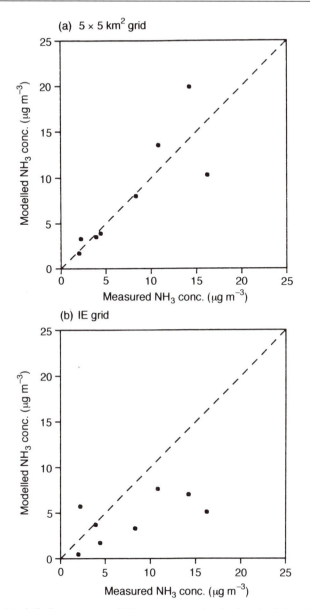

Fig. 28.3. Modelled vs. measured NH_3 concentration in the air of the Netherlands, with emissions on: (a) a 5×5 km^2 grid; (b) an IE grid (75×75 km^2). These figures show that the NH_3 concentrations in the air can be modelled correctly only from an emission inventory with a high spatial resolution, because NH_3 concentrations are largely influenced by local sources of NH_3. (Source: Asman, W.A.H. and Van Jaarsveld, J.A. (1992) A variable-resolution transport model applied for NH_x in Europe. *Atmospheric Environment* 26A, 445–464. Reprinted with kind permission from Elsevier Science Ltd, The Boulevard, Langford Lane, Kidlington OX5 1GB, UK.)

type of error because fluxes can be intermittent and patchily distributed in space.

At a larger scale, Matson and Vitousek (1990) attempted to estimate the fluxes of N_2O from tropical forest by delineating ecosystems. Although the processes and regulating factors of N_2O fluxes are not known in detail, Matson and Vitousek (1990) found a reasonable correlation between N_2O

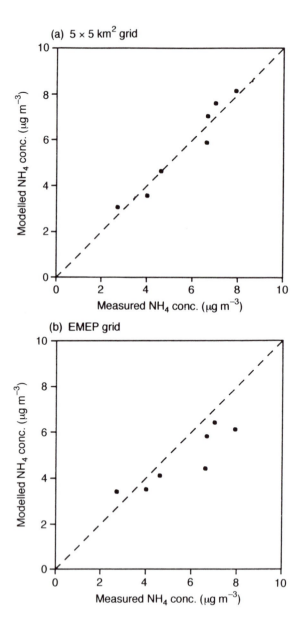

fluxes and soil fertility. For tropical ecosystems this means a total flux of about a factor of two lower than a previous estimate by McElroy and Wofsy (1986), based on a representative flux measurement for all tropical systems.

Measurement Strategy

Measurements should capture the spatial and temporal variability within the site, system or population. Measurements can be used to investigate correlations with certain key factors. In the simplest form of extrapolation, the flux values for that particular site or population are multiplied by the area or number of individuals (e.g. for tropical forests by McElroy and Wofsy, 1986; for animals by Bouwman *et al.*, 1995, 1997). A more advanced approach is to develop measurement strategies for determining a relationship between fluxes and those key variables for which information exists at the next higher scale level.

It is sometimes difficult to develop a measurement strategy because of problems in delineating functional types. For example, the available information on fluxes of NH_3, N_2O and NO_x from savanna and grassland burning (the primary form of global biomass burning) may come from biased observations. Crutzen and Andreae (1990), Andreae (1991), Hegg *et al.* (1990) and Andreae *et al.* (1997) proposed an emission factor as a percentage of the N present in the fuel. Later it appeared that N gas emission factors are biased towards smouldering fires in field measurements (due to high temperatures of the flaming phase), while flaming conditions prevail in aircraft measurements. Inefficient combustion occurs during smouldering, whereas the oxidation grade is higher in flaming conditions. The burning conditions may be more important for the N gas emissions than the composition of the fuel (Griffith *et al.*, 1990). The practical problems of measurements, the extreme variability in time of the burning conditions, and the general uncertainty as to fuel composition in large-scale fires, make it very difficult as yet to propose reliable emission factors.

Fig. 28.4. (opposite) Modelled vs. measured NH_4^+ aerosols concentration in the air of the Netherlands, with emissions on: (a) a 5×5 km^2 grid; (b) an EMEP grid (150×150 km^2). These figures show that the NH_4^+ aerosol concentrations can still be modelled reasonably well from an emission inventory with a relatively coarse resolution. NH_4^+ aerosol concentrations are not influenced much by nearby sources of NH_3 because of the time needed to convert NH_3 to aerosol NH_4^+. (Source: Asman, W.A.H. and Van Jaarsveld, J.A. (1992) A variable-resolution transport model applied for NH_x in Europe. *Atmospheric Environment* 26A, 445–464. Reprinted with kind permission from Elsevier Science Ltd, The Boulevard, Langford Lane, Kidlington OX5 1GB, UK.)

With chamber techniques it is difficult to derive reliable flux values for a plot. Folorunso and Rolston (1984) reported that about 350 chamber measurements were required to estimate the N_2O emission within 10% of the true mean in a 3×36 m plot with a fairly homogeneous soil (see also Chapter 16). It is possible to avoid the problem of within-plot variability by using measurement techniques that cover larger areas, such as those used in micrometeorology and remote sensing. Mosier and Hutchinson (1981) used micrometeorological techniques to determine vertical N_2O flux density from a cropped field, and Hutchinson *et al.* (1982) used them to measure NH_3 and amine flux densities from a cattle feedlot in the USA. Since then these methods have improved substantially, and instruments are now able to analyse a range of different gaseous species with much higher precision than 15 years ago (Fowler and Duyzer, 1989; Choularton *et al.*, 1995; Fowler *et al.*, 1995; Zahniser *et al.*, 1995). Micrometeorological techniques and remote sensing are particularly appropriate for determining fluxes at the field and landscape scale, which are used to validate field scale models or landscape models. Chamber techniques, however, are still invaluable in the study of the processes responsible for gas formation and exchange, and in the development of models and correlation (Stewart *et al.*, 1989: see also Chapter 30).

Model Development

The firmest basis for scaling involves developing an understanding of the mechanisms that regulate spatial and temporal patterns of processes, and describing these mechanisms in models. The most widely used types of models are the descriptive, regression models and the explanatory, process-based models.

Generally, models break down a system into its component parts and describe the behaviour of the system through the interaction of those parts. There are several approaches to building mechanistic models. One of the most productive and satisfying is the individual-based approach, in which one begins with the factors impinging on an individual and uses them as a basis for understanding the behaviour of aggregates of individuals.

To begin the study of any system, the scales of variation of the key variables should be examined and those variables identified that change across scales similar enough to allow some interaction (Levin, 1993). In some studies where feedback mechanisms operate, transposition of scale may occur. Transposition or direct scaling involves the danger of ignoring potential feedbacks at the next higher level of scale. For example, fluxes between the soil surface and the air may not represent the atmospheric flux if there is re-absorption within the canopy (Denmead *et al.*, 1976), a process that operates at the next higher scale level. The type of model used depends on the scale, because the key factors that operate cannot be described similarly at each level of scale.

Langeveld *et al.* (see Chapter 29) describe soil process models for the soil microbe, aggregate and soil column scales, and extrapolate to regression models for the field scale and for the regional and national scale (see also Chapter 30). A discussion follows of a few examples of field-scale models from the literature. At field scale, the basic problem is the translation of field-scale state variables, such as soil moisture, into field-scale rates of processes such as decomposition. Correlations play an important role in suggesting candidate mechanisms for investigation. The effect of major regulating factors on processes such as soil organic matter turn-over are generally included as simple relationships derived from regression between soil moisture and temperature and decomposition (Parton *et al.*, 1987; Schimel *et al.*, 1994).

The first example model is from Mosier and Parton (1985), who developed a regression model of N_2O production from grassland soils. The purpose was to estimate fluxes over large areas, accounting for spatial and interannual variability. Model parameters were developed by relating N_2O flux to soil moisture and temperature for two sites representing much of the variability in the Colorado shortgrass ecosystem. Because no time series data of NO_3^- and NH_4^+ are available at the target scale of the study, the model was simplified with an empirical multiplier representing N availability.

The second example is the process model that Li *et al.* (1992a) developed to simulate N_2O fluxes from decomposition and denitrification in agricultural soils at field level. The model can also be used to describe NO_x fluxes. It is much more complex than the model of Mosier and Parton (1985), as it uses soil, climate and data on management to feed three submodels (thermal-hydraulic, denitrification and decomposition). The management practices it considers include tillage timing and intensity, fertilizer application, irrigation (amount and timing), manure amendment, crop type and rotation. Although detailed descriptions of various processes are included, the results may be dominated by the effects of temperature and moisture, which operate at nearly all levels in the model. This explains the similarity of the results Mosier and Parton (1985) obtained with their simpler approach, to those of Li *et al.* (1992a,b).

The third example is a farm model for understanding NH_3 fluxes from livestock production. At farm level, NH_3 losses can occur in the animal house, during grazing or during and after spreading of animal manure. Therefore a model is required to describe N flows and predict NH_3 volatilization at the farm level, as described by Hutchings *et al.* (1996). Their model describes NH_3 losses from animal housings, stored slurry, application of slurry and urine patches. The model builds on knowledge acquired from other experiments and model studies of animal housing (Janssen and Krause, 1990), waste storage (Olesen and Sommer, 1994) and farming practices (Van der

Molen *et al.*, 1990; Jarvis *et al.*, 1991; Sommer *et al.*, 1991). The model tracks the N input as animal feed until it is lost as NH_3. More examples of N balancing in grasslands are given by Peel *et al.* (see Chapter 33) and further assessment of the N cycle in livestock production systems is discussed by Sapek (see Chapter 34).

Constraining the Results

Testing in a scaling exercise involves determining the boundary conditions and constraints. Ideally, a hierarchical framework for model aggregation is based on the mechanism operating at the subordinate scale level, validated against data of the scale of focus, and constrained by the superior level (Reynolds *et al.*, 1993). The following examples illustrate the top-down scaling approach to setting boundary conditions at higher levels of scale.

Deposition of NH_3 and its product NH_4^+ occurs mainly in the form of dry deposition of NH_3 and wet deposition of NH_4^+. Wet deposition of NH_4^+ can be measured reasonably well and does not show a large spatial variability within 100 km. Dry deposition of NH_3, however, is mainly influenced by very local sources and shows a very large spatial variability. Therefore, the required detail of emission inventories is determined by this variability of NH_3 fluxes (Figs 28.3, 28.4). Asman and Van Jaarsveld (1992) compared results of an atmospheric transport model for emission inventories at different scales, ranging from 5×5 km to 150×150 km in the Netherlands. The results indicate that the model results for NH_3 are consistent with measurements only for the finest grid size and corresponding detail in the emission estimates.

Dentener and Crutzen (1994) used a global atmospheric transport model to simulate deposition. The results for Europe and North America could be constrained to a certain extent by deposition measurements, but the NH_3 budget estimate for large parts of the world remained largely speculative due to lack of deposition measurements. A striking conclusion was that the NH_4^+ concentration in rain cannot be modelled correctly unless there is NH_3 emission from the oceans. Hence, the top-down approach led to identification of a whole source category which was missing in the original inventory.

Inverse modelling derives estimates of flux values by departing from spatial and temporal patterns of atmospheric concentrations. It calculates the emissions needed to produce the observed atmospheric concentrations by reversing atmospheric transport models. For N_2O, Prinn *et al.* (1990) confirmed the conclusions of earlier scaling efforts by Matson and Vitousek (1990), that the tropical ecosystems are dominant global sources, and that conversion of tropical forest to grasslands contributes to the increase in atmospheric N_2O (Luizao *et al.*, 1989). In a different inverse modelling approach, Hartley and Prinn (1993) attempted to determine the best location for observation sites of atmospheric concentrations (Fig. 28.5).

Conclusions

It is hard to separate the concepts of scale at the local level from those at any other level. The importance of such concepts may be manifested differently at each level of scale, but they apply at all scales. The starting point for any

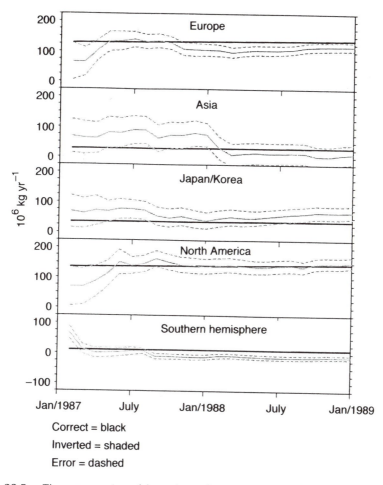

Correct = black
Inverted = shaded
Error = dashed

Fig. 28.5. Time progression of the estimated source strengths during an inversion run at five source regions. The results are based on a 'pseudodata' set that was defined by averaging the observations and modelled concentrations that were derived from emission estimates. Averaging was continued until the hybrid values were within 1 standard deviation of the simulated values. Resolution of the model is 11.5° latitude × 22.5° longitude, with 26 levels between the surface and 72 km altitude. (Source: Hartley and Prinn (1993). Reprinted with permission from the American Geophysical Union.)

scaling of N gas exchange must be embedded in the spatial and temporal data related to environmental resources (e.g. soils, climate) and management practices.

The basis for measurement strategies can be the disaggregation of a population or the delineation of functionally different types. These operations should be done where distinct and easily identified differences in structure and composition of a landscape coincide with the functions or management systems relevant to N-gas fluxes. The aim of defining functional groupings is to improve the understanding of the factors which cause variability within a system. Within the functional groupings, the variability is generally reduced when compared to the landscape or population as a whole.

The aim of the measurements within the functional groupings should be to determine relationships between fluxes and key processes and variables, which can be described and used in models at the next higher scale level. The spatial and temporal variability of the processes responsible for gaseous fluxes should be represented in the measurements. This allows for integration of state variables or controlling factors to the scale of the measurements at the next higher level of scale.

The firmest basis for scaling involves developing an understanding of the mechanisms that regulate spatial and temporal patterns of processes, and describing these mechanisms in models. Modelling plays a crucial role in suggesting candidate mechanisms for investigation. In general, the choice of the key factors or parameters at a certain scale is fairly subjective, making scaling not a science, but an art (Leffelaar, 1990). For livestock production systems, dynamic whole farm approaches are invaluable tools for extrapolating; such approaches are also useful in the assessment of strategies to reduce N-gas fluxes and other adverse environmental effects of livestock production.

Another important tool is the top-down scaling approach, whereby boundary conditions derived from the next higher scale level are used to constrain or test the possible variation in the results of scaling from smaller scales. The interaction between top-down and bottom-up approaches may yield better results more rapidly than can either the bottom-up or top-down approaches alone.

This volume presents various examples of scaling exercises. The use of models to bridge the gap between different scale levels is discussed by Langeveld *et al.* and Müller *et al.* (see Chapters 29 and 30). The supranational scale and the global extrapolation are discussed by Lee *et al.* (see Chapter 31). Two examples describe the N flows and gaseous fluxes at the level of a permanent grassland (see Chapter 34), or of a livestock production system (see Chapters 33 and 35). Reduction of emissions of N and other gases in grassland farm systems are discussed in Peel *et al.* and Kazenwadel and Zeddies (see Chapters 32 and 33).

Acknowledgements

The work of A.F.B. was done as part of RIVM project No. 481508. W.A.H.A. received financial support from the Danish Environmental Research Programme.

References

Andreae, M.O. (1991) Biomass burning: its history, use and distribution and its impact on environmental quality and global climate. In: Levine, J.S. (ed.) *Global Biomass Burning*. MIT Press, Cambridge, pp. 3–28.

Andreae, M.O., Atlas, E., Cachier, H., Cofer III, W.R., Harris, G.W., Helas, G., Koppmann, R., Lacaux, J.P. and Ward, D.E. (1997) Trace gas and aerosol emissions from savanna fires. In: Levine, J.S. (ed.) *Biomass Burning and Global Change*, Vol. 1: *Remote sensing and modeling of biomass burning, and biomass burning in the boreal forest*. MIT Press, Cambridge, MA.

Asman, W.A.H. (1992) Ammonia emission in Europe: updated emission and emission variations. Report 228471008, National Institute of Public Health and the Environment, Bilthoven, The Netherlands.

Asman, W.A.H. and Van Jaarsveld, J.A. (1992) A variable-resolution transport model applied for NH_x in Europe. *Atmospheric Environment* 26A, 445–464.

Bouwman, A.F. (1996) Direct emission of nitrous oxide from agricultural soil. *Nutrient Cycling in Agroecosystems* 46, 53–70.

Bouwman, A.F. (1997) Long-term scenarios of livestock-crop-land use interactions for the assessment of environmental indicators in developing countries. *Land and Water Bulletins* No. 5, Food and Agriculture Organization of the United Nations, Rome.

Bouwman, A.F., Van der Hoek, K.W. and Olivier, J.G.J. (1995) Uncertainties in the global source distribution of nitrous oxide. *Journal of Geophysical Research* 100, 2785–2800.

Bouwman, A.F., Lee, D.S., Asman, W.A.H., Dentener, F.J., Van der Hoek, K.W. and Olivier, J.G.J. (1997) A global high-resolution emission inventory to ammonia. *Global Biogeochemical Cycles* (in press).

Choularton, T.W., Gallagher, M.W., Bower, K.N., Fowler, D., Zahniser, M. and Kaye, A. (1995) Trace gas flux measurements at the landscape scale using boundary layer budgets. *Philosophical Transactions. Royal Society London* 351, 357–369.

Crutzen, P.J. (1983) Atmospheric interactions in homogeneous gas reactions of C, N and S containing compounds In: Bolin, B. and Cook, R.B. (eds) *The Major Biogeochemical Cycles and Their Interactions*. SCOPE Vol. 21. John Wiley & Sons, New York, pp. 67–112.

Crutzen, P.J. and Andreae, M.O. (1990) Biomass burning in the tropics: impact on atmospheric chemistry and biogeochemical cycles. *Science* 250, 1669–1678.

Denmead, O.T., Freney J.R. and Simpson, J.R. (1976) A closed ammonia cycle within a plant canopy. *Soil Biology and Biochemistry* 8, 161–164.

Dentener, F.J. and Crutzen, P.J. (1994) A three-dimensional model of the global ammonia cycle. *Journal of Atmospheric Chemistry* 19, 331–369.

ECETOC (1994) Ammonia emissions to air in Western Europe. *Technical Report*, No. 62. European Centre for Ecotoxicology and Toxicology of Chemicals, Brussels.

FAO/IFA/IFDC (1994) *Fertilizer Use by Crop 2. ESS/MISC/1994/4*. Food and Agriculture Organization of the United Nations, Rome (data on disk International Fertilizer Industry Association, Paris, 1994), 44 pp.

Folorunso, O.A. and Rolston, D.E. (1984) Spatial variability of field measured denitrification gas fluxes. *Soil Science Society of America Journal* 48, 1214–1219.

Fowler, D. and Duyzer, J.H. (1989) Micrometeorological techniques for the measurement of trace gas exchange. In: Andreae, M.O. and Schimel, D.S. (eds) *Exchange of Trace Gases Between Terrestrial Ecosystems and the Atmosphere*. Wiley, Chichester, pp. 189–207.

Fowler, D., Hargreaves, K.J., Skiba, U., Milne, R., Zahniser, M.S., Moncrieff, J.B., Beverland, L.J. and Gallagher, M.W. (1995) Measurement of CH_4 and N_2O fluxes at the landscape scale using micrometeorological methods. *Philosophical Transactions of the Royal Society London* 351, 339–356.

Goudriaan, J. (1993) Model, schaal en aggregatieniveau. Inaugural speech, 30 September 1993, Wageningen Agricultural University, Wageningen, 29 pp.

Griffith, D.W.T., Mankin, W.G., Coffey, M.T., Ward, D.E. and Riebau, A. (1990) FTIR remote sensing of biomass burning emissions of CO_2, CO, CH_4, CH_2O, NO, NO_2, NH_3 and N_2O In: Levine, J.S. (ed) *Global Biomass Burning*. MIT Press, Cambridge, pp. 230–239.

Groffman, P.M. (1991) Ecology of nitrification in soil evaluated at scales relevant to atmospheric chemistry. In: Rogers, J.E. and Whitman, W.B. (eds) *Microbial Production and Consumption of Greenhouse Gases: Methane, Nitrogen Oxides and Halomethanes*. American Society of Microbiology, Washington, DC, pp. 201–217.

Hartley, D. and Prinn, R. (1993) Feasibility of determining surface emissions of trace gases using an inverse method in a three-dimensional chemical transport model. *Journal of Geophysical Research* 20, 5183–5197.

Hegg, D.A., Radke, L.F. and Hobbs, P.V. (1990) Emissions of some trace gases from biomass fires. *Journal of Geophysical Research* 95, 5669–5675.

Hutchings, N.J., Sommer, S.G. and Jarvis, S.C. (1996) A model of ammonia volatilization from a grazing livestock farm. *Atmospheric Environment* 30, 589–599.

Hutchinson, G.L., Mosier, A.R. and Andre, C.E. (1982) Ammonia and amine emissions from a large cattle feedlot. *Journal of Environmental Quality* 11, 288–293.

Janssen, J. and Krause, K-H. (1990) Messung und Simulation von Ammoniak Konzentrationen in Ställen. In: Döhler, H. and Van der Weghe, H. (eds) *Ammoniak in der Umwelt*. KTBL-Schriften-Vertrieb in Landwirtschaftsverlag GMBH, Münster-Hiltrup, pp. 21.1–21.12.

Jarvis, S.C., Hatch, D. and Roberts, D.H. (1991) Micrometeorological studies of ammonia emission from grazed swards. *Journal of Agricultural Sciences* 117, 101–109.

Leffelaar, P.A. (1990) On scale problems in modelling: an example from soil ecology. In: Rabbinge, R., Goudriaan, J., Van Keulen, H., Penning de Vries, F.W.T. and Van Laar, H.H. (eds) *Theoretical Production Ecology: Reflections and Prospects*. Pudoc, Wageningen, Netherlands, pp. 57–73.

Levin, S.A. (1993) Concepts of scale at the local level. In: Ehleringer, J.R. and C.B. Field (eds) *Scaling Physiological Processes. Leaf to Globe.* Academic, London, pp. 7–19.

Li, C., Frolking, S. and Frolking, T.A. (1992a) A model of nitrous oxide evolution from soil driven by rainfall events: I. Model structure and sensitivity. *Journal of Geophysical Research* 97, 9759–9776.

Li, C., Frolking, S. and Frolking, T.A. (1992b) A model of nitrous oxide evolution from soil driven by rainfall events: II. Model applications. *Journal of Geophysical Research* 97, 9777–9783.

Luizao, F., Matson, P., Livingston, G., Luizao, R. and Vitousek, P. (1989) Nitrous oxide flux following tropical land clearing. *Global Biogeochemical Cycles* 3, 281–285.

Matson, P.A. and Vitousek, P.M. (1990) Ecosystem approach for the development of a global nitrous oxide budget. Processes that regulate gas emissions vary in predictable ways. *Bioscience* 40, 667–672.

Matson, P.A., Vitousek, P.M. and Schimel, D.S. (1989) Regional extrapolation of trace gas fluxes based on soils and ecosystems. In: Andreae, M.O. and Schimel, D.S. (eds) *Exchange of Trace Gases Between Terrestrial Ecosystems and the Atmosphere.* Wiley, Chichester, pp. 97–108.

McElroy, M.B. and Wofsy, S.C. (1986) Tropical forests: interactions with the atmosphere In: Prance, G.T. (ed.) *Tropical Rain Forests and the World Atmosphere.* AAAS Selected Symposium 101. Westview Press Inc., Boulder, Colorado, pp. 33–60.

Mosier, A.R. and Hutchinson, G.L. (1981) Nitrous oxide emissions from cropped fields. *Journal of Environmental Quality* 10, 169–173.

Mosier, A.R. and Parton, W.J. (1985) Denitrification in a shortgrass prairie: a modelling approach In: Caldwell, D.E., Brierley, J.A. and Brierley, C.L. (eds) *Planetary Ecology.* Van Nostrand Reinhold, New York, pp. 441–451.

Olesen, J.E. and Sommer, S.G. (1994) Modelling effects of wind speed and surface cover on ammonia volatilization from stored pig slurry. *Atmospheric Environment* 27, 2567–2574.

Parton, W.J., Schimel, D.S., Cole, C.V. and Ojima, D.S. (1987) Analysis of factors controlling soil organic matter levels in Great Plains grasslands. *Soil Science Society of America Journal* 51, 1173–1179.

Pedersen, S., Takai, H., Johnsen, J. and Birch, H. (1996) Ammoniak og støv i kvæg-, svine- og fjerkræstalde II. *Internal Report No. 65, Danish Institute of Animal Science,* Horsens, Denmark, 54 pp.

Poth, M. and Focht, D.D. (1985) ^{15}N kinetic analysis of N$_2$O production by *Nitrosomonas europaea*: an examination of nitrifier denitrification. *Applied Environmental Microbiology* 49, 1134–1141.

Prinn, R., Cunnold, D., Rasmussen, R., Simmonds, P., Alyea, F., Crawford, A., Fraser, P. and Rosen, R. (1990) Atmospheric emissions and trends of nitrous oxide deduced from ten years of ALEGAGE data. *Journal of Geophysical Research* 95, 18369–18385.

Ramanathan, V., Cicerone, R.J., Singh, H.B. and Kiehl, J.T. (1985) Trace gas trends and their potential role in climate change. *Journal of Geophysical Research* 90, 5547–5566.

Reynolds, J.F., Hilbert, D.W. and Kemp, P.R. (1993) Scaling ecophysiology from the plant to the ecosystem: a conceptual framework. In: Ehleringer, J.R. and C.B.

Field (eds) *Scaling Physiological Processes. Leaf to Globe.* Academic, London, pp. 127–140.

Schimel, D.S., Parton, W.J., Adamsen, F.J., Woodmansee, R.G., Senft, R.L. and Stillwell, M.A. (1986) The role of cattle in the volatile loss of nitrogen from a shortgrass steppe. *Biogeochemistry* 2, 39–52.

Schimel, D.S., Braswell, B.H., Holland, E.A., McKeown, R., Ojima, D.S., Painter, T.H., Parton, W.J. and Townsend, A.R. (1994) Climatic, edaphic, and biotic controls over storage and turnover of carbon in soils. *Global Biogeochemical Cycles* 8, 279–293.

Sommer, S.G., Olesen, J.E. and Christensen, T.B. (1991) Effects of temperature, wind speed and air humidity on ammonia volatilization from surface applied cattle slurry. *Journal of Agricultural Science, Cambridge* 117, 91–100.

Stewart, J.W.B., Aselmann, I., Bouwman, A.F., Dejardins, R.L., Hicks, B.B., Matson, P.A., Rodhe, H., Schimel, D.S., Svensson, B.H., Wassmann, R., Whiticar, M.J. and Yang, W-X. (1989) Extrapolation of flux measurements to regional and global scales. In: Andreae, M.O. and Schimel, D.S. (eds) *Exchange of Trace Gases Between Terrestrial Ecosystems and the Atmosphere.* Wiley, Chichester, pp. 155–174.

Van Breemen, N., Burrough, P.A., Veldhorst, E.J., Van Dobben, H.F., De Wit, T., Ridder, T.B. and Reijders, H.F.R. (1982) Soil acidification from atmospheric ammonium sulphate in forest canopy throughfall. *Nature* 299, 548–550.

Van der Molen, J., Beiljaars, A.C., Chardon, W.J., Jury, W.A. and Van Faassen, H.G. (1990) Ammonia volatilization from arable land after application of cattle slurry. 2. Derivation of a transfer model. *Netherlands Journal of Agricultural Sciences* 38, 239–254.

Zahniser, M.S., Nelson, D.D., McManus J.B. and Kebabian, P.L. (1995) Measurement of trace gas fluxes using tunable diode laser spectroscopy. *Philosophical Transactions of the Royal Society London* 351, 371–382.

Modelling Nitrous Oxide Emissions at Various Scales

C.A. Langeveld, P.A. Leffelaar and J. Goudriaan

Department of Theoretical Production Ecology, Wageningen Agricultural University, PO Box 430, 6700 AK Wageningen, The Netherlands

Summary. Soils are a major source of atmospheric nitrous oxide (N_2O), which is partly responsible for global climatic change. The modelling of soil-borne N_2O emissions is the subject of this chapter. Different factors affect soil-borne N_2O emissions and quantification of the effects would be helpful in assessing the impact of, for example, land use change or climatic change. Net N_2O production in soils depends mainly on the balance of two reduction steps in denitrification. Nitrous oxide emission from soil is further determined by transport processes and heterogeneities. An integrative approach using quantitative models is the obvious way to assess the sensitivity of net emission for different factors. Here, models at various spatial scales are discussed. In relation to experimental work, process-based modelling is considered at the microbe and rhizotron scales more closely. The microbe scale model explained the time courses of nitrogenous denitrification intermediates in anaerobically incubated peat well, but underestimated carbon dioxide (CO_2) production. The model at the rhizotron scale is under development. Establishing the relationship between subsoil N_2O gradients and the resulting fluxes at the soil surface is a major challenge with this.

Introduction

Nitrous oxide (N_2O) is involved in both the enhanced greenhouse effect and the depletion of the stratospheric ozone layer (Bouwman, 1995). Its atmospheric concentration of about 310 ppbv (parts per billion by volume) currently rises at a rate of about 0.25% per year. According to the first-order differential equation of Den Elzen *et al.* (1995), the atmospheric concentration will rise to an equilibrium value of about 400 ppbv if current emissions remained unchanged.

About 50% of the gross global emission is attributed to soils. However, considering the various ranges published by Bouwman (1995), the uncertainty in this figure is large. Also, the variability of N_2O fluxes measured in individual monitoring studies is large. In their field study, Syring and Benckiser (1990) attributed a substantial part of the flux variability to random noise. The other part of the variability was attributed to controlling factors. These factors are scale dependent. Their effects must be better quantified, before emission reduction scenarios or climatic change scenarios can be soundly evaluated. The purpose of this chapter is to clarify essential features and achievements of models aiming at identification and quantification of the determining factors at various scales. The relevant literature is discussed, as well as our own modelling activities, the latter in connection with experiments.

Modelling Approaches at Various Spatial Scales

For modelling soil-borne N_2O emission, six integration levels are distinguished, related to the spatial scales of: (i) microbes; (ii) aggregates; (iii) soil columns; (iv) rhizotrons; (v) fields; (vi) regions; and countries. Below, illustrative models from the literature at these scales are discussed.

The microbe scale

Studies at this scale aim at characterizing the dynamics of biological processes underlying N_2O emission. The three phases in the soil are considered to be homogeneous and transformations of dissolved gases and nutrients are described. The transport of these substances is not explicitly considered: a steady state is assumed for the distribution over the phases. A relatively detailed process-based model was proposed by Leffelaar and Wessel (1988) and in their simulations, the time courses of the sequentially appearing nitrogenous denitrification intermediates depended strongly on the initial fraction of denitrifiers in the microbial biomass.

The aggregate scale

In soil mechanics, an aggregate is conceived as a group of soil particles that cohere relatively strongly to each other. In process-based modelling of N_2O emission, a more abstract concept is desired in which an aggregate is a functional unit with respect to transport and transformation of substances.

Such a concept was provided by Rappoldt (1992), who showed the minor relevance of exact geometrical aggregate shape for diffusion properties.

The spatial scale of aggregates ranges from millimetres to centimetres. At this scale, non-homogeneous distributions of water, nutrients and gases can occur because of the interference between biological and transport processes. Arah and Smith (1990) analysed a theoretical process-based aggregate model and identified aggregate radius and oxygen demand as the main factors affecting the ratio $N_2O:(N_2O + N_2)$ evolving from the aggregates in denitrification.

The soil column and the rhizotron scale

In soil columns (0.1–1 m), not only is transport within aggregates important, but so also is that through macropores between aggregates. The model of Rolston and Marino (1976) described nitrate (NO_3^-) dynamics in a soil column well, but not the dynamics of N_2O and molecular nitrogen (N_2). This might be explained by the high sensitivity of N_2O dynamics for transport parameters (diffusion coefficients) above the aggregate scale.

The spatial scale of rhizotrons is in metres (Van de Geijn *et al.*, 1994). In contrast with the soil column scale, the description of vegetation or crop growth processes has also to be included. A rhizotron system could be used in developing and testing a field-scale model but we are not aware of N_2O emission models developed at the rhizotron scale.

The field scale

The field scale extends from 10 to 1000 m. In principle, lateral heterogeneities (soil structure, nutrient availability, etc.) are relevant. Models suitable for describing N_2O emission at this scale have been briefly reviewed (Langeveld and Leffelaar, 1996). In deterministic regression models, simple relationships between fluxes and field-scale variables (N content, temperature, moisture content) typically explained up to 50% of the flux variability. Li *et al.* (1992) and Grant *et al.* (1994) described fluxes after extreme watering events (showers, snow melt) satisfactorily by deterministic process-based modelling. Stochastic models are still in their infancy.

The regional and national scale

The modelling activities at this scale concern extrapolations of experimental observations to estimate regional or national emissions. Kroeze (1994) estimated the total soil-borne N_2O emission from the Netherlands by combining the results of flux monitoring studies with statistical information on soil type, land use and N fluxes. The scarcity of suitable experimental data and the restricted justification of underlying assumptions should be translated into uncertainties in the emission estimates.

Two Case Studies

This work forms part of the Dutch Integrated N_2O and methane (CH_4) Grassland Project, one of the aims of which is the modelling of N_2O emissions from grasslands at the above-mentioned scales. The microbe and rhizotron scale have been focused on, and preliminary results are discussed below.

Case I: process-based modelling at the microbe scale

Agricultural drained peat soils emit relatively large amounts of N_2O (Kroeze, 1994). To investigate the dynamics of the biological processes involved, peat soil samples were incubated in gas-tight covered petri dishes (Langeveld and Hofman, 1994). To simulate these processes under anaerobic conditions, the model of Leffelaar and Wessel was adapted (1988). The major changes to this model were:

1. Glucose, added in the experiments described by Leffelaar and Wessel, was replaced by endogenous total water soluble carbon (WSC); it is assumed that WSC plays a comparable role to glucose with respect to the processes described.

2. A pH effect was imposed on the rate constants for the successive steps in denitrification ($NO_3^- \rightarrow NO_2^- \rightarrow N_2O \rightarrow N_2$), according to Li *et al.* (1992), to account for the pH effect on nitrite (NO_2^-) accumulation (Bernet *et al.*, 1995).

In Fig. 29.1(a), the measured and simulated time courses of nitrogenous compounds, for one of the treatments, are presented. The correspondence is satisfactory (high amounts of N_2 at days 3 and 5 are attributed to experimental imperfections). Contrary to the model predictions, no substantial amounts of NO_2^- were found (data not shown). The modelled pH effect on the parameters reduced, but did not stop, NO_2^- accumulation. The *quality* of the organic matter, not considered in the model, could be relevant for NO_2^- accumulation (Minzoni *et al.*, 1988). The model under-estimates carbon dioxide (CO_2) production (Fig. 29.1(b)). In anaerobically incubated peat, CO_2 must have been produced by processes other than denitrification. This is consistent with findings of Yavitt and Lang (1990). The exact nature of these processes is, however, unclear. Since they will also require organic compounds, they may affect denitrification, and thus N_2O dynamics.

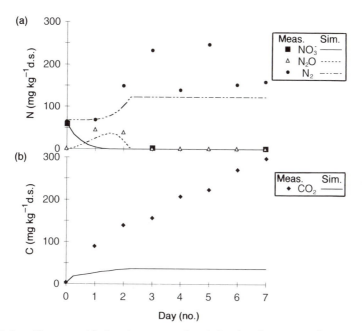

Fig. 29.1. Changes with time in measured and simulated amounts of NO_3^-, N_2O and N_2 (mg N kg^{-1} dry soil (d.s.)) (a) and CO_2 (mg C kg^{-1} d.s.) (b), in anaerobic incubation of peat soil at 20°C. Samples were from the 2–8-cm layer of a drained plot (average ground water table: −0.5 m) with no fertilizer N added. Simulation parameters are from the literature compilation of Leffelaar and Wessel (1988), but the maximum relative effects of growth rates of bacteria on NO_3^- and N_2O is adapted according to Li *et al.* (1992), and the initial mass fraction of denitrifiers in the microbial biomass (free parameter) set at 0.033.

Case II: process-based modelling at the rhizotron scale

For the development and testing of a model to describe N_2O emission at the field scale, a 2-year experiment was conducted in the Wageningen Rhizolab. Here, the conditions were better controlled and monitored than in field situations. N_2O emissions were monitored in four compartments (1.25 × 1.25 × 1.00 m), each containing sandy soil under grass, and subsoil profiles of several gases, nutrients, temperature, soil moisture content and the pressure head were observed.

In order to describe this system, the model of Leffelaar (1988), which operates satisfactorily at the aggregate scale, will be adapted. Processes in the top soil deserve special attention. In agreement with Benckiser (1994), a relationship was not found between N_2O concentrations in the gas phase at 10 cm depth and N_2O fluxes at the soil surface (data not shown).

High microbiological activities in the top soil and/or changes in transport properties in time might explain this. The model could be used to test this hypothesis.

Discussion and Conclusions

When considering validation and justification of assumptions, models at the microbe and aggregate scale are in a relatively advanced stage of development. Validation studies on deterministic process-based models at the field scale have focused thus far on their behaviour under extreme moisture conditions events. The first case study showed that in peat other biological processes consuming organic compounds interfere with denitrification, which influences N_2O dynamics. The second case study showed the importance of the upper soil layer with respect to N_2O emission in a rhizotron.

Although the determining variables and parameters generally vary with scale, the fundamental processes underlying N_2O emission (production, consumption and transport) are the same at all scales. Therefore, the models operating at different scales should be consistent. Establishing this consistency via sensitivity analyses is an important future task.

Acknowledgements

We thank R. Segers and A.M. van Dam for comments on drafts of this paper, and the Dutch National Research Programme on Global Air Pollution and Climate Change for financial support.

References

Arah, J.R.M. and Smith, K.A. (1990) Factors influencing the fraction of the gaseous products of soil denitrification evolved to the atmosphere as nitrous oxide. In: Bouwman, A.F. (ed.) *Soils and the Greenhouse Effect.* John Wiley, Chichester, pp. 475–480.

Benckiser, G. (1994) Relationships between field-measured denitrification losses, CO_2 formation and diffusional constraints. *Soil Biology and Biochemistry* 26, 891–899.

Bernet, N., Bizeau, C., Moletta, R., Cornier, J.C. and Deguin, A. (1995) Study of physicochemical factors controlling nitrite build-up during heterotrophic denitrification. *Environmental Technology* 16, 165–172.

Bouwman, A.F. (1995) Compilation of a global inventory of emission of nitrous oxide. PhD thesis, Wageningen Agricultural University, Wageningen, The Netherlands, 143 pp.

Den Elzen, M., Beusen, A. and Rotmans, J. (1995) *Modelling Global Biogeochemical Cycles.* National Institute of Public Health and the Environment (RIVM), Bilthoven, The Netherlands, 104 pp.

Grant, R.F., Nyborg, M. and Laidlaw, J.W. (1993) Evolution of nitrous oxide from soil: I. Model development. *Soil Science* 156, 259–265.

Kroeze, C. (1994) *Nitrous Oxide (N2O). Emission Inventory and Options for Control in the Netherlands.* National Institute of Public Health and the Environment (RIVM), Bilthoven, The Netherlands, 163 pp.

Langeveld, C.A. and Hofman, J.E. (1994) Nitrous oxide production and consumption in peat soils. In: Van Ham, J., Janssen, L.J.H.M. and Swart, R.J. (eds) *Non-CO2 Greenhouse Gases: why and how to control?* Kluwer Academic Publishers, Dordrecht, pp. 433–438.

Langeveld, C.A. and Leffelaar, P.A. (1996) Approaches in field scale modelling of nitrous oxide emission from grassland soils. In: Diekkrüger, B., Heinemeyer, O. and Nieder, R. (eds) *Transactions of the 9th Nitrogen Workshop, Braunschweig, Germany, 9–12 September 1996.* Technische Universität Braunschweig and FAL Völkenrode, Braunschweig, pp. 153–156.

Leffelaar, P.A. (1988) Dynamics of partial anaerobiosis, denitrification and water in a soil aggregate: Simulation. *Soil Science* 146, 427–444.

Leffelaar, P.A. and Wessel, W. (1988) Denitrification in a homogeneous, closed system: experiment and simulation. *Soil Science* 146, 335–349.

Li, C., Frolking, S. and Frolking, T.A. (1992) A model of nitrous oxide evolution from soil driven by rainfall events: 1. Model structure and sensitivity. *Journal of Geophysical Research* 97, 9759–9776.

Minzoni, F., Bonetto, C. and Golterman, H.L. (1988) The nitrogen cycle in shallow water sediment systems of rice fields. Part 1: The denitrification process. *Hydrobiologia* 159, 189–202.

Rappoldt, C. (1992) Diffusion in aggregated soil. PhD Thesis. Wageningen Agricultural University, Wageningen, The Netherlands, 162 pp.

Rolston, D.E. and Marino, M.A. (1976) Simultaneous transport of nitrate and gaseous denitrification products in soil. *Soil Science Society of America Journal* 40, 860–865.

Syring, K.M. and Benckiser, G. (1990) Modeling denitrification losses from arable land. *Mitteilungen der Deutschen Bodenkundlichen Gesellschaft* 60, 403–406.

Van de Geijn, S.C., Vos, J., Groenwold, J., Goudriaan, J. and Leffelaar, P.A. (1994) The Wageningen Rhizolab – a facility to study soil–root–shoot–atmosphere interaction in crops. I. Description of main functions. *Plant and Soil* 161, 275–287.

Yavitt, J.B. and Lang, G.E. (1990) Methane production in contrasting wetland sites: response to organic-chemical components of peat and to sulfate reduction. *Geomicrobiology Journal* 8, 27–46.

Application of a Mechanistic Model to Calculate Nitrous Oxide Emissions at a National Scale

30

C. Müller[1,*], R.R. Sherlock[1], K.C. Cameron[1] and J.R.F. Barringer[2]

[1]*Department of Soil Science, Lincoln University, New Zealand;*
[2]*Manaaki Whenua–Landcare Research, Lincoln, New Zealand;*
*Present address: Institute of Applied Microbiology,
Justus-Liebig University, Gießen, Germany

Summary.   The development of a mechanistic nitrous oxide (N_2O) emission model and its potential application in a Geographic Information System (GIS) for the estimation of soil-derived N_2O emissions on a national scale is reported. The mechanistic model proposed is based on Michaelis–Menten kinetics using data collected during a field experiment in Canterbury, New Zealand. Soil temperature, moisture and mineral nitrogen (N) concentrations were measured or interpolated on a daily time step for the 0–10-cm soil depth. Total N_2O and N_2O resulting from nitrification (F_{nit}) and denitrification (F_{den}) were quantified in the field using an acetylene blocking technique. Relationships between the potassium chloride (KCl) extractable concentrations of ammonium (NH_4^+) and nitrate (NO_3^-) with F_{nit} and F_{den}, respectively, were developed. This enabled the calculation of F_{nit} and F_{den} for situations where only total N_2O was measured. The special influence of rainfall events on stimulating N_2O emissions was also recognized through the use of two sets of Michaelis–Menten equations: one set being used for the 10-day period immediately following rainfall events exceeding 20 mm day^{-1} and the other set being used for all other occasions. The practical application of a mechanistic model on a national scale requires a careful consideration of the input parameters. Such parameters must either be readily accessible from climate records (e.g. rainfall and temperatures) or from soils inventory databases before the model can be implemented in a GIS. To achieve this, the time step was increased by using monthly mean temperatures and rainfall data and a simple soil water model developed to calculate monthly mean soil water suction values for a wide spectrum of soil types. At this stage in the model's development, typical mean annual NH_4^+-N and NO_3^--N values for a wide range of ecosystem classes are accessed by the GIS from a 'look-up table'. An initial comparison of the GIS model output with an 8-month data set of measured N_2O emissions at a single pasture site showed good agreement between the total measured and modelled amounts of N_2O released.

Introduction

Upon signing the 1992 framework convention on Climate Change in Rio de Janeiro, 154 nations committed themselves to publishing national greenhouse gas inventories, including anthropogenic emissions of nitrous oxide (N_2O) (Subak *et al.*, 1993). The Intergovernmental Panel on Climate Change (IPCC) has since undertaken to provide guidelines for national greenhouse gas inventories with the aim of harmonizing estimates (Subak *et al.*, 1993). The IPCC has recently distributed the Phase II methodology (IPCC, 1996) where anthropogenic N_2O emissions are estimated from nitrogen (N) inputs through synthetic fertilizer, animal excreta, biological N fixation, crop residues and cultivation of organic soils (Histosols). In addition, indirect N_2O emissions are estimated from N deposition, N leaching and sewage. Nitrous oxide emissions are calculated by multiplying the N inputs by emission factors. This approach is very rough and generalized due to the underlying assumption that agricultural systems across the globe behave similarly. No effect of climatic variables, soil effects and management systems on N_2O emissions are taken into account and the need to apply process-based models for estimates of N_2O emissions is recognized (IPCC, 1996).

Other methods to produce national inventories are based on an ecosystem approach by multiplying flux measurements made at several sites within an ecosystem by the area of that ecosystem (Matson and Vitousek, 1990). The underlying assumption of this approach is that the N_2O flux from the whole ecosystem is equal to the mean flux obtained from the measurements made. Such an approach was used in an earlier N_2O inventory for New Zealand (Sherlock *et al.*, 1992).

However, N_2O emitted by soil originates from extremely dynamic processes, with both spatial and temporal variability governed by a number of factors and processes and their interactions (Granli and Bøckman, 1994). Therefore the use of only a single or a few flux values can never be representative of the emissions from a whole ecosystem. Matson and Vitousek (1990) recognized this difficulty in their revised estimate of N_2O emissions from humid tropical forests by dividing the whole ecosystem into areas of low, intermediate and high fertility and assigning N_2O loss factors to each fertility level. Ideally, the effects of other variables such as soil moisture and soil temperature should also be included in N_2O emission models to refine such estimates even further (Matson and Vitousek, 1990). Thus, more accurate estimates of trace gas emissions have to integrate available data sets and known relationships of N_2O emissions with vegetation, soils, meteorological data and other factors (IPCC, 1996).

The use of a Geographical Information System (GIS), which can hold spatial information of possible input parameters, could be an ideal platform to calculate inventories with such process-based models for whole countries.

Matthews (1993) demonstrated this for estimates of global methane (CH_4) emission. It was concluded that geographical databases provide an ideal framework for coherent and systematic estimates of trace gas emissions and for analysing the global distribution of the sources of these gases (Matthews, 1993). A recent N_2O estimate for New Zealand combined available rainfall information, soil drainage class and a simple temperature relationship for an N_2O inventory calculated within a GIS (Carran *et al.*, 1993). However, those estimates were based on somewhat arbitrary emission factors obtained from three N_2O flux studies in the North Island of New Zealand and not on process-oriented relationships. Important driving variables for N_2O emissions, such as inorganic N concentrations (Granli and Bøckman, 1994), were not included in the model (Carran *et al.*, 1993).

In order to formulate a process-based emission model, the main driving variables as well as the main processes for N_2O emission from soils should be included. The N_2O emission model developed in this study is GIS based and, in contrast to that of Carran *et al.* (1993), is developed on mechanistic relationships observed from a field study. It combines the effect of the most important driving variables for N_2O emissions, such as soil temperature, soil moisture, inorganic N (NO_3^- and NH_4^+) and rainfall intensity, as well as the main mechanisms for N_2O emission being both nitrification and denitrification (Müller, 1996). The mechanistic basis of this model should allow it to be used in different climatic regions as well as in different ecosystems and it could therefore be applied for a whole country, in this case New Zealand.

Materials and Methods

Full details of the field experiments used for the development of the mechanistic N_2O emission model are presented elsewhere (Müller, 1996). However, some of the more important experimental aspects are detailed below.

Experimental design, soils, treatments

Nitrous oxide was measured for 1 year from intensive grassland on Templeton silt loam (Udic Ustochrept; USDA Soil Taxonomy) sown with rye-grass (*Lolium perenne* L.) and white clover (*Trifolium repens* L.) in New Zealand's South Island. The treatments were synthetic sheep urine applied at a rate of $4.073 \, l \, m^{-2}$ (N: 500, K: 400, Cl: 100, S: 15 kg ha^{-1}) and a control (no N applied).

Gas collection and analysis of N₂O

A modified version of the soil cover technique described by Hutchinson and Mosier (1981) was used to determine N_2O emissions from the treatments. Analysis of gas samples for N_2O was carried out on a gas chromatography system with a stainless steel column (length 3 m) packed with Porapak Q (80/100 mesh), similar to that described in Mosier and Mack (1980). Nitrous oxide fluxes were calculated using the non-linear equation given in Hutchinson and Mosier (1981).

To quantify N_2O emission resulting from nitrification and denitrification, the acetylene (C_2H_2) inhibition technique of Klemedtssson *et al.* (1988) was applied during a separate field experiment on the same site. Field incubations of soil cores under two C_2H_2 concentrations (0 Pa = I_{tot} and 5 Pa = I_{den}; $I_{nit} = I_{tot} - I_{den}$) were performed in glass jars (Agee; volume approx. 1100 ml) inserted into the ground and covered with a thin wooden plate to track field temperature conditions as closely as possible. After 5 h incubation, gas samples were taken from the jars and analysed for N_2O as described above. During the incubation time in the field, N_2O was also measured via the cover technique (F_{tot}) as described above. Nitrous oxide emissions via nitrification, F_{nit} and denitrification, F_{den} were calculated by multiplying the two fractions:

$$\frac{I_{den}}{I_{tot}}, \frac{I_{nit}}{I_{tot}}$$

obtained from the incubation by F_{tot} making the assumption that the fractions determined from the jar incubation were equal to the average daily fractions in the plots where N_2O was measured with the soil cover technique.

Soil moisture status

Soil volumetric water content in the top 15 cm (in 5-cm increments) was determined gravimetrically (Marshall and Holmes, 1988; equation 1.11) and converted with the power function (Buchan and Grewal, 1990) to soil water suction with the previously determined moisture characteristic.

NO₃-N, NH₄⁺-N and water-soluble carbon

Soil NH_4^+ and NO_3^- were extracted immediately after sampling in 2 M KCl (10 g moist soil in 50 ml KCl), according to the procedure given in Maynard and Kalra (1993), and analysed colorimetrically by flow-injection analysis (Tector Flow Injection Analyser). Water-soluble carbon (C) analyses were performed as described in Burford and Bremner (1975). Inorganic N (kg N ha⁻¹) and water-soluble C (kg C ha^{-1}) were adjusted to the water content and the bulk density of the soil.

Soil and air temperature

Soil and air temperature were measured with thermistors (Campbell Scientific, Inc.) and logged half-hourly with a datalogger (21X, Campbell Scientific, Inc.).

Model Description

Description of original model

The process-oriented model described here is based on a model (KNOM), which was developed to explain and predict N$_2$O emissions from urine-affected intensive grassland in New Zealand (Müller et al., 1997). The two main processes for N$_2$O emission, nitrification and denitrification were modelled separately with functions based on Michaelis–Menten kinetics in the form:

$$F = \frac{F_m \alpha \cdot N}{F_m + \alpha \cdot N} \qquad (1)$$

where F represents the N$_2$O flux either from denitrification or nitrification and N is the inorganic N concentration (N_{NH4} = N H$_4$-N for nitrification or N_{NO3} = NO$_3^-$-N for denitrification). F_m and α are process specific Michaelis–Menten parameters describing the relationship between N and F. The parameters F_m and α were estimated from a data set obtained from a field study on intensive grassland in Canterbury, New Zealand. The data set (*DATA1*) contained daily total N$_2$O emissions (F_{tot}), together with the main driving variables for N$_2$O emissions: soil temperature (T), soil water suction (S) and inorganic N concentrations N_{NH4} and N_{NO3}, from the top 5 cm of soil (Müller et al., 1995). The application of results from a separate acetylene field incubation study on the same site enabled the daily total N$_2$O emission values in *DATA1* to be partitioned into N$_2$O emitted via nitrification (F_{nit}) and denitrification (F_{den}) (Müller et al., 1997). In the acetylene field incubation study, relationships for F_{nit} and F_{den} against N_{NH4} and N_{NO3}, respectively, were developed for three soil water suction classes, S ($S_1 < 55$ cm of water; $S_2 = 55$–150 cm of water and $S_3 > 150$ cm of water) by incubating soil under field conditions in the presence of 0 Pa, 5 Pa and 10 kPa C$_2$H$_2$ (Klemedtsson et al., 1988; Müller et al., 1997). These relationships were then used to obtain F_{nit} and F_{den} for the whole data set *DATA1* (Müller et al., 1997). Different Michaelis–Menten parameters (F_m and α) for F_{nit} and F_{den} were estimated from *DATA1* by non-linear curve fitting (SigmaPlot, Jandel Scientific) of N versus F under various soil temperature–soil water suction regimes using equation (1) (Müller *et al.*, 1997). In addition, separate Michaelis–Menten parameters were estimated for times up to 10 days after rainfall intensities (R) higher than 20 mm day^{-1}. This was done because different N_2O emission characteristics were observed under those conditions compared to the rest of the observations when only a gradual soil water change occurred (Müller *et al.*, 1995).

In the original implementation of this model, calculations for both F_{nit} and F_{den} were made with equation (1) on a daily time step by using Michaelis–Menten parameters chosen according to the appropriate factor combination for that day (i.e. combination of R, S, T and N). Total N_2O emitted daily (F_{tot}) was calculated as the sum of $F_{nit} + F_{den}$ (Müller *et al.*, 1997). Model calculations of F_{tot} were in good agreement with N_2O emissions measured daily from a site on intensive grassland in Canterbury, New Zealand (Müller *et al.*, 1997).

Application of the model in GIS

The practical application of the mechanistic model described above on a national scale required a careful consideration of the number and complexity of the input variables. Such variables must either be readily accessible from climate records (e.g. rainfall, temperatures) or should be available in soils inventory databases before the model can be implemented. To handle the enormous amounts of data and perform calculations, the model had to be rewritten using a monthly rather than a daily time step.

From the input variables needed, only rainfall intensities and soil temperatures were readily accessible from climatic databases on a national scale. Inorganic N concentrations and soil water suctions were not available for New Zealand and had to be derived from other available information. Inorganic N concentrations (NH_4^+-N and NO_3^--N) for the top 10 cm of soil were estimated from available observations from different ecosystems. The estimates were grouped into the ecosystem classification of Newsome (1987), which is available in a GIS format. For those ecosystems where no data were available a best estimate was made (Goh, McMillan, Stewart, Monaghan, Lincoln, 1995, personal communications).

All inputs to the model, except inorganic-N, were stored as raster layers which have been derived from data sources held by Landcare Research. The base soil spatial layer was from the New Zealand Land Resource Inventory (NZLRI). The original vector layer of soil type was converted to soil attribute raster layers via a relational join to the National Soils Database (NSD). The NSD holds a wide range of soil properties, including texture, mineralogy, chemistry and soil water. Inorganic-N concentrations were derived from a 'look up' table based on the ecosystem classification of Newsome (1987), which is linked to the NZLRI vegetation attribute to generate NH_4^+-N and NO_3^--N raster layers. Rainfall and temperature layers were from 1-km resolution national climate layers which were derived by empirical interpolation (D.J. Giltrap, 1996, personal communication) from the National Institute of Water and Atmosphere's climate station database (CLIDB) and the national digital elevation model derived from the Department of Survey and Land Information's digital 100 m contour coverage. The model was programmed in Arc Macro Language (AML).

To obtain monthly mean soil water suctions, a simple water model using only inputs, which were available in spatial databases, had to be developed. The main requirement for the model calculations was the soil water suction in the top 10 cm. However, the water content in the top soil layer was not independent of the conditions below 10 cm and therefore a two-layered soil water model was developed based on relationship (2):

$$\Delta\theta_A = R - ET_A - D_A \ (units \ in \ mm) \tag{2}$$

where $\Delta\theta_A$ is the change in volumetric water content, R is the total monthly rainfall, ET is the total monthly evapotranspiration and D is the drainage from the A into the B layer. The subscript denotes the layer (in this case A for the top layer). The thickness of the A layer was 0.1 m and the B layer 0.1–1.0 m, depending on the drainage characteristic of the soil (poorly drained soil = 0.1, well-drained soil = 1.0 m). No exchange of water occurred on the bottom boundary of the B layer. Starting conditions for the model calculations were in mid-winter and assumed that the soil was at field capacity ($S = 70$ cm of water). Evapotranspiration (ET) was taken from the A and B layers in fractions of 0.25 and 0.75, respectively, when the soil water suction in the A layer was < 2000 cm of water and in fractions of 0.09 and 0.91, respectively, when the soil water suction in the A layer was > 2000 cm of water. This somewhat arbitrary relationship was determined from correlations of soil water content change in different soil layers with evapotranspiration, using data from Müller (1996). It was assumed that the ET fraction from the B layer originates always from a 1-m-deep layer. This condition takes into account situations where a high water table leads to slow drainage, but ET may still be high from the water table itself.

Total monthly rainfall (R_i) was added to the A layer and remained there if the volumetric water content was lower than field capacity (i.e. $\theta_A < \theta_{FC}$). Under conditions when the volumetric water content was higher than field capacity (i.e. $\theta_A > \theta_{FC}$) water moved from the A layer into the B layer at a maximum rate given by the unsaturated hydraulic conductivity (K_A), assuming that downward movement of water was driven by a unit hydraulic gradient. Unsaturated hydraulic conductivity was predicted according to Vereecken *et al.* (1990), who found relationships for sand, silt, clay, organic matter contents and bulk density against saturated and unsaturated hydraulic conductivities for a range of temperate soils in Belgium. These inputs are in soil inventories for the whole of New Zealand and therefore can be used to predict unsaturated hydraulic conductivities for temperate New Zealand soils.

Drainage from the A into the B layer continued until the B layer was saturated. Thereafter drainage stopped and filled the A layer up to saturation (θ_{Sat}) The new calculated soil water content was then converted to soil water suction via the power function (Buchan and Grewal, 1990) (3):

$$S = A\theta^B \qquad (3)$$

where S is the soil water suction and θ the volumetric water content. A and B are parameters from a linear regression of ln S versus ln θ. The determination of A and B parameters is usually done with a whole range of S-θ data pairs. However, because of limitations of this kind of information for the whole country, the parameters were determined from S-θ pairs at field capacity and permanent wilting point only. These moisture limits are available for most soils and if unavailable can easily be interpolated from available data. Comparison of this '2-point' approach with a full data set showed that the moisture release curve could be predicted satisfactorily using the simplified approach. Comparison of measured and modelled soil water suctions, using the water model described above, are shown in Fig. 30.1. The N$_2$O emission model does not require the exact value of the soil water suction, only that the soil be assigned to one of three soil water suction classes ($S_1 < 55$ cm of water;

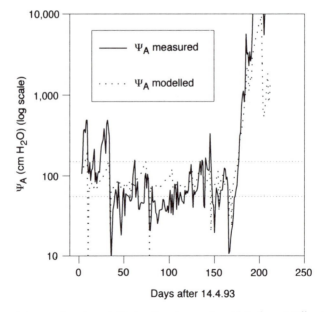

Fig. 30.1. Measured and modelled soil water suction (data from Müller, 1996).

S_2 = 55–150 cm of water and S_3 > 150 cm of water). From the model output it can be concluded that the soil water suction class can be assigned satisfactorily using the simple water model (data from Müller, 1996).

Results and Discussion

Initial testing of the GIS model outside the GIS environment

The original model (KNOM, Müller, 1996) was developed from data collected during a field experiment on urine-affected intensive grassland where N_2O emission as well as inorganic N, volumetric water content, soil temperature and meteorological data were obtained. Figure 30.2 shows the output obtained if the model, modified as described above for GIS implementation, is used to predict monthly N_2O emissions from urine-affected intensive grassland in Canterbury, New Zealand.

Although the model was unable to predict the exact amount of N_2O lost each month (Fig. 30.2), the overall trend in the N_2O emission as well as the total N_2O lost were represented well. In addition to the modelled N_2O output, all inputs to the model are presented in Fig. 30.2. This simple illustration shows that average monthly inputs appear to be sufficient to model total N_2O emission from the same period. The danger in using monthly rather than daily time steps lies in the reduction of extreme factor combinations which may lead to high N_2O emissions for only short periods but not longer, e.g. a month (Müller *et al.*, 1997). Müller (1996) showed that most of the N_2O over a year was emitted during a few days, when key factors reached a critical combination which prompted high N_2O fluxes. Extreme factor combinations are most likely eliminated by using monthly inputs. However, this does not seem to be a big problem, at least for the situation examined here, as demonstrated by the good agreement between measurements and model output achieved using monthly input values (Fig. 30.2). Indeed, in spite of this apparent contradiction, the total amount of N_2O predicted by the model over 8 months (3759 g N_2O-N ha^{-1}) is very close to the total measured N_2O emission during the same time period (3740 g N_2O-N ha^{-1}), which provides an additional indication that the model is in good agreement with measured data.

Initial testing of the GIS model within the GIS environment

A 'pilot' application of the full emission model has been implemented in a GIS using ARC/INFO's raster modelling environment ARC/GRID (Environmental Systems Research Institute). This 'pilot' implementation was carried out for a 10 × 10 km area centred on Lincoln University, Canterbury New Zealand. The maximum data resolution was 25 m.

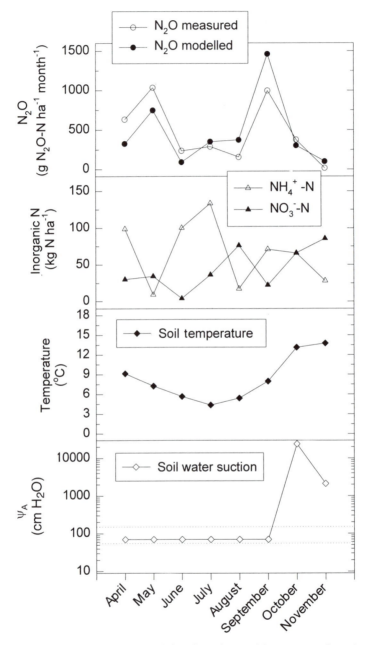

Fig. 30.2. Application of a GIS model to data obtained from urine-affected intensive grassland (Müller, 1996).

The emission values predicted are consistent with values expected for this region for that period of the year. This demonstrates that the algorithms used and the specialist GIS procedures employed can be combined successfully without the introduction of significant errors, either mathematical or procedural. What is required now is verification of the efficacy of the model for other climatic regions and soil types within New Zealand which differ significantly from those used for the original model development. This will require comparisons of GIS-predicted N_2O fluxes with other datasets of direct measurements. Such comparisons may identify unanticipated problems with the GIS implementation approach taken here or highlight possible limitations within the original model itself. For example, the original model was developed with data from a pasture system in which water-soluble metabolizable-C was shown to be non-limiting for the rate of denitrification (Müller, 1996). It must be acknowledged that this underpinning assumption may not necessarily hold for all soils throughout the country. In addition, the use of a 'look-up' table for the assignment of mineral-N inputs is a necessary expedient at this stage in the model's development and future work should attempt to provide a more robust basis for the values used. Nevertheless, the original N_2O emission model and its simplified implementation through the GIS, provides a usable, mechanistic framework for further development and for the testing of future scenarios: e.g. predicting a change in N_2O emission from a region where a shift in land use alters its 'ecosystem classification'. Similarly, it should be possible to predict the likely changes in N_2O emissions resulting from anticipated effects of climate change (Mullan, 1990).

Conclusion

The results obtained thus far for the application of a process-oriented N_2O emission model in GIS lead to the following conclusions:

- It is possible to implement simple mechanistic models in GIS to obtain spatial outputs for trace gas emission.
- GIS appears to be an ideal platform to hold spatial data sets which are needed in model calculations.
- More detailed spatial databases are needed of basic soil characteristics which can be used as inputs in mechanistic models.
- To further develop and validate the model, comprehensive datasets are needed which combine actual N_2O emissions with the most important factors and processes from various climatic regions and ecosystems.

Acknowledgements

The authors would like to thank The Crop and Food Crown Research Institute, Lincoln and MAF Policy, Wellington for their funding support. We also wish to thank Sheryl Hendrickson of the Lincoln Soil Quality Research Centre for her assistance.

References

Buchan, G.D. and Grewal, K.S. (1990) The power-function model for the soil moisture characteristic. *Journal of Soil Science* 41, 111–117.

Burford, J.R. and Bremner, J.M. (1975) Relationships between the denitrification capacities of soils and total, water soluble and readily decomposable soil organic matter. *Soil Biology and Biochemistry* 7, 389–394.

Carran, R.A., Willoughby, E.J., Giltrap, D.J. and Newsome, P.F.J. (1993) *Nitrous Oxide Inventory Information for New Zealand.* A report prepared for the Ministry of the Environment. Manaaki Whenua, Landcare Research New Zealand Ltd, 17 pp.

Granli, T. and Bøckman, O.C. (1994) Nitrous oxide from agriculture. *Norwegian Journal of Agricultural Science* (suppl. 12), 1–128.

Hutchinson, G.L. and Mosier, A.R. (1981) Improved soil cover method for field measurement of nitrous oxide fluxes. *Soil Science Society of America Journal* 45, 311–316.

IPCC (1996) *Nitrous Oxide and Carbon Dioxide in Agriculture OECD/IPCC/IEA Phase II Development of IPCC Guidelines for National Greenhouse Gas Inventory Methodology.* Workshop report December 4–6, 1995, 47 pp.

Klemedtsson, L., Svensson, B.H. and Rosswall, T. (1988) A method of selective inhibition to distinguish between nitrification and denitrification as sources of nitrous oxide in soil. *Biology and Fertility of Soils* 6, 112–119.

Marshall, T.J. and Holmes, J.W. (1988) *Soil Physics.* Cambridge University Press, Cambridge.

Matthews, E. (1993) Global geographical databases for modelling trace gas fluxes. *International Journal of Geographical Information Systems* 7, 125–142.

Matson, P.A. and Vitousek, P.M. (1990) Ecosystem approach to a global nitrous oxide budget. *BioScience* 9, 667–672.

Maynard, D.G. and Kalra, Y.P. (1993) Nitrate and exchangeable ammonium nitrogen. In: Carter, M.R. (ed.) *Soil Sampling and Methods of Analysis.* Lewis Publishers, Boca Raton, pp. 25–38.

Mosier, A.R. and Mack, L. (1980) Gas chromatographic system for precise, rapid analysis of nitrous oxide. *Soil Science Society of America Journal* 44, 1121–1123.

Mullan, A.B. (1990) Calculating the climate. In: *New Zealand Climate Change Report 1990. A Report Prepared by the New Zealand Climate Change Committee of the Royal Society of New Zealand.* Royal Society of New Zealand, pp. 44–52.

Müller, C. (1996) Nitrous oxide emission from intensive grassland in Canterbury, New Zealand. Edition Wissenschaft Rehe Biologie, Vol. 65. Tectum Verlag, Marburg.

Müller, C., Sherlock, R.R. and Williams, P.H. (1995) Direct field measurements of annual nitrous oxide emissions from urine-affected and urine-unaffected pasture in Canterbury. In: Currie, L.D. and Loganathan, P. (eds) *Fertilizer Requirements of Grazed Pasture and Field Crops: macro- and micro-nutrients. Occasional Report no. 8.* Fertilizer and Lime Research Centre, Massey University, Palmerston North, pp. 243–247.

Müller, C., Sherlock, R.R. and Williams, P.H. (1997) Mechanistic model for nitrous oxide emission via nitrification and dentrification. *Biology and Fertility of Soils* 24, 231–238.

Newsome, P.F.J. (1987) *The Vegetative Cover of New Zealand.* National Water and Soil Conservation Authority, Wellington, New Zealand.

Sherlock, R.R., Müller, C., Russell, J.M. and Haynes, R.J. (1992) *Inventory Information on Nitrous Oxide. Report prepared for the Ministry of the Environment.* Soil Science Department, Lincoln University, 72 pp.

Subak, S., Raskin, P. and Hippel, D. von (1993) National greenhouse gas accounts: current anthropogenic sources and sinks. *Climate Change* 25, 15–58.

Vereecken, H., Maes, J. and Feyen, J. (1990) Estimating unsaturated hydraulic conductvity from easily measured soil properties. *Soil Science* 149, 1–12.

Emissions of Nitric Oxide, Nitrous Oxide and Ammonia from Grasslands on a Global Scale

31

D.S. Lee[1], A.F. Bouwman[2], W.A.H. Asman[3], F.J. Dentener[4], K.W. van der Hoek[2] and J.G.J. Olivier[2]

[1]AEA Technology plc, National Environmental Technology Centre, Culham Laboratory, Oxfordshire OX14 3DB, UK; [2]National Institute of Public Health and the Environment (RIVM), PO Box 1,NL3720 BA Bilthoven, The Netherlands; [3]Assensvej 6, 4000 Roskilde, Denmark; [4]Department of Air Quality, Biotechnion, Bomenweg 2, Wageningen Agricultural University, 6700 EV Wageningen, The Netherlands

Summary. Quantification of both natural and man-made emissions of many trace gas species is needed for global modelling studies in the context of ecosystem modification and climate change. Emissions of nitric oxide (NO), nitrous oxide (N_2O) and ammonia (NH_3) represent the major sources of nitrogen (N)-containing gases to the atmosphere, and these gases participate either directly, or indirectly, in processes which result in both positive and negative radiative forcing. Emissions from natural grasslands, grasslands fertilized by animal excreta and biomass burning of savannas are considered. Estimates for emissions of NO, N_2O and NH_3 are given and shown on a $1° \times 1°$ grid. The calculated global emission of N from grasslands is 6, 2.5 and 24 Tg year^{-1} from NO, N_2O and NH_3, respectively. These emissions constitute 13, 18 and 40% of the estimated global source strength of these gases.

Introduction

Quantification of emissions of nitric oxide (NO), nitrous oxide (N_2O) and ammonia (NH_3) is required for input data in global and regional atmospheric chemistry/transport models. Both NO and NH_3 interact with acidifying pollutants and are important on global scales (Graedel *et al.*, 1995). Nitrous oxide is a 'greenhouse gas' in its own right and has been estimated to have a globally averaged radiative forcing of approximately 0.14 W m^{-2} (Schimel *et al.*, 1996). Emissions of nitrogen oxides, NO_x (NO_x = NO_2 + NO), are precursors of tropospheric ozone (O_3) which also causes radiative forcing. The globally averaged radiative forcing from tropospheric O_3 is more uncertain than that from N_2O, but has been estimated to be of the order of 0.4 ($\pm$0.2) W m^{-2} (Schimel *et al.*, 1996). Atmospheric NH_3 plays a role in the oxidation of $S^{(IV)}$ to $S^{(VI)}$ and is therefore of potential importance, as sulphate aerosols have been shown to have a net negative radiative forcing (Charlson *et al.*, 1990, 1991). Sulphate aerosol exerts a negative forcing both directly, i.e. by scattering and absorption of solar radiation, and indirectly, by altering the optical properties of clouds and their lifetimes. Neutralization of acid sulphate droplets has the effect of allowing further oxidation of sulphur dioxide (SO_2) to sulphate (SO_4^{2-}) by O_3 in solution (when hydrogen peroxide is limited) as this reaction is pH limited, and thus the presence of NH_3 may perturb the life cycle of SO_2, effectively increasing its atmospheric lifetime. There is also a limited amount of evidence that the acid sulphate aerosol, H_2SO_4, has different optical properties to that of its fully neutralized form, $(NH_4)_2SO_4$ (Horvath, 1992; Kiehl and Briegleb, 1993), although this is very uncertain.

Grasslands and their soils are an important source of NO, N_2O and NH_3 (Lee *et al.*, 1996, unpublished results; Dentener and Crutzen, 1994; Bouwman *et al.*, 1995). In this chapter, global estimates of emissions of these trace gases from grasslands are presented, the relative importance in global budgets quantified, and their emissions shown on a 1° × 1° grid.

Grasslands, Distribution, Importance and Definition of Emissions

For the purposes of quantifying emissions of N from grasslands, a definition is required. Three components have been identified for this definition: 'natural' undisturbed grasslands, 'fertilized' grasslands (i.e. grasslands which are used for agricultural grazing purposes) and the burning of savannas. The complex of grasslands used here is from Bouwman *et al.* (1995), who defined a grassland complex including 2390 Mha of mediterranean grazing areas, warm grass/shrub and cool grass/shrub complexes given in the land cover database of Olsen *et al.* (1983). In addition, the 757-Mha difference between 2195 Mha of total agricultural land and 1438 Mha of arable land (including fallow land) was assumed to be part of the grassland complex. This yielded a total area of 3147 Mha of grasslands, which is somewhat less than the global area of grasslands given by FAO (1991) for 1990, but correction was not considered feasible because of the uncertainties in estimates of grasslands (Bouwman, 1996).

Many estimates of the global magnitude of biomass burning have been made (Crutzen *et al.*, 1979; Seiler and Crutzen, 1980; Andreae, 1991), but these are based mostly upon estimates of burning frequency and locations made in the 1970s and 1980s. The spatial disaggregation of such emissions has been made using the tropical biomass inventory of Hao *et al.* (1990). However, it is known that the extent of natural and prescribed fire extends well into temperate zones (Levine, 1991). The biomass inventory is taken from Bouwman *et al.* (1995), which covers tropical and temperate zone burning of savannas, agricultural waste burning and deforestation.

The amount and distribution of tropical savanna burning, 2375 Tg dry matter (DM) year^{-1}, was taken from Hao and Liu (1994) and extended for Australia, using the inventory of Walker (1981) of 300 Tg DM year^{-1}. This was distributed on the basis of the relationship between structural vegetation types and vegetation types for Australia listed in Galbally *et al.* (1992), using the $1° \times 1°$ vegetation/land use land-use database of Matthews (1983) and is shown in Fig. 31.1.

Global Emissions of NO, N$_2$O and NH$_3$ from Grasslands

NO

Nitric oxide is produced from soils as a result of microbial denitrification and nitrification, and chemodenitrification, although the most important mechanism is considered to be nitrification (Bouwman, 1990). Many studies have been made of NO emissions from soil surfaces in various environments, using mostly chamber techniques. Such measurements show clearly variable but significant fluxes from various soil and ecosystem types.

In areas where O$_3$ is available, the conversion of NO to NO$_2$ takes place over a time scale of the order of minutes. The principal terrestrial sink for NO$_2$ is thought to be stomatal (Hargreaves *et al.*, 1992), so that some of the emissions from soils may not escape the canopy, especially within dense vegetation. In order to account for canopy deposition of NO$_2$, the global total is often reduced by an arbitrary factor so that fluxes to the boundary layer can be estimated (Müller, 1992; Lee *et al.*, 1996, unpublished results). Utilizing such a method, Lee *et al.* arrived at a global estimate of 4 Tg N year^{-1} from all soil types. However, this method can be improved upon by using some estimate of the upward flux reduction arising from in-canopy deposition. A simplification of the global land-use database of Matthews (1983) was made to encompass the following biome types: tropical rainforest, other forest, woodland, shrubland, grassland, tundra and cultivated land. For these categories, emission factors were reviewed from the literature (Lee *et al.*, 1996, unpublished results) which were mostly chamber studies, and applied to

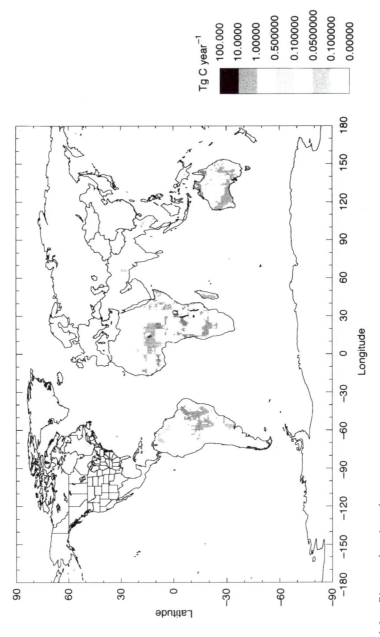

Fig. 31.1.　Biomass burning of savannas.

these biomes with an estimate of the potential canopy reduction by dry deposition. This resulted in a global emission of 6.9 Tg N year^{-1}.

Losses of NO from fertilized agricultural grasslands are important, and studies report a wide range of emission rates. The rate used here is 10 ng N m^{-2} s^{-1}, based upon a review of chamber measurements (Lee *et al.*, 1996, unpublished results). The area to which this emission rate was applied was the portions of cells allocated to agricultural land, which are not cropland according to Olsen *et al.* (1983): globally, this amounts to 760 Mha. This is assumed to be intensively used grassland, since it is not fallow land (which is accounted for in the 1430 Mha cropland). It is recognized that there will be some overestimation caused by not accounting for forests, trees, bush, waste land, infrastructure and farms/buildings, etc. The canopy reduction factor used was 20%, resulting in a global emission from intensively grazed grasslands of 2 Tg N year^{-1} (Fig. 31.2). The fertilization of these grasslands is assumed to be from animal wastes, rather than synthetic fertilizers, as a trivial amount of fertilizer is applied to such grasslands (Bouwman *et al.*, 1995), but there are no quantitative data on relationships between emissions of NO and fertilization from animal waste. However, it has been observed that fertilized grasslands emit more NO than natural grasslands.

The distribution of natural grasslands was assumed to be equal to the 2376 Mha of mediterranean grazing, cold and warm grass complexes from Olsen *et al.* (1983). These areas do not cover all ecosystems with grass in it: e.g. savannas, tundras and Siberian parks are not included. An emission rate of 1 ng m^{-2} s^{-1} was used from a review of chamber measurements and a canopy reduction factor of 20% used (Lee *et al.*, 1996, unpublished results). The estimated emission of NO originating from natural grasslands is 0.6 Tg N year^{-1}, and is shown in Fig. 31.3.

For savanna burning an emission factor for NO$_x$ of 1.2 g N kg^{-1} (Lee *et al.*, 1996) can be combined with the savanna burning inventory described above and results in a global emission of 3.2 Tg N year^{-1}.

N$_2$O

Nitrous oxide emissions may arise from animal wastes and synthetic fertilizers. The amount of synthetic N fertilizers applied to intensively grazed grasslands is very small on the global scale and results in an emission of 0.1 Tg N year^{-1} (Bouwman *et al.*, 1995). In grazing areas, N$_2$O formation is thought to occur primarily in urine patches by hydrolysis of urea to NH$_4^+$ and conversion to NO$_2^-$ and NO$_3^-$ by nitrifying bacteria, and subsequent denitrification to N$_2$O. Global animal population distributions were taken from Lerner *et al.* (1988) and Bouwman *et al.* (1995), and updated using FAO statistics for 1990. Annual N excretion rates were calculated for dairy cattle, non-dairy cattle, sheep, goats, pigs, buffalo, camels and poultry for developed and

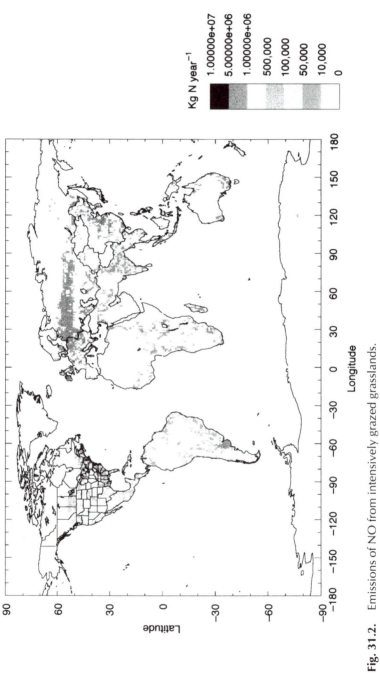

Fig. 31.2. Emissions of NO from intensively grazed grasslands.

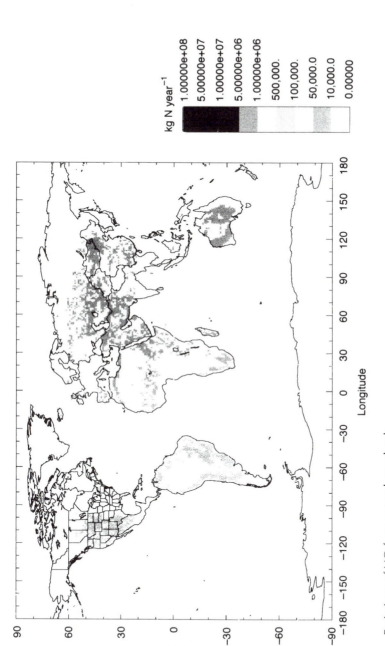

Fig. 31.3. Emissions of NO from natural grasslands.

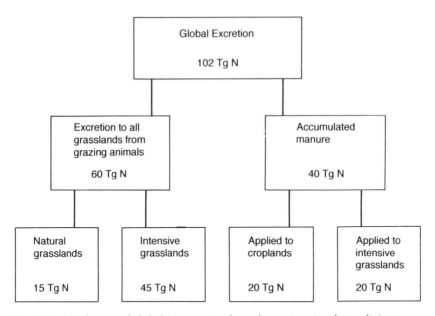

Fig. 31.4.   Pathways of global N excretion from domestic animal populations.

undeveloped agricultural systems (Bouwman *et al.*, 1995). Not all the animal waste from these populations is applied to grasslands, as some is applied to croplands. The global fate of N excretion from domestic animal populations is shown in Fig. 31.4. Based upon a global N excretion of about 100 Tg N year^{-1} from all domesticated animals and an emission factor for N$_2$O of 1%, the combined emission from these animal populations is 0.8 Tg N year^{-1} (Fig. 31.5). In addition to this term, soils emit naturally N$_2$O, and it is estimated that the 'background' emission from soil processes in intensively grazed grasslands is 0.4 Tg N year^{-1} (Bouwman *et al.*, 1995). Thus, the global emission of N$_2$O from intensively grazed grasslands is estimated to be 1.1 Tg N year^{-1}.

Natural soils are known to emit N$_2$O, and the emissions were calculated using the simple global model of Bouwman *et al.* (1993). This model describes the variation in space and time of the major factors that regulate N$_2$O production in soils. The basis of the model is the strong relationship between N$_2$O fluxes and the amount of N cycling through the soil–plant–microbial biomass system (Matson and Vitousek, 1987). The model includes indices for five regulators of N$_2$O production. The first index ranks the organic matter input into the soil. The second index, soil fertility, indicates quality and N content of the organic matter. Soil moisture and temperature rank the decomposition, mineralization and nitrification rates. The fifth index, soil

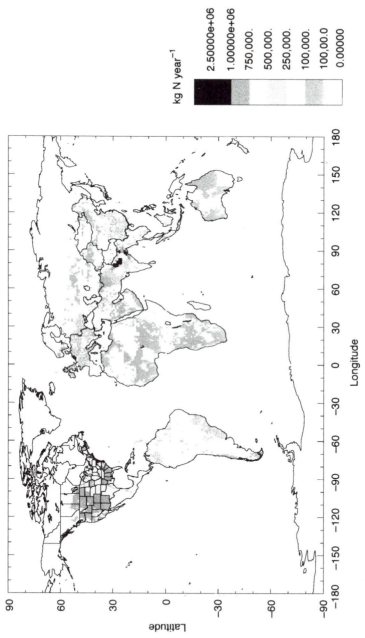

Fig. 31.5. Emissions of N_2O from intensively grazed grasslands.

oxygen status, describes conditions for denitrification. The potential for N_2O production, derived from these five indices, was compared with measurements reported in the literature. The resulting regression equation ($r^2 = 0.6$) was used to calculate emissions at a $1° \times 1°$ resolution. The resulting N_2O emission for grasslands is 1.0 Tg N year^{-1}, added to which is 0.15 Tg N year^{-1} from excretion to natural grasslands giving a total of 1.2 Tg N year^{-1} (Fig. 31.6).

For savanna burning, the N : C ratio was assumed to be 0.006 (Crutzen and Andreae, 1990) and an emission factor for N_2O of 7 g N kg^{-1} was used (Bouwman *et al.*, 1995). This resulted in a global emission of 0.05 Tg N year^{-1} from savanna burning. In addition to the emission of N_2O caused by savanna burning, burning is known to promote fluxes from the soil (Anderson *et al.*, 1988). However, the source strength of this has not been calculated because of the paucity of data.

NH₃

Emissions of NH_3 from fertilized grasslands are principally the result of microbial decomposition of urea in animal waste to NH_4^+ and NH_3 in solution. Various factors including temperature, the physicochemical status of the soil and wind speed affect the amount of NH_3 volatilized. Animal waste may be either dropped in the field during grazing, or in more intensive agricultural systems, emissions arise from housing units, waste storage units and from field spreading of accumulated waste. Emissions of NH_3 have been calculated on a global basis for animal populations described in the inventory for N_2O, taken from Lerner *et al.* (1988) and Bouwman *et al.* (1995), and updated using FAO statistics for 1990. Emission factors for each animal type were calculated for developed and non-developed agricultural systems (Bouwman *et al.*, 1995). The combined emission from these animal populations is 21.6 Tg year^{-1} (Fig. 31.7). Not all of the emissions from domestic animal populations originate from grasslands, as some of the waste is applied to croplands. It is estimated that the amount originating from an intensively grazed 'grassland landscape' is approximately 17 Tg N year^{-1}. This includes the natural 'background' emission from these areas. The amount of synthetic fertilizer aplied to grasslands is small on a global basis and results in an estimated emission of 0.6 Tg N year^{-1}.

There is a paucity of measurements of NH_3 emission from natural soils. Previously, attempts to quantify this source were made by extrapolation of flux measurements (Schlesinger and Hartley, 1992), or by allocating emission factors to continents and using the 'compensation point' concept to calculate emissions using a global atmospheric transport chemistry model (Dentener and Crutzen, 1994). Schlesinger and Hartley (1992) used emission factors which were clearly inappropriate to 'natural' soils, and Dentener and Crutzen

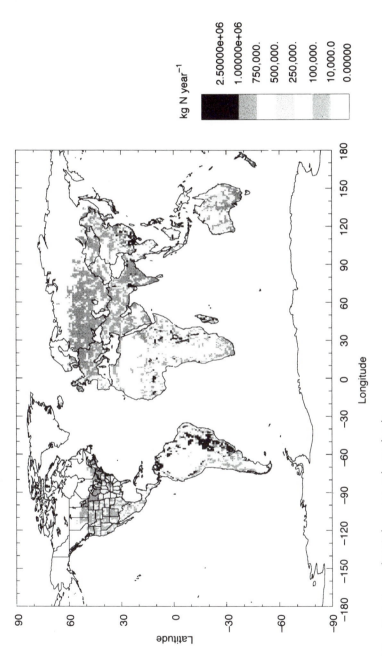

Fig. 31.6.   Emissions of N₂O from natural grasslands.

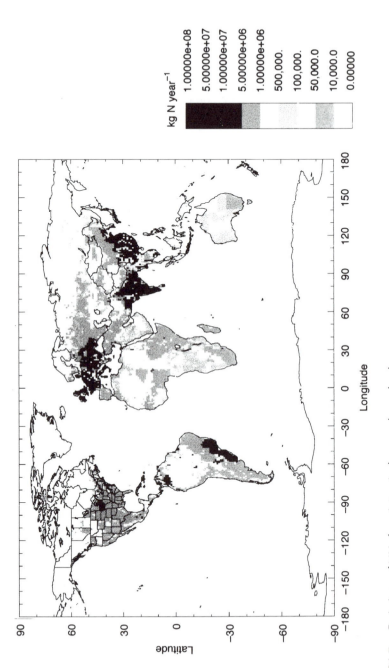

Fig. 31.7. Emissions of NH$_3$ from intensively grazed grasslands.

(1994) pointed out that their own calculations were 'highly speculative'. Unfortunately, the emission term calculated by both Schlesinger and Hartley (1992) and Dentener and Crutzen (1994) showed that this source was important, and thus cannot be ignored.

A method devised by Bouwman *et al.* (1996, unpublished results) is presented here as an alternative. The global net primary production (NPP) distribution of Klein-Goldewijk *et al.* (1994) was used, assuming that this includes both above-ground and below-ground NPP. The net N mineralization was calculated using the C : N ratios given by Melillo *et al.* (1993) in combination with NPP estimates of Klein-Goldewijk *et al.* (1994). This resulted in a net global N mineralization of 1160 Tg N year^{-1}, which is consistent with the estimate of Nadelhoffer *et al.* (1991). It was assumed that the N cycled in the top 0–10 cm of the soil could emit NH_3, and that 50% of the N in NPP was mineralized and is present in the form of NH_4^+. It was assumed that < 4% of the N mineralization escapes as NH_3 to the atmosphere from the surface 0–10 cm (Bouwman *et al.*, 1995). Thus, the potential NH_3 volatilization to the air, as an upper limit, is 20.2 Tg NH_3-N year^{-1} of the total 1010 Tg year^{-1} net N mineralization. The NH_3 emitted from the soil surface may be absorbed by leaves within the plant canopy (Denmead *et al.*, 1976). To account for this, canopy absorption coefficients are used which range from 0.8 for tropical rainforest, 0.5 for other forests and woodlands and 0.2 for all other vegetation types, including tundras, grasslands, shrublands and arable land.

The amount of NH_3 escaping to the atmosphere is thus reduced to 9.7 Tg N year^{-1}. The assumed canopy reduction coefficients may not be correct, particularly in temperate deciduous and tropical seasonal forests, where part of the NH_3 volatilizes from the soil in autumn and spring, during which periods canopy reduction is minimal. However, the other uncertainties are such that it would be fallacious to attempt to refine this component of the calculation. The component of these natural soil emissions from the global 3140 Mha of grasslands is calculated to be 3.3 Tg N year^{-1} (Fig. 31.8). In addition to this, Bouwman *et al.* (1996, unpublished results) have calculated a contribution of 0.1 Tg N year^{-1} from wild animal excreta.

Emissions of NH_3 from savanna burning are calculated on the basis of the biomass inventory described above. Emissions have been found to vary by flame type and vegetation type. However, no consistent emission factors have emerged, thus the estimate of emissions from biomass burning must be considered very uncertain. The emission factor used was 1.5 g NH_3-N kg^{-1} dry matter (Bouwman *et al.*, unpublished results), which is consistent with figures reported by Andreae (1991) and Andreae *et al.* (1997). In the calculations presented here, it has been assumed that all vegetation types contain 45% by dry mass of C. This results in a global emission from savanna burning of 1.8 Tg N year^{-1}.

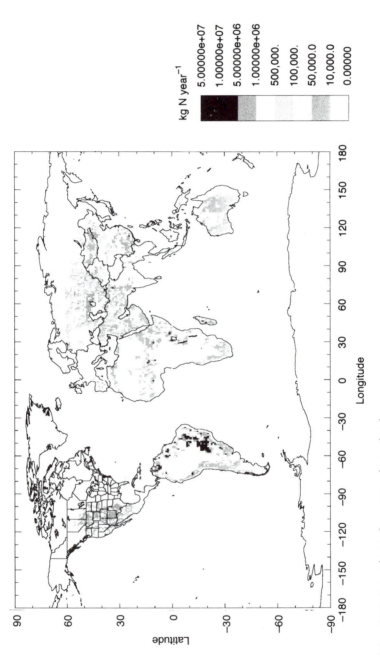

Fig. 31.8. Emissions of NH₃ from natural grasslands.

Discussion

In common with most emission calculations, the data presented here are, at best, estimates and rely upon quantification of the 'activity' statistics (e.g. grassland extent, animal populations) and the emission factor for the activity and particular trace gas. In many cases, statistics on activities are better quantified than the emission factor, and uncertainties arising from the activity are often easier to estimate. Uncertainties are not considered here but are presented elsewhere for all sources of NO, N_2O and NH_3 (Lee *et al.*, 1996, unpublished results; Bouwman *et al.*, 1995).

Global emissions from all sources of NO, N_2O and NH_3 are presented in Table 31.1, to put into context the emissions from intensively grazed grasslands, natural grasslands and savanna burning, and are quantified in Table 31.2. Of the global emissions of NO, grasslands represent a relatively small but significant source. It is of note that the global source strength of 6.9 Tg N year^{-1} from soils presented here is not in agreement with that of IPCC (1995), which suggests an emission term of 12 Tg N year^{-1}. Yienger and

Table 31.1. Global emissions of NO, N_2O and NH_3 from all sources (Tg N year^{-1}).

Sources	NO	N_2O	NH_3
Fossil fuel combustion	22	0.3	0.1
Other industrial and combustion sources	0.85 (aircraft)	0.5 (adipic acid production)	0.2 (fertilizer, NH_3 and HNO_3 production)
Animal excreta (fertilized grasslands)	—	1.0	21.6
Biomass burning			
Savanna	3.2	0.05	1.8
Deforestation	2.5	0.07	1.5
Agricultural waste	1.5	0.03	0.5
Post-clearing/burning-enhanced flux	1.7	0.4	—
Biofuel combustion	—	0.07	2.2
Humans	—	—	2.7
N-fertilizer usage	—	1.0	9.0
Croplands	—	0.9	3.6
Soils under natural vegetation incl. grasslands	6.9	5.7	9.7
Lightning	5	—	—
Oceans	—	3.6	8.2
Other sources	0.9 (NH_3 oxidation)	0.3–1.2	1.6 (other animals)
Stratospheric flux to troposphere	0.6	—	—
Total	45	14	61

Table 31.2. Global emissions of N from grasslands from NO, N_2O and NH_3 (Tg N year^{-1}).

Emission source	NO	N_2O	NH_3	Totals
Background emissions from intensively used grasslands	2.0	0.4	—	2.4
Animal wastes	—	0.7*	16.9[†]	17.6
Synthetic fertilizer	—	0.1	0.6	0.7
Savanna burning	3.2	0.1	1.8	5.1
Natural grasslands	0.6	1.2[‡]	4.9[§]	6.7
Total N emission	6	2.5	24	32.5
Grassland contribution to total N emissions (%)	13	18	40	27

*Excluding 0.2 Tg N year^{-1} from manure application to croplands.
[†]Excluding 3.2 Tg N year^{-1} from manure applied to croplands, and including 'background' emission from decomposition of grass.
[‡]Including 1.0 Tg N year^{-1} 'background' emission and 0.2 Tg N year^{-1} from 15 Tg N year^{-1} excretion in natural grasslands and emissions from wild animals.
[§]Including 3.3 Tg N year^{-1} soil emission from decomposition of grass, 1.5 Tg N year^{-1} from ~15 Tg N year^{-1} excretion in natural grasslands based on average emission coefficient of 10%, and the contribution from other ('wild') animals of 0.1 Tg N year^{-1}.

Levy (1995) recently presented an emission calculation suggesting a global emission term of 5.5 Tg N year^{-1} (range = 3.3–7.7 Tg N year^{-1}), which is consistent with the calculations presented here. For N_2O, grasslands represent a more signficant source, approximately 17% of the global total. However, NH_3 emissions from grasslands represent a significant fraction, 36%, of the global emission term. This is the result of the large emissions of NH_3 from animal waste.

Emissions from livestock production systems contribute about 40% to the global emission of NH_3, and 8% to the global budget of N_2O. As the atmospheric residence time of N_2O is very long (> 100 year), expression of the contribution of animal production systems to the atmospheric *increase* of perhaps 50% (Bouwman *et al.*, 1995) is more meaningful. No quantitative relationships between NO emission and grassland fertilization from livestock production exist. In future, the emission of N gases associated with animal production may increase considerably, particularly in developing countries. The population of non-dairy cattle and associated waste production and emissions may increase by some 40% between now and 2025 in developing countries including China (Bouwman, 1996). In addition, it is envisaged that the use of feedstuffs to support animal production will have to increase considerably. At present, 40% of global cereal production and 25% of global production of starchy foods is fed to animals and a major part of oilseed production is also used to produce animal feed (Bouwman, 1996). Indirect effects caused by feedstuff production result in the increased use of synthetic fertilizers and forest clearing for expansion of grasslands and croplands. A shift towards less meat and milk consumption in rich countries may help to reduce emissions. However, as a result of the fast population-driven growth in demand for livestock products in developing countries, livestock production will continue to be an important source of acidifying and radiatively active trace gases in the coming decades.

Acknowledgements

We would like to thank the following institutes for financial suppport of this work: the Hadley Centre for Climate Prediction and Research and the National Institute of Public Health and the Environment (Project No. 481508).

References

Anderson, I.C., Levine, J.S., Poth, M.A. and Riggan, P.J. (1988) Enhanced biogenic emissions of nitric oxide and nitrous oxide following surface biomass burning. *Journal of Geophysical Research* 93, 3893–3898.'

Andreae, M.O. (1991) Biomass burning: its history, use and distribution and its impact on environmental quality and global climate. In: Levine, J.S. (ed.) *Global Biomass Burning: atmospheric, climatic, and biospheric implications.* MIT Press, Cambridge, MA, pp. 3–21.

Andreae, M.O., Atlas, E., Cachier, H., Cofer III, W.R., Harris, G.W., Helas, G., Koppmann, R., Lacaux, J.P. and Ward, J.E. (1997) Trace gas and aerosol emissions from savanna fires. In: Levine, J.S. (ed.) *Biomass Burning and Global Change.* MIT Press, Cambridge, Massachusetts, pp. 278–295.

Bouwman, A.F. (1990) Exchange of greenhouse gases between terrestrial ecosytems and the atmosphere. In: Bouwman, A.F. (ed.) *Soils and the Greenhouse Effect.* John Wiley, Chichester, pp. 62–127.

Bouwman, A.F. (1996) *Long-term Scenarios of Livestock–Crop–Land Use Interactions for the Assessment of Environmental Indicators in Developing Countries. Land and Water Bulletins.* Food and Agriculture Organization of the United Nations, Rome (in press).

Bouwman, A.F., Fung, I., Matthews, E. and John, J. (1993) Global analysis for the potential for N_2O production in natural soils. *Global Biogeochemical Cycles* 7, 557–597.

Bouwman, A.F., van der Hoek, K. and Olivier, J.J.G. (1995) Uncertainties in the global source distribution of nitrous oxide. *Journal of Geophysical Research*, 100, 2785–2800.

Charlson, R.J., Langner, J. and Rodhe, H. (1990) Sulphate aerosol and climate. *Nature* 348, 22.

Charlson, R.J., Langner, J., Rodhe, H., Leovy, C.B. and Warren S.G. (1991) Perturbation of the Northern Hemisphere radiative balance by backscattering from anthropogenic sulphate aerosols. *Tellus* 43A, 152–163.

Crutzen, P.J. and Andreae, M.O. (1990) Biomass burning in the tropics: impact on atmospheric chemistry and biogeochemical cycles. *Science* 250, 1669–1678.

Crutzen, P.J., Heidt, L.E., Krasnec, J.P., Pollock, W.H. and Seiler, W. (1979) Biomass burning as a source of atmospheric gases CO, H_2, N_2O, NO, CH_3Cl and COS. *Nature* 282, 253–256.

Denmead, O.T., Freney, J.R. and Simpson, J.R. (1976) A closed ammonia cycle within a plant canopy. *Soil Biology and Biochemistry* 8, 161–164.

Dentener, F.D. and Crutzen, P.J. (1994) A three-dimensional model of the global ammonia cycle. *Journal of Atmospheric Chemistry* 19, 331–369.

FAO (1991) *Agrostat PC, Computerized Information Series 1/3: land use.* FAO Publications Division, Food and Agriculture Organization of the United Nations, Rome.

Galbally, I.E., Fraser, P.J., Meyer, C.P. and Griffith, D.W.T. (1992) Biosphere–atmosphere exchange of trace gases over Australia. In: Gifford, R.M. and Barson, M.M. (eds), *Australia's Renewable Resources: sustainability and global change.* BRR Proceedings No. 14, Bureau of Rural Resources and CSIRO Division of Plant Industry, pp. 118–148.

Graedel, T.E., Benkovitz, C.M., Keene, W.C., Lee, D.S. and Marland, G. (1995) Global emissions inventories of acid-related compounds. *Water, Air and Soil Pollution* 85, 25–36.

Hao, W.M., Liu, M.-H. and Crutzen, P.J. (1990) Estimates of annual and regional releases of CO_2 and other trace gases to the atmosphere from fires in the tropics, based on the FAO statistics for the period 1975–1980. In: Goldammer, J.G. (ed.) *Fire in the Tropical Biota, Ecol. Stud.*, vol. 84, Springer-Verlag, New York, pp. 440–462.

Hao, W.M. and Liu, M.-H. (1994) Spatial and temporal distribution of tropical biomass burning. *Global Biogeochemical Cycles* 8, 495–503.

Hargreaves, K.J., Fowler, D., Storeton-West, R.L. and Duyzer, J.H. (1992) The exchange of nitric oxide, nitrogen dioxide and ozone between pasture and the atmosphere. *Environmental Pollution* 75, 53–59.

Horvath, H. (1992) Effects on visibility, weather and climate. In: Radojecic, M. and Harrison, R.M. (eds) *Atmospheric Acidity Sources, Consequences and Abatement.* Elsevier Applied Science, London, pp. 435–466.

IPCC (1995) *Climate Change 1994. Radiative Forcing of Climate Change and An Evaluation of the IPCC IS92 Emission Scenarios,* Cambridge University Press, Cambridge.

Kiehl, J.T. and Briegleb, B.P. (1993) The relative roles of sulfate aerosol and greenhouse gases in climate forcing. *Science* 260, 311–314.

Klein-Goldewijk, J.G., Van Minnen, J.J.G., Kreileman, M., Vloedbeld, M. and Leemans, R. (1994) Simulating the carbon flux between the terrestrial environment and the atmosphere. *Water, Air and Soil Pollution* 76, 199–230.

Lerner, J., Matthews, E. and Fung, I. (1988) Methane emissions from animals: a global high resolution database. *Global Biogeochemical Cycles* 2, 139–156.

Levine, J.S. (ed.) (1991) *Global Biomass Burning: Atmospheric, Climatic, and Biospheric Implications.* MIT Press, Cambridge, Massachusetts.

Matson, P.A. and Vitousek, P.M. (1987) Cross-ecosystem comparisons of soil nitrogen and nitrous oxide flux in tropical ecosystems. *Global Biogeochemical Cycles* 1, 163–170.

Matthews, E. (1983) Global vegetation and land use: new high resolution data bases for climate studies. *Journal of Climate and Applied Meteorology* 22, 474–487.

Melillo, J.M., McGuire, A.D., Kicklighter, D.W., Moore III, B., Vorosmarty, C.J. and Schloss, A.L. (1993) Global climate change and terrestrial net primary production. *Nature* 363, 234–240.

Müller, J.F. (1992) Geographical distribution and seasonal variation of surface emissions and deposition velocities of atmospheric trace gases. *Journal of Geophysical Research* 97, 3787–3804.

Nadelhoffer, K.J., Giblin, E.A., Shaver, G.R. and Linkins, A.E. (1991) Microbial processes and plant nutrient availability in arctic soils. In: Chapin, F.S. *et al.* (eds) *Arctic Ecosystems in a Changing Climate.* Academic, San Diego, California, pp. 281–300.

Olsen, J.S., Watts, J.A. and Allison, L.J. (1983) *Carbon in Live Vegetation of Major World Ecosystems.* ORNL 5862. Environmental Sciences Division, Publication No. 1997. Oak Ridge National Laboratory, Oak Ridge, Tennessee, National Technical Information Service. US Dept. Commerce.

Schimel, D., Alves, D., Enting, I., Heimann, M., Joos, F., Raynaud, D., Wigley, T., Prather, M., Derwent, R., Ehhalt, D., Fraser, P., Sanhueza, E., Zhou, X., Jonas, P., Charlson, R., Rodhe, H., Sadasivan, S., Shine, K.P., Fouquart, Y., Ramaswamy, V., Solomon, S., Srinivasan, J., Albritten, D., Isaksen, I., Lal, M. and Wuebbles, D. (1996) Radiative forcing of climate change. In: Houghton, J.T., Meiro Filho, L.G., Callander, B.A., Harris, N., Kattenberg, A. and Maskell, K. (eds) *Climate Change 1995: The Science of Climate Change. Contribution of Working Group I to the Second Assessment Report of the Intergovernmental Panel on Climate Change.* Cambridge University Press.

Schlesinger, W.H. and Hartley, A.E. (1992) A global budget for atmospheric NH_3. *Biogeochemistry* 15, 191–211.

Seiler, W. and Crutzen, P.J. (1980) Estimates of gross and net fluxes of carbon between the biosphere and the atmosphere from biomass burning. *Climatic Change* 2, 207–247.

Walker, J. (1981) Fuel dynamics in Australian vegetation. In: Gill, A.M., Groves, R.H. and Noble, J.R. (eds) *Fire and the Australian Biota.* Australian Academy of Science, Canberra.

Yienger, J.J. and Levy, H. (1995) Empirical model of global soil-biogenic NO_x emissions. *Journal of Geophysical Research* 100, 11447–11464.

Balancing Greenhouse Gases and Reduction of Emission in Grassland Farming Systems

32

G. Kazenwadel, J. Zeddies and K. Löthe

Institute of Farm Management, University of Hohenheim, D-70593 Stuttgart, Germany

Summary. One cause of the greenhouse effect is emission of the greenhouse gases carbon dioxide (CO_2), methane (CH_4) and nitrous oxide (N_2O) into the atmosphere. Particularly with CH_4 and N_2O pollution, agriculture is responsible for a significant share of the emissions. Agriculture is unequivocally the main source of ammonia (NH_3) emission. This paper focuses on costs of reducing nitrogen (N) leaching, NH_3 and greenhouse gas emissions. The results supply decision aids as to which technical strategies are economically and ecologically efficient and what relationships exist between them. Furthermore, the results can be used as a basis for assessing compensation payments for ecological performances in reducing greenhouse gases. The relationships between land use and environment are shown and optimized with the help of linear programming. Based on extensive data collection from research, the calculations are used to examine possibilities in existing farms. Different strategies to reduce NH_3 and greenhouse gases are analysed.

Several technical possibilities to reduce emission of greenhouse gases from grassland farming systems are shown. Measures to reduce the intensity of grassland use and to increase milk yields are particularly efficient and are, to some extent, profitable at the farm level. Under the defined conditions, covering manure stores, use of manure additives, manure injection and the installation of biogas plants are not economic at the farm scale. In order to introduce these measures, it would be necessary either to establish energy taxes or to grant subsidies for protection measures and for the use of renewable resources. However, agriculture could substantially contribute to reductions in environmental stress, particularly with respect to the greenhouse effect.

Introduction

Emissions of carbon dioxide (CO_2) and methane (CH_4) are major contributors to the greenhouse effect and, with nitrous oxide (N_2O), are related to the intensification of food production. Agriculture is at the same time both a victim and a causal agent of greenhouse effects. That part of the total emissions of these gases resulting from agriculture in Germany is shown in Table 32.1.

Table 32.1. Contribution to greenhouse gas emission by agriculture in Germany. (Source: Umweltbundesamt, 1992.)

	NH_3 (1000 t)	CO_2 (10^6 t)	CH_4 (1000 t)	N_2O (1000 t)
Emissions from agriculture	384	17	1400	59.1
Total emissions	402	705	4700	164
Proportion from agriculture	96%	2.4%	30%	36%

Methods

Comparative balancing-system delineation

The limitation of a reference system should follow criteria that allow a clear distinction between those processes included in the balancing and those not to be considered. In this system, the complete N cycle and the production and consumption of intermediate goods (e.g. fertilizers and pesticides) are examined. Production of capital items (machines and buildings) is not included. For the N balance, the nitrate (NO_3^-) in the soil is considered as an emission when it is leached from the top soil. The reference system for this study is an ecosystem undisturbed by man. It is naturally conditioned and the sinks and sources of the different gases are almost in balance. Therefore, the system emits the same amount of greenhouse gases that the vegetation of the system has already removed from the atmosphere.

The feed of a dairy cow removes CO_2 from the atmosphere and part of this is emitted as CH_4 from the rumen. Because of the higher greenhouse gas potential of CH_4 compared to CO_2, this process causes a net increase in greenhouse effects. Thus these emissions are aggregated with the global warming potential (GWP) to CO_2 equivalents, which also includes consideration of their atmospheric life span (Table 32.2). In this study, a period of 100 years has been considered.

Table 32.2. Lifetime and greenhouse warming potential (GWP) of different gases. (Source: Deutscher Bundestag, 1990, 1994.)

	CO_2	CH_4	N_2O
Average lifetime (years)	120	10.5	132
GWP (100 years)	1	21.0	290

The linear programming farm models

The basic data for the models have been gathered in a survey covering 131 dairy farms in an intensively farmed grassland region in Southern Germany (Allgäu). In this sample, different farming systems have been identified. The model used is a static linear programming model with a time span of 1 year, but the dynamic aspects of grassland growth, lactation, feeding and manure use have been integrated by dividing the year into 37 smaller periods. A detailed description of the models can be found in Trunk (1995). When considering environmental effects, the models calculate NH_3 losses from livestock (housing, manure storage and spreading), the amount of NO_3^- leached and the N_2O emissions. Ammonia losses from manure depend on the weather conditions at spreading and the technique (Horlacher and Marschner, 1990). The N cycle is characterized in the models by the whole-farm N emission. The major source of CH_4 is the rumen fermentation in cattle and the amounts emitted depend on the diets of the animals (Kirchgessner *et al.*, 1991). In addition, manure storage causes CH_4 emission, because organic matter in the manure is fermented when the waste is stored anaerobically (Heyer, 1994). Carbon dioxide emissions from fossil energy, concentrates and mineral fertilizer are also considered. The time interval for the measurement of the output of greenhouse gases is 1 year.

The following strategies are analysed with the models: (i) reduction of NH_3 emissions by manure treatment such as covering liquid manure stores, use of liquid manure additives, injection of liquid manure; (ii) reduction of N_2O emissions by less intensive grassland use; (iii) increased milk yield in order to reduce the emission of greenhouse gases per unit of production; and (iv) biogas production in order to reduce CH_4 emissions from manure stores and to substitute for fossil energy.

Results

Manure treatment

The covering of manure stores costs, for different grassland farming systems, approximately 3 German marks per kg reduction in NH_3 emission, taking into account the lower N fertilizer input (Trunk, 1995).

Due to the high protein contents in the basic ration, the manure of grassland dairy farms contains much ammonium. Additives are available to reduce the NH_3 emissions. Table 32.3 shows that the use of additives causes a decrease in the gross margin of about 2% and reduces the amount of mineral N-fertilizer from 70 to 50 kg ha^{-1}. The NH_3 emissions decrease by 25%. The influence of additive use on greenhouse gases is marginal: the reduction of mineral fertilizer reduces the CO_2 emission caused by fertilizer production.

Injection of manure is shown to be a very efficient measure to reduce NH_3 emissions, but, compared to the reference farm, the gross margin decreases by more than 12,000 German marks $year^{-1}$. If the opportunity costs for labour and reduced machinery needs are considered, this decreases to about 7000 German marks $year^{-1}$. Manure injection allows a reduction of mineral fertilizer input to 25 kg ha^{-1}. The NH_3 losses are reduced considerably and the farm inputs and outputs become better balanced. The CO_2 emissions from fertilizer production are reduced, but this effect is over-compensated for by the higher input of fossil energy, so the total greenhouse gas output increases slightly.

Table 32.3. Effects of manure additives and liquid manure injection (stanchion barn, 300 t milk quota). (Source: Trunk and Zeddies, 1996.)

	Number of farm model		
	1	2	3
Manure treatment	Standard	Additives	Injection
Grassland (ha)	34	34	34
Gross margin (DM)	153,576	150,494	141,421
Labour (h $year^{-1}$)	5046	5040	4778
Cows (no.)	50	50	50
N-fertilizer bought (kg ha^{-1})	70	54	25
NH_3-N loss during spreading (kg)	1805	1264	524
Sum of NH_3-N loss (kg LU^{-1})	23.4	17.4	9.3
NH_3-N loss (g kg^{-1} milk)	4.7	3.5	1.9
NO_3-N leaching (kg ha^{-1})	38.4	38.4	38.8
N_2O-N emission (kg ha^{-1})	5.5	5.5	5.6
Sum of N emissions (kg)	3619	3079	2355
CH_4 emission (kg)	9843	9842	9847
CH_4 emission (g kg^{-1} milk)	21.7	21.7	21.7
CO_2 emission from fossil energy (kg)	37,837	37,757	42,187
CO_2 emission from concentrate production (kg)	16,649	16,649	16,742
CO_2 emission from fertilizer production (kg)	6171	4773	2185
CO_2 equivalent (100 years) (kg)	329,144	327,708	330,817
CO_2 equivalent (100 years) (kg kg^{-1} milk)	0.73	0.72	0.73

Reduction of intensity in grassland use

One possibility to reduce greenhouse gas emissions in agriculture is to reduce the intensity of grassland use. In Table 32.4, a 34-ha dairy farm with a milk quota of 300 t, 50 cows and silage feeding reduces the intensity significantly by using less N fertilizer. In farm model 5 with intermediate intensity, the gross margin is decreased by only 1000 German marks year^{-1}. Young stock numbers are reduced and any feed deficit is compensated for by concentrates. Altogether the reduction in intensity has an influence on emissions. Ammonia per Livestock Unit (LU) and per kg milk are decreased by about 10%, because the protein content of the grass is lower and the cows excrete less urea. Nitrate leaching and N_2O emission decrease more than proportionally. The total N emission decreases by one-third. Methane emission is

Table 32.4. Reduction of intensity in grassland use (cubicle house, 300 t milk quota). (Source: Trunk and Zeddies, 1996.)

	Number of farm model			
	4	5	6	7
Intensity of grassland use	High	Intermediate	Low	Low
Grassland (ha)	34	34	34	44
Gross margin	153,259	152,289	148,100	156,657
Labour (h year^{-1})	3985	3736	3431	3905
Cows (no.)	50	50	50	50
Young stock (no.)	42	37	27	42
N fertilizer bought (kg ha^{-1})	85	42	0	0
Net-yield grassland (t dry matter)	94	83	75	72
Concentrate bought (t)	56.5	70.9	76.4	62
Protein surplus in ration (%)	19	9	4	5
Nitrogen segregation per cow (kg N cow^{-1})	112	100	94	95
Sum of NH_3-N loss (kg LU^{-1})	29.5	26.8	24.2	22.1
NH_3-N loss (g kg^{-1} milk)	5.9	5.4	4.9	4.4
NO_3-N leaching (kg ha^{-1})	38.7	11.1	1.8	1.8
N_2O-N emission (kg ha^{-1})	5.6	3.3	1.9	1.9
Sum of N emissions (kg)	4193	2815	2066	2171
CH_4 emission (kg)	9846	8777	7980	9280
CH_4 emission (g kg^{-1} milk)	21.7	20.3	20.3	20.7
CO_2 emission from fossil energy (kg)	37,946	30,822	34,923	31,901
CO_2 emission from concentrate production (kg)	19,499	20,684	22,331	18,159
CO_2 emission from fertilizer production (kg)	7525	3676	0	0
CO_2 equivalent (100 years) (kg)	331,113	269,040	235,317	260,395
CO_2 equivalent (100 years) (kg kg^{-1} milk)	0.73	0.62	0.59	0.58

reduced, because the share of concentrates in the diet is higher, and the emission caused by digestion of concentrates is lower than the emissions from roughage. The sum of the greenhouse gases is reduced by 18% LU^{-1} and by 15% kg^{-1} milk.

A further reduction in the intensity (farm model 6) of management decreases the gross margin to a greater extent (5000 German marks $year^{-1}$). The number of young stock is further reduced and the diminished roughage production requires more concentrates to be purchased. Ammonia emissions, N_2O emissions and NO_3^- leaching are sharply decreased. Compared with the highly intensive management, the N emissions are reduced to one-half. Only CH_4 emission kg^{-1} milk remains constant because the roughage includes more crude fibre.

In model 7, the farm has 44 ha of grassland instead of 34 ha. As a result the farm can work at the lowest intensity and the gross margin is 3000 German marks higher than it is in the highly intensive variant with 34 ha. The environmental effects are again reduced.

Increase in milk yield

When the milk yield rises from 5000 to 6000 l per cow, the gross margin increases by 4500 German marks and the NH_3 emission decreases by 13% and the CH_4 emissions by 9% (Table 32.5). The CO_2 equivalent output of the whole farm is reduced by 10%. Because with an increasing milk yield the share of the food needed for maintenance is reduced, an increase in milk yield has positive effects on the income and on the environmental effects up to certain level. Above this level, the gross margin can only be increased if the resources, i.e. labour and housing, can be used for other purposes or if additional milk quota can be bought or leased in but the environmental effects continue to be positive. If this is not possible the farm will no longer use all its grassland.

Biogas production in dairy farms

With a biogas fermenter, 30–40% of the organic matter in the manure can be transformed into biogas to be used for production of heat or electricity. Biogas fermenters are important in two respects: first, to prevent CH_4 emission from the manure store and, second, the production of biogas can substitute for fossil energy. Table 32.6 shows the effect of biogas production in a dairy farm. In model 12, an on-farm fermenter is compared with a 'normal' system (11) and with the use of a central fermenter (model 13). Both variants decrease greenhouse gas emission significantly.

Table 32.5.   Effects of different milk yield levels (cubicle house, 272 t milk quota).
(Source: Trunk and Zeddies, 1996.)

	Number of farm model		
	8	9	10
Milk yield	5000	6000	7000
Grassland (ha)	34	34	32
Gross margin (DM)	140,038	144,551	143,080
Labour (h year^{-1})	3948	3564	3073
Cows (no.)	55	46	39
N-fertilizer bought (kg ha^{-1})	33	40	35
NH$_3$-N loss during spreading (kg)	1805	1264	524
Sum of NH$_3$-N loss (kg LU^{-1})	24.8	26.2	27
NH$_3$-N loss (g kg^{-1} milk)	6.1	5.3	4.7
NO$_3$-N leaching (kg ha^{-1})	32.8	33.0	32.0
N$_2$O-N emission (kg ha^{-1})	3.9	3.9	3.6
Sum of N emissions (kg)	3552	3419	3074
CH$_4$ emission (kg)	8849	8583	7706
CH$_4$ emission (g kg^{-1} milk)	23.1	21.0	19.1
CO$_2$ emission from fossil energy (kg)	31,779	29,194	25,420
CO$_2$ emission from concentrate production (kg)	15,976	14,003	15,249
CO$_2$ emission from fertilizer production (kg)	2923	3564	2925
CO$_2$ equivalent (100 years) (kg)	275,221	266,570	240,080
CO$_2$ equivalent (100 years) (kg kg^{-1} milk)	0.72	0.65	0.59

Conclusions

There is a wide range of technical measures available to reduce NO$_3^-$ leaching as well as the emission of greenhouse gases from grassland farming systems by up to 40% over the short and medium term. Measures to reduce the intensity of grassland use and to increase milk yields are particularly efficient and to some extent profitable at the farm level. Under the stated conditions, covering manure stores, use of manure additives, manure injection and the installation of biogas fermenters are not economic farm options. In order to introduce these measures, it would be necessary either to establish energy taxes or to grant subsidies for protection measures and for the use of renewable resources. Such subsidies could be part of other agri-environmental programmes. Suitable changes in agricultural management could contribute substantially to reductions in environmental stress, particularly with regard to greenhouse effects.

Table 32.6. Biogas production in dairy farms (cubicle house, 272 t milk quota). (Source: Trunk and Zeddies, 1996.)

	Number of farm model		
	11	12	13
Biogas fermenter	Without	On farm	Central
Investment subsidy (%)		30	30
Gas production organic matter (l kg^{-1})		220	300
Use of heat energy (%)		50	80
Gross margin (DM)	144,551	138,532	140,347
Labour (h)	3564	3554	3728
N fertilizer bought (kg ha^{-1})	40	26	26
NH$_3$-N loss (g kg^{-1} milk)	5.3	4.9	4.9
NO$_3$-N leaching (kg N ha^{-1})	33	32	32
N$_2$O emission (kg ha^{-1})	3.9	3.8	3.8
N emissions total (kg)	3419	3258	3259
CH$_4$ emission manure storage (kg)	1574	0	0
CH$_4$ emission total (kg)	8583	6998	6996
CH$_4$ emission (g kg^{-1} milk)	21.0	17.2	17.2
CO$_2$ emission from fossil energy (kg)	29,194	29,048	30,733
CO$_2$ emission from concentrate production (kg)	14,003	14,309	14,259
CO$_2$ emission from fertilizer production (kg)	3564	2308	2305
Electricity from biogas (kWh)	0	45,848	62,645
Heating energy (l oil equivalent)	0	2709	5923
CO$_2$ equivalent (100 years) total (kg)	266,572	266,225	208,356
CO$_2$ equivalent (100 years) (kg kg^{-1} milk)	0.65	0.55	0.51

References

Deutscher Bundestag (1990) *Schutz der Erde. Eine Bestandsaufnahme mit Vorschlägen zu einer neuen Energiepolitik.* Band 1. Bonn.

Deutscher Bundestag (1994) *Schutz der grünen Erde. Klimaschutz durch umweltgerechte Landwirtschaft und Erhalt der Wälder.* Enquete-Komission 'Schutz der Erdatmosphäre' des Deutschen Bundestages, Bonn.

Heyer, J. (1994) Methan. In: Deutscher Bundestag (ed.) *Studienprogramm Landwirtschaft,* Band 1, Teilband 1. Bonn.

Horlacher, D. and Marschner, H. (1990) Schätzrahmen zur Beurteilung von Ammoniakverlusten nach Ausbringung von Rinderflüssigmist. *Zeitschrift für Pflanzenernährtung und Bodenkunde* 153, 107–115.

Kirchgessner, M., Windisch, W., Müller, H.L. and Kreuzer, M. (1991) Release of methane and of carbon dioxide by dairy cattle. *Agribiological Research* 44, 2–3, 92–103.

Trunk, W. (1995) Ökonomische Beurteilung von Strategien zur Vermeidung von Schadgasemissionen bei der Milcherzeugung – dargestellt für Allgäuer

Futterbaubetriebe. Dissertation Hohenheim. Schriftenreihe 'Studien zur Agrarökologie', Band 15.

Trunk, W. and Zeddies, J. (1996) Ökonomische Beurteilung von Strategien zur Vermeidung von Schadgasemissionen bei der Milcherzeugung. *Agrarwirtschaft* 45(2), pp. 111–120.

Umweltbundesamt (1992) *Daten zur Umwelt* 1991/1992. Berlin.

Reducing Nitrogen Emissions from Complete Dairy Farm Systems

33

S. Peel[1], B.J. Chambers[2], R. Harrison[3] and S.C. Jarvis[4]

[1]ADAS Bridgets, Martyr Worthy, Winchester, Hampshire SO21 1AP, UK; [2]ADAS Gleadthorpe, The Grange, Meden Vale, Mansfield, Nottinghamshire NG20 9PD, UK; [3]ADAS Boxworth, Boxworth, Cambridge CB3 8NN, UK; [4]IGER, North Wyke, Okehampton, Devon EX17 2SB, UK

Summary. In the UK, dairy farms are seen as potentially major sources of nitrogen (N) emissions. Typically, total N inputs are 300–400 kg ha^{-1} compared with offtakes of 60–80 kg ha^{-1}. The MIDaS (Minimal Impact Dairy Systems) project, at ADAS Bridgets, was set up to develop systems incorporating a package of measures to reduce emissions to an acceptable level whilst remaining economically viable. System 1 represents commercial practice, system 2 is partially improved with no manure spreading in September–November and system 3 is fully improved, no manure spreading in September–January and injection of slurry. In 1994/95, leaching losses from grassland were greatly reduced on the improved systems. Ammonia (NH$_3$) losses have been measured over 7-day periods following slurry applications. Losses as % of NH$_4^+$-N applied have averaged 25% when broadcast, 18% when shallow injected, and 5% when rapidly ploughed in before maize. When expressed for whole systems, NH$_3$ losses from manures represent only 12 kg N ha^{-1} on System 1 and 3 kg ha^{-1} on System 3. Limited measurements of denitrification have been made. Conceptual models have been constructed for each system. These predict leaching losses reasonably well, but tended to overestimate NH$_3$ losses. Two years of milk production have now been completed; results so far indicate that the partially improved system may achieve acceptable levels of leaching from grassland, if not from maize. Financial margins per hectare may be reduced by perhaps 10% compared to current commercial practice. The reduction will be greater if substantial extra manure storage is constructed.

Introduction

In the UK, dairy farms are seen as potentially major sources of nitrogen (N) emissions. For example, Pain *et al.* (1996) estimate ammonia (NH_3) loss from dairy excreta at 62 kt year^{-1}. Fertilizer inputs on dairy farms, based on a random sample of sites, are shown in Table 33.1.

The economic optimum N input for grass on most soils in the UK is *c.* 300 kg ha^{-1} (MAFF, 1994) and much grassland on specialist dairy farms receives this amount. Little, if any, allowance is made in fertilizer policies for the N value of organic manures. Maize is still a relatively small crop in the UK (105,000 ha compared with grass at 6.7 million ha) but is increasing rapidly. It often receives heavy dressings of manures in addition to fertilizer-N. Manure stores are not covered, and manure is often applied in the autumn and early winter, invariably by broadcasting. There are voluntary codes (MAFF, 1992) but no legislation governing these practices in the UK.

Table 33.1. Fertilizer use on dairy farms, England and Wales, 1995*.

	Fertilizer applied (kg ha^{-1})			Area receiving organic manure (%)
	N	P	K	
Grassland	189	11	38	62
Forage maize	86	29	47	83

*1265 grass fields all on dairy farms, 40 maize fields mainly on dairy farms (Burnhill *et al.*, 1996).

Methods

The MIDaS Project: objective, structure and management

The MIDaS (Minimal Impact Dairy Systems) project was set up in 1994 to develop systems incorporating a package of measures to reduce emissions to an acceptable level whilst remaining economically viable. The project is based at ADAS Bridgets in Southern England, which is predominantly on free draining shallow soils overlying chalk where the water table depth is *c.* 25 m. Long-term mean excess winter rainfall is 260 mm, and thus to reduce nitrate (NO_3^-) levels in water to < 50 mg l^{-1}, leaching must be < 30 kg N ha^{-1}. Meeting this target is a high priority. The structure of the project is shown in Table 33.2.

No replication is possible because of the large land area required. Each system has its own dedicated land area, cows, silos and slurry facilities. The land use and managements are shown in Table 33.3. No manures are applied from September to November on system 2 and from September to January

Table 33.2. MIDaS: Minimal Impact Dairy Systems study.

Three self-contained systems, each with 36 cows, yield > 6000 litres cow^{-1}:
  System 1 – Control. Good commercial practice, high output
  System 2 – Reduced loss, high output
  System 3 – Minimal loss, reduced intensity

Table 33.3. MIDaS Land use (ha) and management.

	System 1	System 2	System 3
Land use			
Permanent grass	13	13	17
Italian rye-grass	6	—	—
Maize	—	6	6
Total area	19	19	23
Management			
Slurry application			
Grass	Broadcast	Broadcast (diluted)	Injected (open slit)
Maize	—	Rapid incorporation	Rapid incorporation
Slurry storage	1 month	3 months	5 months
Fertilizer			
Grass	Full economic optimum	Tactical reduction	Planned reduction
Maize	—	Total N supply 180 kg ha^{-1}	Total N supply 150 kg ha^{-1}
Feed	Least cost min. 18% CP. Normal P levels	No surplus ERDP Moderate P levels	No surplus ERDP Low P

ERDP = effective rumen degradable protein.

inclusive on system 3. Maize is cropped on the same areas continuously but is undersown with an Italian rye-grass cover crop which is destroyed in February prior to application of manure in late March, which is ploughed in within 30–60 min. On systems 2 and 3, diets are formulated using the UK Metabolisable Protein system (AFRC, 1993) and any surplus Effective Rumen Degradable Protein (ERDP) is minimized. Studies of N balance in the dairy cow have been made in parallel with the MIDaS project (Metcalf *et al.*, 1996).

Modelling the fate of nitrogen

Jarvis (1993) described N flows on a typical UK dairy farm using experimental data, model calculations and best estimates. Using the same approach, but including recent data on leaching from maize (Chalmers, 1996, personal communication), model calculations of each of the three MIDaS systems are shown in Table 33.4.

Table 33.4. MIDaS fate of nitrogen (kg ha^{-1}): modelling. (Compiled by S.C. Jarvis and A. Davies, Institute of Grassland and Environmental Research, based on Jarvis, 1993.)

	System 1	System 2	System 3
Surplus (external inputs minus output)	320	190	121
Losses			
Leaching	67	47	25
Denitrification			
Slurry spreading	11	12	10
Grazing + fertilizer	20	12	5
Volatilization			
Slurry spreading	22	18	11
Grazing + fertilizer	7	5	1
Buildings/stores	48	21	5
Total losses	175	115	57
Immobilized in soil	75	75	75
Unaccounted for	70	0	−11

Results and Discussion

Modelled outputs

Model calculations were based on the planned inputs and outputs of each system. In the base system 1, each of the three major pathways of loss was predicted to be substantial (including denitrification, even on this well-drained soil) and to be reduced on systems 2 and 3. Because of a shortage of UK data, volatilization from buildings and stores was calculated as the difference between N excreted and that contained in typical slurry. The values for this component of systems 2 and 3 were therefore questionable. The estimate of N immobilized in soil organic matter was based on a long-term experiment on a similar soil type (Tyson *et al.*, 1990). Clearly, the calculations do not balance and there was a substantial amount unaccounted for in system 1, but in systems 2 and 3 there may be substantial overaccounting if actual estimates of loss from buildings were higher.

Actual mineral balance

The balance of external N inputs and outputs is shown in Table 33.5. Inputs of N, and hence surpluses, were higher than those planned, and greater than those used in modelling, for all systems. This was partly from extra feed purchased due to a shortage of forage, and partly from extra fertilizer due to insufficient manures being available in the introductory winter before the systems were fully established.

Leaching

Nitrate leaching was measured using a total of 150 porous ceramic cups on each system. In each field unit, 30 cups were installed, arranged in three lines of 10, with the cups 2 m apart. Table 33.6 shows that in both years, leaching from grassland was substantially less in systems 2 and 3 than in system 1. Losses from maize were relatively high, particularly in the first year; for the second year fertilizer input was reduced. Whole system leaching losses met the target of 30 kg N ha^{-1}, but there was great variation between fields. For example on system 3 in 1994/95 the loss ranged between 2 and 30 kg N ha^{-1} on two grazed fields.

Table 33.5. Overall N balance sheet (kg ha^{-1}).

	System 1	System 2	System 3
Inputs			
Fertilizer	325	190	146
Feed	117	116	90
Atmosphere (estimated)	30	30	30
Total	472	336	266
Outputs			
Milk	66	64	54
Liveweight*	6	7	4
Total	72	71	58
Surplus	400	265	208

*N content of calves, and of cull cows, purchased cows and heifers, assumed to be 2.53%.

Table 33.6. Nitrogen leached (kg N ha^{-1}).

	1994/95	1995/96*
System 1		
Permanent grass	71	64
Italian rye-grass†	40	40
Whole system	61	56
System 2		
Permanent grass	16	27
Maize	50	40
Whole system	27	31
System 3		
Permanent grass	11	13
Maize	79	36
Whole system	29	19

*Provisional data.
†Estimates based on soil mineral N measurements.

Table 33.7.   MIDaS NH₃ loss following slurry application, 1994/95: mean results.

	Broadcast	Injected	Rapid incorporation
No. of measurement periods	5	2	1
Application rate (m^3 ha^{-1})	27.2	39.0	47.0
Dry matter (%)	7.4	10.0	16.0
NH_3-N loss (kg N ha^{-1})	17.1	13.7	3.8
NH_3-N loss as % NH_4^+-N applied	26	17	5

Field losses of ammonia

The mass balance micrometeorological technique, with rotating passive samplers (Leuning *et al.*, 1985), was used over 7-day periods following manure applications to measure NH_3-N emissions. Results are shown in Table 33.7. Losses from broadcast applications, at 26% of the ammonium-N (NH_4^+-N) applied, were much lower than the 50% predicted from the relationship established by Smith and Chambers (1995). This relationship is based largely on IGER and ADAS data obtained using the wind-tunnel technique which protects the sward from rainfall and is operated at a constant wind speed. The loss following injection of slurry in February and March was less than from broadcasting, but was higher than expected, possibly due to the application rate being set relatively high. This was to reduce the need to make applications in summer when conditions at ADAS Bridgets are often very dry and there is a risk of poor efficiency of utilization and sward damage. Nevertheless, based on these measurements, whole-system estimates of annual NH_3-N loss following slurry applications were low: 16, 10 and 3 kg ha^{-1} on systems 1, 2 and 3, respectively.

Nitrous oxide (N_2O)

A robust field method of measuring N_2O fluxes has recently been developed, (De Klein *et al.*, 1996) and limited measurements on the MIDaS systems have been made. These show that fluxes peaked immediately after slurry applications, and further peaks occurred during or immediately following rainfall events. Rates of emission were doubled when slurry was applied on top of dung pats, and was higher following injection compared with broadcast slurry (Harrison and De Klein, 1996, personal communication).

Conclusions

- Model estimates of N cycling on UK dairy farms suggest that losses via all major pathways can be reduced by a package of practical measures.

There were imbalances in the model calculations and more information, e.g. on soil organic N immobilization, is required to enable more reliable balances to be obtained.

- The MIDaS systems study so far demonstrates that N leaching from grass can be greatly reduced, but is very variable between fields. Nitrogen leaching from maize is relatively high even using a cover crop. Ammonia losses from broadcast slurry have been substantially less than predicted, and hence the benefit of injection may be less than anticipated.

- It is too early to draw conclusions, but it seems possible that large quantities of N will remain unaccounted for. This will raise the need for re-targeting the measurement programme to include more effort on denitrification, NH_3 loss from buildings and stores, N fluxes to and from soil organic matter, and a better understanding of the manipulation of nutrient flow through the dairy cow.

Acknowledgements

The MIDaS Project is funded through LINK by the following: ADAS, BOCM Pauls, Hydro Agri UK, Borregaard LignoTech UK, Linbury Trust, Livestock Systems UK, Maize Growers Association, Ministry of Agriculture, Fisheries and Food, Trident Feeds and Wessex Water.

References

AFRC (Agriculture and Food Research Council) (1993) *Energy and Protein Requirements of Ruminants*. CAB International, Wallingford.

Burnhill, P.M., Chalmers, A.G. and Fairgrieve, J. (1996) *The British Survey of Fertiliser Practice: Fertiliser Use on Farm Crops 1995*. HMSO, Edinburgh.

De Klein, C.A.M., Harrison, R. and Lord, E.I. (1996) Comparison of N_2O flux measurements using gas chromatography and photo-acoustic infra-red spectroscopy. In: van Cleemput, O. *et al.* (eds) *Progress in Nitrogen Cycling*. Kluwer Academic Publishers, Dordrecht, pp. 533–536.

Jarvis, S.C. (1993) Nitrogen cycling and losses from dairy farms. *Soil Use and Management,* 9, 99–105.

Leuning, R., Freney, J.R., Denmead, O.T. and Simpson, J.R. (1985) A sampler for measuring atmospheric ammonia flux. *Atmospheric Environment,* 19, 1117–1124.

MAFF (Ministry of Agriculture, Fisheries and Food) (1992) *Code of Good Agricultural Practice for the Protection of Air*. MAFF, London.

MAFF (Ministry of Agriculture, Fisheries and Food) (1994) *Fertiliser Recommendations* (RB209). HMSO, London.

Metcalf, J.A., Mansbridge, R.J. and Blake, J.S. (1996) Potential for increasing the efficiency of nitrogen and phosphorus use in lactating dairy cows. *Animal Science* 62, 636.

Pain, B.F., Van der Weerden, T.J., Chambers, B.J., Phillips, V.R. and Jarvis, S.C. (1997) A new inventory for ammonia emissions from UK agriculture. *Atmospheric Environment* (in press).

Smith, K.A. and Chambers, B.J. (1995) Muck from waste to resource-utilization: the impacts and implications. *Agricultural Engineer,* Autumn, 33–38.

Tyson, K.C., Roberts, D.H., Clement, C.R. and Garwood, E.A. (1990) Comparison of crop yields and soil conditions during 30 years under annual tillage or grazed pasture. *Journal of Agricultural Science, Cambridge* 115, 29–40.

Nitrogen Balances in Permanent Grassland

34

A. Sapek

Institute for Land Reclamation and Grassland Farming at Falenty, 05-090 Raszyn, Poland

Summary. The aim of this investigation was to recognize the potential nitrogen (N) losses on permanent grasslands where the water supply was adequate to meet herbage needs under Polish conditions. Different levels of N fertilizer rates have been taken into account. The method of investigation was based on determination of N balances. The results showed that most of the N applied in fertilizers could be accounted for in the grass sward. However, at rates of fertilizer greater than 120 kg N ha^{-1}, there were significant concentrations of nitrates (NO$_3$) in leachates. Simulated estimates of denitrification losses showed that these were also significant and increased proportionally with N input. The N balance approach provided a useful mechanism for forecasting N losses and threats to water quality.

Methods

A field experiment was established on a permanent meadow in 1987. It was situated at the Falenty Experimental Station, located near Warsaw. The soil was loamy, coarse sand with a organic carbon content of 2.0%. Reseeding of the grass sward was undertaken in 1994. Rates of nitrogen (N) fertilizer as ammonium nitrate and equal to 120, 240, and 360 kg N ha^{-1} were applied each year to the 80 m^2 plots in four replications. Three cuts were harvested at the beginning of the experiment and four cuts have been taken since 1991. No legumes were observed in the sward during this time. Each plot was equipped with a 9 m^2 fibre glass drain installed in the soil at a depth of 70–80 cm. This device allowed collection of the percolating water. From

©CAB INTERNATIONAL 1997. *Gaseous Nitrogen Emissions from Grasslands*
(eds S.C. Jarvis and B.F. Pain)

1988, the meadow was irrigated by a sprinkler system to ensure an adequate water supply for the growing plants. Equipment for collecting the precipitation water was set up near to the experimental site.

Nitrate (NO_3^-) and ammonium (NH_4^+) contents were determined in samples of percolated water, in precipitation as well as in soil samples taken up to 1 m depth, in spring before the fertilizer-N dressings and after each cut. The N balance for the experiment included measurement of N inputs with the commercial fertilizer, in irrigation water and precipitation, and of the outputs in harvested crops and the NO_3^- leached. The denitrification rate was not measured but the N losses due to denitrification were simulated by means of the CREAMS model (Version 1.8/PC, 1985: see Knisel, 1980). Together, the precipitation and the applied sprinkler irrigation produced from 45 to 200 mm of water per year percolating to below the depth of 70 cm in the soil profile (Table 34.1). The N inputs with precipitation were between 13 and 25 kg N ha^{-1} year^{-1} (Tables 34.2–34.4). The mean N concentration in rain water was c. 3 mg N dm^{-3}. More than 50% of the N in precipitation was in the form of NH_4^+ salts. The total sprinkler irrigation ranged between 120 and 240 mm per year and was applied in doses of between 20 and 30 mm. The mean concentration of N in irrigation water was c. 5 mg N dm^{-3}.

The N uptake with crops from plots fertilized with 120 kg N ha^{-1} was usually greater than that applied in fertilizer, but usually lower than total inputs when N in precipitation and irrigation water was considered (Table 34.2). The N uptake from 240 kg N ha^{-1} plots was usually lower than that applied in fertilizer (Table 34.3) and that from the 360 kg N ha^{-1} plots was always lower than the amount of N applied with fertilizer (Table 34.4).

The measured amounts of NO_3^- leached were proportional to the N applied in commercial fertilizers in each treatment. The mean values of N leached during 9 years of observation were equivalent to c. 10% of the N applied in fertilizers. Similar relationships were found between the fertilizer rates and NO_3^- concentrations in leachate. The NO_3^--N concentrations

Table 34.1. Water balance.

Year	Precipitation (mm)	Irrigation (mm)	Percolation (mm)
1987	495.0	0	128
1988	494.7	140	151
1989	468.5	140	99
1990	451.8	130	85
1991	459.6	260	126
1992	445.3	275	203
1993	451.8	150	45
1994	628.6	160	197
1995	545.1	155	125

in leachate from plots fertilized with 120 kg N ha^{-1} were close to the Polish (10 mg NO$_3$-N dm^{-3}) and the European Union standards for drinking water (11.3 mg NO$_3$-N dm^{-3}). Those concentrations from plots fertilized with

Table 34.2. Nitrogen balance on treatments fertilized with 120 kg N ha^{-1}.

Balance factors	kg N ha^{-1}								
	1987	1988	1989	1990	1991	1992	1993	1994	1995
Fertilizer N	120	120	120	120	120	120	120	100	120
Rain N	17.3	25.7	20.1	17.8	14.5	20.1	13.2	16.3	14.7
Irrigation	0.0	6.2	5.9	5.7	12.8	13.9	6.7	7.8	7.3
Total inputs	137.3	151.9	146.0	143.5	147.2	153.9	139.9	114.1	142.0
N uptake	111.8	145.9	132.8	128.6	122.2	111.6	128.8	73.4	130.0
N leached	25.7	18.1	12.1	5.5	8.7	13.8	2.7	16.3	15.6
Denitrification	16.7	18.6	18.2	9.0	11.0	16.3	4.0	25.4	16.5
Total outputs	154.2	182.6	163.0	143.0	141.9	141.7	135.5	115.1	162.2
Inputs − outputs	−16.9	−30.7	−17.0	0.5	5.4	12.3	4.4	−1.0	−20.1
Soil NO$_3$-N difference	20.5	11.9	4.3	2.4	2.3	0.8	−2.2	−4.8	8.0
Total difference	3.6	−18.8	−12.7	2.9	7.7	13.1	−2.2	−4.8	8.0
Mean NO$_3$-N concentration (mg dm^{-3} in leachate)	20.1	12.0	12.1	6.5	6.9	6.8	6.1	8.3	12.5

Table 34.3. Nitrogen balance on treatments fertilized with 240 kg N ha^{-1}.

Balance factors	kg N ha^{-1}								
	1987	1988	1989	1990	1991	1992	1993	1994	1995
Fertilizer N	240	240	240	240	240	240	240	200	240
Rain N	17.3	25.7	20.1	17.8	14.5	20.1	13.2	16.3	14.7
Irrigation	0.0	6.2	5.9	5.7	12.8	13.9	6.7	7.8	7.3
Total inputs	257.3	271.9	266.0	263.5	267.2	274.0	259.9	174.1	262.0
N uptake	199.0	246.3	237.6	223.4	228.8	218.0	231.1	98.8	225.3
N leached	34.1	29.8	19.8	10.7	19.3	23.8	8.1	29.8	33.8
Denitrification	23.7	31.2	28.7	19.6	24.0	25.3	13.3	43.2	36.1
Total outputs	256.8	307.3	286.0	253.6	272.0	267.1	252.5	171.8	295.3
Inputs − outputs	0.6	−35.3	−20.0	9.9	−4.8	6.9	7.4	2.3	−33.3
Soil NO$_3$-N difference	6.2	14.2	7.3	−5.8	11.4	5.6	9.6	−21.3	34.5
Total difference	6.8	−21.1	−12.7	4.1	6.6	12.5	17.0	−19.0	1.2
Mean NO$_3$-N concentration (mg dm^{-3} in leachate)	26.8	19.9	21.2	13.1	15.4	11.9	10.6	15.3	27.1

Table 34.4. Nitrogen balance on treatments fertilized with 360 kg N ha^{-1}.

Balance factors	kg N ha^{-1}								
	1987	1988	1989	1990	1991	1992	1993	1994	1995
Fertilizer N	360	360	360	360	360	360	360	300	360
Rain N	17.3	25.7	20.1	17.8	14.5	20.1	13.2	16.3	14.7
Irrigation	0.0	6.2	5.9	5.7	12.8	13.9	6.7	7.8	7.3
Total inputs	377.3	391.9	386.0	383.5	387.2	394.0	379.9	234.1	382.0
N uptake	276.6	340.7	315.7	327.9	306.2	320.5	342.0	146.3	312.7
N leached	40.4	47.4	31.7	21.7	34.0	44.9	6.5	29.6	47.5
Denitrification	29.3	48.8	47.7	34.7	40.9	44.2	10.2	46.3	50.6
Total outputs	346.4	436.9	395.1	384.3	381.1	409.6	358.7	222.1	410.9
Inputs – outputs	30.9	–45.0	–9.1	–0.8	6.1	–15.6	21.2	12.0	–28.9
Soil NO$_3$-N difference	–20.6	21.3	–2.6	4.9	1.0	26.1	3.3	5.8	–2.8
Total difference	10.3	–23.7	–11.7	4.1	7.1	10.5	24.5	17.8	–31.7
Mean NO$_3$-N concentration (mg dm^{-3} in leachate)	31.9	31.4	35.3	25.7	27.3	22.4	14.5	15.4	39.9

higher N rates were obviously higher than these standards, occasionally exceeding 50 mg N dm^{-3}. The mean NO$_3$ concentrations in leachate on the same treatment differed subsequently between the different years of the experiment, and were always higher in the samples taken during the autumn.

The simulated N losses due to denitrification were higher than the measured losses due to NO$_3$ leaching, but were also proportional to the N inputs. The model also simulates the denitrification process on the basis of the NO$_3$ and carbon content of the soil during the time when the soil water content was above field capacity. The other major factor controlling the denitrification is the soil temperature. The simulated data should be regarded as a trend rather than providing exact quantification of this form of loss. In spite of this, the simulated denitrification losses of N are realistic and fit sensibly into the calculated N balance.

The N balances show that the differences between inputs and outputs were generally smaller than 20% for all three fertilizer rates. These differences were much smaller when the differences between the NO$_3$ content in soil before the beginning of the vegetative season each year were also taken into account. The greatest differences were observed in the year after changing the management system, i.e. introducing the sprinkler irrigation in 1988 and reseeding the plots in 1994. Such changes would have influenced the N mobilization and/or immobilization processes, which were not considered in the present calculations.

Discussion

The experiment described does not conform with the real situation on productive Polish grasslands in many circumstances. The water supply is not sufficient for continuous growth on many meadows. Also, the application of N with commercial fertilizer varies from 0 to 120 kg N ha^{-1}. Higher rates are seldom used. There are only a few Polish farms where grassland covers 100% of farm area, which means that the animal manure is mostly used on arable land and not on grassland. The productive dairy farms are often located on shallow organic soils drained only a few decades ago and where some accelerated mineralization of organic matter is observed and the production of herbage dry matter is enhanced by N from this irreversible supply. Even when the water supply is not adequate, high rates of N fertilizer are often not needed to obtain good yields.

The results obtained and the calculated balances show that even with satisfactory water supply, grasses are not able to remove all the NO_3^- from soil and some N losses due to NO_3^- leaching and denitrification should be expected. The N leaching from loamy coarse sand soil used in the present experiment was high enough even from the lowest N rate to pose a genuine threat to groundwater with regard to the drinking water standards.

The method of N balance on grassland, taking account of inputs with fertilizers and removal in crops, is a useful tool in forecasting the N losses and the threat to water quality. The N inputs with precipitation and different irrigation water are significant only on grassland with no or only low rates of fertilizer, which is a frequent case observed in current Polish agriculture.

Conclusion

1. The N balance on irrigated grassland shows that more than 90% of the N from fertilizers was taken up by growing plants.

2. In the case of N rates greater than 120 kg N ha^{-1}, the concentration of NO_3^- in leachate water can be higher than the Polish or European Union standards for drinking water.

3. The simulation of N losses due to denitrification by the CREAMS model provided a fair estimate of the effects of fertilizer addition.

Reference

Knisel, W.G. (ed.) (1980) *CREAMS, A Field Scale Model for Chemicals, Runoff, and Erosion from Agricultural Management Systems.* US Department of Agriculture, Science and Education Administration, Conservative Research Report 26.

Environmentally Responsible Management of Grassland in Organic Farming Systems

35

C.A. Watson[1] and L. Philips[2]

[1]Land Resources Department, SAC, Craibstone Estate, Aberdeen AB21 9TQ, UK; [2]Elm Farm Research Centre, Hamstead Marshall, Newbury, Berkshire RG20 0HR, UK

Summary. One of the major challenges facing the agricultural industry is to maintain acceptable levels of productivity whilst minimizing losses to the environment, either in gaseous forms or by leaching. Organic farming is a production system which offers the opportunity to meet both production and environmental objectives. Organic farming systems in the UK are typically based on ley/arable rotations, where up to 70% of the farmed area is under grass/clover ley. Research has demonstrated that losses to the environment can be manipulated by altering management practices. This paper addresses key issues in the management of N losses from organic systems.

Introduction

Increased concern over the environmental, economic and social impacts of agriculture have led to a search for farming systems which sustain food production and the rural environment with minimal demand on non-renewable resources. Maintaining plant nutrient supply whilst minimizing dependence on agrochemicals demands a shift away from chemical solutions, to a greater understanding of, and reliance on, management of biologically regulated processes. The enhancement of biological cycles involving microorganisms, soil fauna, plants and animals is the basis of organic agriculture (Lampkin, 1990). Organic production systems are defined by the EC Regulation 2092/91 (EC, 1991) and in the UK are administered by the United Kingdom Register of Organic Food Standards (UKROFS, 1991). Since 1994 the UK Government has supported those converting to organic farming with payments under the Organic Aid Scheme (MAFF, 1994). Currently there are over 800 registered organic farmers in the UK, accounting for more than 48,000 ha of land (UKROFS, 1996, personal communication).

By definition, organic systems include the rational use of organic manure, the use of appropriate cultivation techniques, the avoidance of soluble fertilizers, the prohibition of agrochemical pesticides and, most importantly, the employment of balanced rotations which inherently requires gaseous and leaching losses to be minimized. Organic farming systems in the UK are traditionally based on ley/arable crop rotations. Up to 70% of the farmed area comprises mixed grass and legume leys (NRA, 1992). These leys offer a powerful mechanism for supplying nitrogen (N) through their potential to harvest biologically fixed N to support both animal production and a subsequent phase of arable cropping.

Research on nutrient cycling in organic farming systems has largely concentrated on measuring leaching losses, particularly those associated with cultivation of leys (NRA, 1992; Watson *et al.*, 1993). Little emphasis has, as yet, been placed on measuring gaseous losses from these systems and the only published work known to the authors relates to measurement of denitrification losses following incorporation of green manures and cover crops (Baggs *et al.*, 1996). Ongoing work on organic farms at SAC aims to measure denitrification losses from both organically managed grassland and cereals. This chapter deals with management opportunities for minimizing gaseous losses from organically managed systems. Emphasis has been placed on those management decisions which have a particular relevance to organically managed systems, although many of the principles can be transferred to other husbandry systems.

Management Practices

Decision-making processes relating to optimizing the balance between productivity and environmental impact are inherently complex in all agricultural systems. However, within organic systems, where the use of readily soluble fertilizers is prohibited, an unusually high management input may be required. Table 35.1 summarizes the major opportunities for management of N losses within organically managed grassland and key practices are discussed in more detail below.

Rotation design

The choice of crops in a ley/arable rotation largely determines both the economic viability and environmental impact of organic systems: the balance between the ley and arable phases is critical (Cuttle and Bourne, 1992;

Table 35.1.   Summary of opportunities for N management for organically managed grassland.

Practice	Decision	Options example
Rotation	Proportion ley : arable	3 years ley : 3 years arable
Grassland establishment	Purpose of ley	Grazing, cutting or mulching
	Legume species	Red or white clover
	Ley mixture	Proportion of legume : grass
	Duration of ley	1–5 years
	Method of establishment	Drilled or undersown
Stocking	Class of stock	Cattle, sheep, mixed grazing
	Stocking rate	Range 0.9–2.2 LSU ha^{-1}*
	Grazing period	Early and late season
Cutting	Conservation	Silage or hay
	Mulching	Height and frequency
Manure	Type	Slurry, manure or compost
	Phase of rotation	Conserved or grazed ley or root crops
	Application	Rate, frequency, timing
Cultivation	Season	Spring, summer, autumn, winter
	Timing	Early or late
	Method	Ploughed or rotavated
	Number	Single or multiple

*Fowler *et al.*, 1993.

Johnston *et al.*, 1994). The immobilized N which builds up during the grassland phase is utilized during the following arable phase. Johnston *et al.* (1994) suggested that neither the ley nor the arable period should exceed 3 years, as the N stored in a 3-year ley would have been used up in the subsequent 3 years of arable cropping. Cuttle and Bourne (1992) suggested that there was an advantage in keeping the ley period short, as this ensured net immobilization during the grassland phase, with the result that N is withdrawn from the actively cycling pool and is thus unavailable for loss by leaching or gaseous processes which in turn limits losses. Francis *et al.* (1995) found that leaching losses were very low after 3–4 years of arable cropping after ley, which could be interpreted to mean that N was becoming limiting.

During the arable phase of the rotation, cover cropping, the practice of growing a crop over winter between two arable crops in a rotation, is frequently used with the aim of minimizing both leaching and denitrification losses, since the cover crop will take up N in autumn, effectively lowering the nitrate (NO_3^-) pool. While the ability of cover crops to maintain N within the plant soil system is widely accepted (Martinez and Guiraud, 1990; Jackson *et al.*, 1993), it is not clear how best to optimize use of that N. Rayns and Lennartsson (1995), amongst others, have suggested that while the use of cover crops solves an immediate problem, it may simply move the problem forward in time. Further information is needed on the dynamics and the practical control of N release from incorporated plant material.

Establishment

Grass leys are established either by undersowing into a spring cereal or by direct drilling into a seedbed prepared after harvest of the previous arable crop. Undersowing has the advantage of both avoiding autumn cultivation and the ensured establishment of winter groundcover, both of these will minimize the pool of NO_3^- available for autumn denitrification.

Grazing and cutting strategy

There are a number of considerations to be made in managing organic grassland for grazing. It is a compromise between supplying fodder for the grazing animal and providing fertility for subsequent arable crops. The productivity of organic grassland (8.8 t DOMD ha^{-1} $year^{-1}$ (Newton, 1995)) and the requirement that the majority of the forage is produced on-farm, leads to lower stocking rates for organic farms, typically 1.6 livestock unit (LSU) ha^{-1} (range 0.9–2.2) (Fowler *et al.*, 1993). This results in a reduction in the quantity of N cycling within organic farm units (Watson and Younie, 1995). Since ammonia losses during grazing are directly related to stocking rate (Ball and Ryden, 1984), losses may be lower than from conventional systems.

Stocking

Organic farms frequently have diverse livestock enterprises to allow a 'clean grazing' system to be employed as a means of parasite control and an efficient grazing strategy. However, these mixed grazing systems complicate the N fluxes and have the effect of extending the grazing season. Removal of stock in the autumn has been shown to reduce the leaching losses in fertilizer-based farming systems (Titchen *et al.*, 1989; Cuttle and Bourne, 1992).

Stockless systems

There is increasing interest in developing stockless rotations for organic farming. The use of grass and legume leys is unlikely to be economic in these situations. One option to increase N inputs to stockless organic rotations is the use of short-term leys or green manures, such as red clover. Figure 35.1 shows the above-ground accumulation of N in white and red clover leys compared with rye-grass. Whilst the N input is valuable, it must be weighed up against the potential volatilization losses if this high N material is cut and mulched.

Ley cultivation

The timing of the ley cultivation has been identified as the critical point in the organic rotation where high leaching losses may occur (Watson *et al.*, 1993). Depending on climatic and soil conditions, it is likely that significant denitrification losses may also occur at this time. Goulding *et al.* (1993) found larger losses by denitrification from fertilized wheat following a grass and clover ley than in a continuous arable rotation There are a number of possible strategies for reducing losses following the cultivation of the ley. These include reducing the tillage intensity and altering timing of cultivation (Philipps *et al.*, 1995). Spring cropping may have an advantage over autumn cropping. The N content and the C : N ratio of the ploughed in material, will alter losses following incorporation. Figure 35.1 illustrates the high loss of NO_3^- by leaching following a red clover ley compared with rye-grass. Baggs *et al.* (1996) found lower denitrification fluxes following spring incorporation of leguminous green manures than from soil left bare over winter. This suggests enhanced immobilization following residue incorporation.

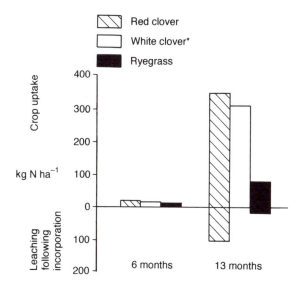

Fig. 35.1. Uptake of N by green manure species and leaching loss following their incorporation. *Leaching not measured following incorporation of white clover (data from Stopes *et al.*, 1996).

Conclusions

Clearly there are a number of possible management strategies for minimizing gaseous losses from organic farming systems. However, further research is needed before the practical, economic and environmental benefits can be fully assessed. Both SAC and Elm Farm Research Centre have extensive ongoing research programmes investigating N cycling in these systems. Additionally, both these organizations are responsible for the provision of advice to farmers, which means that research results are quickly and effectively disseminated to the agricultural community.

Acknowledgements

SAC receives financial support from the Scottish Office Agriculture, Environment and Fisheries Department. The authors wish to acknowledge the financial support of the NRA and MAFF. We are also grateful for the contribution and support of the many staff at Elm Farm Research Centre who have participated in the nitrogen leaching research programme.

References

Baggs, E.M., Watson, C.A., Rees, R.M. and Smith, K.A. (1996) The fate of nitrogen from incorporated crop residues. In: Younie, D. (ed.) *Legumes in sustainable farming systems, Occasional Symposium of the British Grassland Society,* No. 30. British Grassland Society, Reading, pp. 113–118.

Ball, P.R. and Ryden, J.C. (1984) Nitrogen relationships in intensively managed temperate grasslands. *Plant and Soil* 76, 23–33.

Cuttle, S.P. and Bourne, P.C. (1992) Nitrogen immobilization in pasture soils. *Proceedings of the Fertilizer Society* 325, 1–31.

EC (1991) Council Regulation (EEC) No 2092/91 of 24th June 1991 on organic production of agricultural products and indication referring thereto on agricultural products and foodstuffs. *Official Journal of the European Communities* L. 198/1 22 July 1991.

Fowler, S.M., Watson, C.A. and Wilman, D. (1993) N, P and K on organic farms: herbage and cereal production, purchases and sales. *Journal of Agricultural Science, Cambridge* 120, 353–360.

Francis, G.S., Haynes, R.J. and Williams, P.H. (1995) Effect of the timing of ploughing-in temporary leguminous pastures and two winter cover crops on nitrogen mineralization, nitrate leaching and spring wheat growth. *Journal of Agricultural Science, Cambridge* 124, 1–9.

Goulding, K.W.T., Webster, C.P., Powlson, D.S. and Poulton, P.R. (1993) Denitrification losses of nitrogen fertilizer applied to winter wheat following ley and arable rotations as estimated by acetylene inhibition and ^{15}N balance. *Journal of Soil Science* 44, 63–72.

Jackson, L.E., Wyland, L.J. and Stivers, L.J. (1993) Winter cover crops to minimize nitrate losses in intensive lettuce production. *Journal of Agricultural Science, Cambridge* 121, 55–62.

Johnston, A.E., McEwen, J., Lane, P.W., Hewitt, M.V., Poulton, P.R. and Yeoman, D.P. (1994) Effects of one to six year old ryegrass-clover leys on soil nitrogen and on the subsequent yields and fertilizer nitrogen requirements of the arable sequence winter wheat, potatoes, winter wheat, winter beans (*Vicia faba*) grown on a sandy loam soil. *Journal of Agricultural Science, Cambridge* 122, 73–89.

Lampkin, N. (1990) *Organic Farming.* Farming Press, Ipswich, 701 pp.

MAFF (1994) *Organic Aid Scheme Regulations,* MAFF Publications, London.

Martinez, J. and Guiraud, G. (1990) A lysimeter study of the effects of a ryegrass catch crop, during a winter wheat/maize rotation, on nitrate leaching and on the following crop. *Journal of Soil Science* 41, 5–16.

Newton, J.E. (1995) Herbage production from organic farms. *Journal of the Royal Agricultural Society of England* 156, 24–34.

NRA (1992) *Nitrate Reduction for Protection Zones: the role of alternative farming systems.* Bristol, 97 pp.

Philipps, L., Stopes, C.E. and Woodward, L. (1995) The impact of cultivation practice on nitrate leaching from organic farming systems. In: Cook, H.F. and Lee, H.C. (eds) *Soil Management in Sustainable Agriculture.* Wye College Press, Wye, pp. 488–496.

Rayns, F.W. and Lennartsson, E.K.M. (1995) The nitrogen dynamics of winter green manures. In: Cook, H.F. and Lee, H.C. (eds) *Soil Management in Sustainable Agriculture.* Wye College Press, Wye, pp. 308–311.

Stopes, C.E., Millington, S. and Woodward, L. (1996) Dry matter and nitrogen accumulation by three leguminous green manure species and the yield of following wheat crop in an organic production system. *Agriculture, Ecosystems and Environment* 57, 189–196.

Titchen, N.M., Wilkins, R.J., Philipps, L. and Scholefield, D. (1989). Strategies of fertilizer nitrogen application for beef: effects on production and dynamics. In: *XVI International Grassland Conference.* The French Grassland Society, Nice, pp. 183–184.

UKROFS (1991) *UKROFS Standards for Organic Agriculture.* Food from Britain, London.

Watson, C.A. and Younie, D. (1995) Nitrogen balances in organically and conventionally managed beef systems. In: Pollot, G.E. (ed.) *Grassland into the 21st Century.* Occasional Symposium of the British Grassland Society, Reading, pp 197–199.

Watson, C.A., Fowler, S.M. and Wilman, D. (1993) Soil inorganic-N and nitrate leaching on organic farms. *Journal of Agricultural Science, Cambridge* 120, 361–369.

Posters 16–22 $\qquad$ V

Mineral Balance and Farming Systems

A. Farruggia, L. Pichot and C. Perrot

Institut de l'Elevage, 149 rue de Bercy, 75595 Paris Cedex 12, France

The mineral accounting system is a farm-level management tool to assess potential pollution by nitrogen (N), phosphorus (P) and potassium (K). In this system, a farm's nutrient inputs (bought in fertilizers, feeds, animals, etc.) and outputs (sale of animal products such as milk and meat, and plant products, etc.) are recorded, the farm itself being considered as a black box. All quantities are multiplied by the N, P, and K contents of the various components in order to compute the mineral balance. In order to provide useful advice to livestock producers, it makes no sense to express a mineral balance in absolute terms. Explanations on the mineral balance become meaningful when a comparison is given between the farm surplus and the results from a 'reference' farm that uses a similar operating system and is considered to make optimal use of organic manure, mineral fertilizers and animal feed. Thus, the purpose of this study is to contribute to the collection of information on which to base the reference farm.

Data from the livestock networks (Chambers of Agriculture and the Institut de l'Elevage) were used to build up the reference materials and information. System descriptions are presented as 'standard cases' that have been derived from models of the diverse functions of the pilot farms, with emphasis on the functions that are technically and economically the most efficient. For each 'standard case', data is provided for mineral fertilizers, animal feed, bought straw, milk and meat products which are used to calculate the mineral balance. In this study, the standard cases serve as a reliable basis for categorizing and ranking livestock production systems in terms of their mineral balance, and for calibrating the mineral balance according to system type. The national sample will comprise 96 'standard cases' (for cattle), thereby covering the diversity of systems in all the major cattle-producing regions. This chapter will only present the results for N.

There was a wide range of variability in mineral balances and input-output figures throughout the samples. Nitrogen surpluses vary between 2 kg ha^{-1} year^{-1} in a milk-herbage system in Franche Comté and 218 kg ha^{-1} year^{-1} in an intensively reared pork/suckler system in Brittany, which may not seem very high in comparison to Dutch dairy farms. Nitrogen inputs through mineral fertilizers may reach a maximum of 205 kg ha^{-1} year^{-1}, which is twice the peak found in feed (110 kg ha^{-1} year^{-1}). By way of comparison, N output from animal products levels off at a maximum of only 50 kg ha^{-1} year^{-1} in milk, which corresponds to a high level of production, and 45 kg ha^{-1} year^{-1} in meat, which corresponds to a very intensive fattener standard case. On the other hand, output through plant products can reach 105 kg ha^{-1} year^{-1}, i.e. double that of animal products.

Considering the relatively low removal through the sale of animal products, fertilizer and feed inputs will have a decisive influence on the balance. The same is true for the percentage of commercial crops in each crop rotation since removal through plants is high. A multiple linear regression that only includes these three inputs can be used to estimate the N balance with an error of no more than 11 kg^{-1} ha^{-1} (standard deviation with $r^2 = 0.95$): i.e. N ha^{-1} balance = 0.92 N fert. ha^{-1} + 0.58 units N concen. ha^{-1} – 1.18%, where N fert. ha^{-1} = kg N ha^{-1} input by mineral fertilizer and N concen. ha^{-1} = kg N ha^{-1} input by bought in concentrates.

Figure P16.1 shows the standard case in terms of total N input on the x-axis and total N output on the y-axis. The diagonal lines indicate the levels of surplus: 50–100–150 kg ha^{-1} year^{-1}. This shows the difference by group : (i) systems with low inputs and outputs, i.e. the suckler, milk-herbage, suckler-fattener herbage system; (ii) systems of high input and low output, i.e. intensive suckler-fattener systems; (iii) systems with high input and medium (40–50 kg ha^{-1} year^{-1}) output, i.e. specialized intensive milk system or systems with beef units; and (iv) systems with medium (100–150 kg ha^{-1} year^{-1}) inputs and high outputs because of removal through commercial crops: i.e. milk or suckler herds with commercial crops. Figure P16.2 shows the range of variation per system, including the median, minimum and maximum. This can be used at a later stage to identify the position of a 'real' farm in relation to one of the 'standard cases' studied.

The mineral balance seems to be effective as a first level diagnosis of the mineral pollution and risks from waste in stock-crop production units. Ranking and calibrating systems throughout the livestock regions can be started, using farm models and standard cases constructed with reference to the

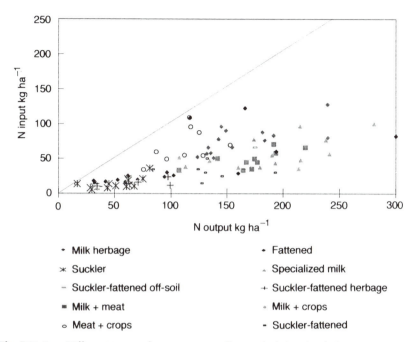

Fig. P16.1. Different types of systems according to their levels of nitrogen input and output.

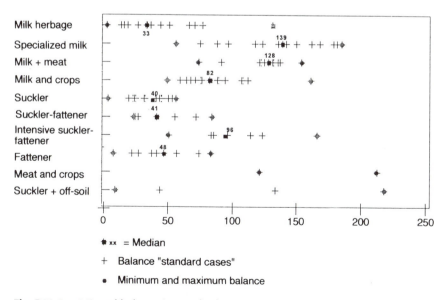

Fig. P16.2. Mineral balance in standard cases per system type.

livestock farm networks. The second part of this study will provide further detailed work on the standard cases. The work on the 'real' farms in these networks will especially emphasize 'benchmark values' in the mineral accounting system per farm type and per region. These values will have been selected because they seem to represent an acceptable risk for the environment and a reasonable, achievable goal for farmers willing to improve the mineral management in their system.

The Renaissance of Mixed Farming Systems: A Way Towards Sustainable Agriculture

E.A. Lantinga and R. Rabbinge
Department of Theoretical Production Ecology, Agricultural University, PO Box 430, 6700 AK Wageningen, The Netherlands

Over the last few decades agricultural production systems have developed in north-western Europe so that waste inputs are suboptimal in biotechnical and environmental terms. In the near future this will lead not only to unacceptable environmental and ecological impacts, but also economical and social effects (Rabbinge, 1992). Therefore, there is a need to develop and test alternative systems which are acceptable in the long term. One of the possibilities for reducing the negative effects of the increased specialization and intensification, which are characterized by too narrow crop rotations and an overuse of external inputs like fertilizers and biocides, is a renaissance of mixed farming systems at farm or regional levels in which products and services are exchanged between the different production systems. The main advantage of a mixed farming system is a reduction in the use of external inputs and an increase in their efficiency through: (i) the use of homegrown concentrates (less purchased concentrates); (ii) more efficient application of animal manure (less waste of nitrogen (N) and minerals); and (iii) broadening the crop rotation (less use of biocides and higher yields due to less problems with soil-borne pests and diseases). There is also better utilization of the available labour and spreading of income risks.

On the Minderhoudhoeve, the experimental farm of Wageningen Agricultural University in Oostelijk Flevoland, two different prototypes of mixed farming systems are being developed, optimized and tested: an integrated farm (135 ha; 90 dairy cows, 60 young cattle, 60 sheep) and an ecological farm (90 ha; 55 dairy cows, 60 young cattle and bulls, 40 sheep, 200 laying hens). Both farms have their own sets of goals and constraints. The production target per ha at the ecological farm is 80% of that on the integrated farm as an average for milk, potatoes and cereals. The location is characterized by a good loam soil with a high nutrient use efficiency and low irrigation needs. Measurements at farm level will start in autumn 1996 when both farms are fully operational. In the foregoing years, the two prototypes were designed and the transition to the present farms was initiated. The integrated type is described here according to its targets and constraints. Nitrogen surplus is used as an example for its perspectives.

The main targets and constraints on the integrated mixed farm are as follows:

1. Minimization of the nitrogen (N) surplus per unit product.

2. Minimization of the use of biocides per unit product under the fertilization regime resulting from target 1 and with the constraint of a good product quality at harvest.

3. In the system there is a variety of crops more or less corresponding with the 'average' Dutch cattle and arable farms: grassland, maize, seed and ware potatoes, sugar beets, winter and summer cereals, vegetables (onions, peas, green beans, etc.).

4. No bare fields until late autumn to prevent nitrate (NO_3^-) leaching.

5. Cultivation of potatoes and sugar beets on a certain field up to a maximum of only once in every 6 years to reduce the risks of soil-borne pests and diseases.

6. Application of slurry only between late winter and mid summer to reduce nutrient losses.

7. Amount of purchased concentrates less than 0.10 kg kg^{-1} milk, i.e. less than about 800 kg cow^{-1} year^{-1}, to restrict nutrient inputs under the constraint of a milk production of about 8000 kg cow^{-1} year^{-1} and about 11,000 kg ha^{-1} of forages (grass, clover, maize, wheat).

8. With the exception of 4 ha permanent grassland surrounding the farm buildings, the grass in rotation is ploughed after 2 or 4 years to prevent nutrient accumulation in the soil.

9. A stock of 60 ewes is kept to increase pasture utilization and condition (consumption of grass rejected by dairy cows, winter grazing, 'biological' weeding in sown pastures).

10. Sufficient phosphorus (P) status of the soil (Pw-value about 25).

11. Weeding, in principle, first through mechanical measures.

The N and phosphorus (P) surpluses per unit of acreage and the N surplus per ton of milk are shown in Table P17.1. This illustrates the possibilities of mixed farming to decrease environmental side effects and to increase profit. This was also concluded by De Koeijer *et al.* (1995) in an environmental economic analysis of mixed crop-livestock farming. The contribution on a country scale is considerable, as the dairy sector is responsible for about two-thirds of the N surplus in Dutch agriculture. The negative P balance results from the aim of achieving a sufficient P status of the soil. Current fertility in most of the fields is far beyond this level.

The calculated results illustrate that nutrient losses per unit product and per ha may be reduced considerably by sound integration of the different production components. It is interesting to note that when the results are extrapolated to the Netherlands as a whole, total milk production is almost

Table P17.1. Calculated nitrogen and phosphorus surpluses excluding deposition on the integrated mixed farm (1996–2000; 50% forage land) compared with the reference year 1993 (56% forage land) and the average of Dutch cattle and arable farms, 1985–86 (65% forage land).

	kg N ha^{-1} year^{-1}	kg P ha^{-1} year^{-1}	kg N ton^{-1} milk
The Netherlands (1985–86)	217	11	37
Minderhoudhoeve (1993)	124	10	25
Integrated mixed farm (1996–2000)	33	−12	6

the same as the current Dutch production volume (11 million tonnes on 2 million ha agricultural land, i.e. 5500 kg ha^{-1}). On the integrated mixed farm, the milk quota equals 5300 kg ha^{-1} of farmland of which only 50% is used for growing forages. This confirms both the good production situation at this site and the prospects for mixed farming systems.

References

de Koeijer, T.J., Renkema, J.A. and Mensvoort, J.J.M. van (1995) Environmental-economic analysis of mixed crop-livestock farming. *Agricultural Systems* 48, 515–530.

Rabbinge, R. (1992) Options for integrated agriculture in Europe. In: van Lenteren, J.C., Minks, A.K. and Ponti, O.M.B. (eds) *Proceedings of an International Conference Organized by the IOBC/WPRS, Veldhoven, The Netherlands, 8–13 September 1991.* Pudoc Scientific Publishers, Wageningen, pp. 211–218.

Effect of Maize/Grassland Balance: Development of Dairy Fodder Systems to Minimize Environmental Impact

A. Le Gall[1], J. Legarto[1] and M.M. Cabaret[2]
[1]Institut de l'Elevage-Rennes, 149 rue de Bevay, 75595 Paris, Cedex 12, France; [2]Chambre d'Agriculture de Bretagne-Saint-Brieuc, France

Dairy systems in the French Atlantic Arc (western and south western parts of the country) are intensive, i.e. 1.8–2.2 livestock units (LU) ha^{-1} produced (6000–8000 kg of milk cow^{-1} year^{-1}). Because of the high nitrogen (N) inputs from the concentrates (40–80 kg N ha^{-1}), the fertilizers and the symbiotic fixation by white clover (120–200 kg N ha^{-1}), the N balance reaches levels of 150–220 kg ha^{-1} (Pflimlin *et al.*, 1996). This remains far below the surplus reached in the Dutch dairy farms (*c.* 380 kg N ha^{-1}) (Mandersloot *et al.*, 1995). Morever, these dairy systems have been strongly affected by an increase in maize silage production. The latter represents at the moment between 30 and 60% of the fodder area and results in a high proportion of bare soils after the maize harvest. This, in an oceanic climate, is a risky situation at this period because of intense organic N mineralization. This phenomenon can be further amplified by regular applications of manures and by the late effect of sward ploughing. Grasslands are mainly grazed with a high grazing pressure (600–900 grazing days ha^{-1} year^{-1}) and are consequently vulnerable to nitrate (NO$_3^-$) leaching. Hence these dairy systems are a potential risk to the environment through ammonia (NH$_3$) emissions, NO$_3^-$ leaching, phosphate accumulation and pesticides, but they can be improved.

Research combines studies involving modelling, experimentation on complete systems and observations in commercial farms. The means used to optimize these systems are: storage and spreading of the animal manures at the optimal stage for the crop or the pasture, rational fertilization, decrease in the N content of the concentrates, reduced bare soils during winter by sowing a cover crop under maize at the stage of four to eight leaves and a mixed mechanical/chemical weed control system for maize. Modelling enabled us to evaluate the expected effects of these different techniques on the N balance and on the NO$_3^-$ leaching, and also to study the effect of the grassland to maize balance. For similar stocking rates, increasing maize silage in the fodder system leads to a decrease in the N outputs per cow and in the N balance, but to an increase in the amount of the bare soil in winter, of the amount of manure handling and to stable NO$_3^-$ leaching according to our hypotheses. These results are consistent with those obtained by Wilkins (1993) in a dairy farm in the south-west of England, where the increase in the proportion of maize in the fodder system from 0 to 25% modified the N losses and reduced the N balance.

This modelling enabled selection of the most interesting systems to test in practice. This is why two intensive dairy fodder systems were simultaneously investigated on two experimental farms in contrasting situations. The first farm was in Ognoas, in south-western France, where the weather is warm with moderate rainfall and irrigation is possible on silt-sandy soils (drainage water: 300–500 mm year^{-1}). A feeding system based only on maize silage (90% of the fodder growing area, stocking rate: 2.2 LU ha^{-1}, over 6 t dry matter (DM) silage cow^{-1} year^{-1}) was compared to a system with a lower proportion of maize (35% of the fodder area, 1.9 LU ha^{-1} and 30 t of ensiled DM cow^{-1} year^{-1}). The second trial was conducted in Crécom, with a wet climate, in Brittany, on granite soils (drainage water: 500 mm year^{-1}). A classical Breton fodder system with 50% of maize silage, 1.8 LU ha^{-1}, 2.9 t of ensiled DM cow^{-1} year^{-1} was compared with a system based on a lower use of fodder (25% of the fodder area in maize, 1.8 LU ha^{-1}, and 1.9 t of ensiled DM cow^{-1} year^{-1}) (Table P18.1). Each system is fertilized by its own manures and slurries (separate pits). Measurements were made on animal intake and production, animal waste, manure storage, handling and spreading, forage production and nutrient losses, especially NO$_3^-$ to the soil and ground water.

The results enabled a comparison between the different systems in terms of N losses per cow, global and semi-global N balances, NO$_3^-$ leaching, and also determination of the whole N, phosphorus and potassium cycles in the system. The first results show (Table P18.1) that the N mineral balance on the

Table P18.1. Main characteristics of the experimental design and preliminary results.

	Experimental farm			
	Ognoas		Crécom	
Situation	South west – silt sandy soils (1.5% OM)		Brittany – sandy loam soils (6% OM)	
Soil capability (t DM ha^{-1})				
Yield herbage	8–10		8–10	
Yield maize silage	14–16		8–12	
System in test*	'All maize'	'Maize and pasture'	'Maxi maize'	'Maxi pasture'
% of crops	50	30	0	0
% of maize silage	90	30	50	25
Stocking rate (LU ha^{-1})[†]	2.2	1.9	1.8	1.8
Ensiled DM cow^{-1} (t)	6.0	3.0	2.0	3.0
Milk production target (kg cow^{-1})	8000	7500	8000	7000–8000
Concentrates (kg cow^{-1})	1500	1000	1300	1300
Management of manure and slurry	Manure and slurry on maize with cover crop	Manure on maize, slurry on pasture and maize	Manure on maize, slurry on pasture	Compost on pasture, slurry on maize
Fertilizers (N chemical ha^{-1})				
Yield herbage	—	100	180	180
Yield maize silage	180	180	0	0
Weed control for maize	Chemical	Chemical	Mixed	Mixed
Cover crop between maize	Only fields with organic fertilizers		Yes	Yes
Nitrogen balance (N ha^{-1})[‡]	+124	+99	+116	+136
Conversion rate	40	43	39	37
% of bare soils	74	38	17	25

*LU ha^{-1} of main fodder area.
[†]Each system has 35 or 40 cows.
[‡]Results from 1994/1995 at Ognoas and 1995/1996 at Crécom.

whole farm was improved in relation to commercial farms. The results for NO$_3^-$ leaching are being analysed and must be compared to whole-farm balances. At Crécom in 1995, the mixed mechanical and chemical weed control for maize gave good results, with a high fodder yield (12 t DM ha^{-1}). The cover crop (rye-grass) sown in the same operation was well developed (between 0.5 and 1 t DM ha^{-1} at 1 February 1996 with 2.9% N).

References

Mandersloot, F., Schreuder, R. and Van Scheppingen, A.T.J. (1995) Reduction of nitrogen and phosphorus surpluses, Proceeding of the PR symposium. *Applied Research for Sustainable Dairy Farming* 48–52.

Pflimlin, A., Le Gall, A., Farruggia, A. and Hacala, S. (1995) More efficient use of manures and nutrients on dairy farms in France, Proceedings of the PR symposium. *Applied Research for Sustainable Dairy Farming* 85–89.

Wilkins, R.J. (1993) Environmental constraints to production systems. In: *The Places for Grass in Land Use Systems*. British Grassland Society, Reading, pp. 19–30.

Nitrogen Budgets for an Organic Dairy Farm

S.P. Cuttle

Institute of Grassland and Environmental Research, Plas Gogerddan, Aberystwyth, Dyfed SY23 3EB, UK

A 3-year project to monitor the physical, environmental and financial implications of converting conventional dairy farms to organic milk production has demonstrated the viability of conversion (IGER, 1995). As part of this study, whole-farm nitrogen (N) budgets were estimated for the Tŷ Gwyn dairy farm at Trawsgoed, near Aberystwyth to provide an indication of potential N losses during conversion of the farm to organic production.

The farm lies at an altitude of between 60 and 110 m with an average annual rainfall of 1202 mm. Soils are a mixture of brown earths (61%) and gleys (39%). Prior to the start of conversion in 1992, the farm was managed as an intensive, conventional dairy farm. Of the total field area of 63 ha, 24 ha are permanent pasture. The remaining fields are in a 7-year rotation of white and red clover/rye-grass leys, together with cereal crops which are harvested for whole-crop silage. Overall stocking rates are equivalent to 1.5–1.6 livestock units (LU) ha^{-1}. Feed concentrates and straw are purchased from outside the farm but there are no additional inputs of animal manures or slurries from external sources.

The potential N loss for each year between 1992 and 1995 was calculated as the difference between the total input of N to the farm and that recovered in milk and livestock. Quantities of N fixed by clover were estimated for each field, assuming fixation of 54 and 40 kg N ha^{-1} per 1000 kg dry matter yield of white and red clover, respectively (van der Werff *et al.*, 1994). Other inputs were calculated from the quantities of purchased feed and straw and their measured N contents, supplemented where necessary by values obtained from the literature. The input in rainfall was estimated from previous measurements at the Institute of Grassland and Environmental Research, Aberystwyth. Outputs were calculated from the volume of milk sales and measured protein content and from estimated body weights of stock sold from the farm, together with published values of N content.

Table P19.1. Annual N budget for Tŷ Gwyn (mean 1992–95) averaged over the whole farm area.

Inputs	(kg N ha^{-1})	Outputs	(kg N ha^{-1})
N fixation	71	Milk	29
Concentrates	42	Livestock	2
Straw	2		
Rain	7		
Total	122	Total	31

Total input to farm – output in products = 91 kg N ha^{-1}.

The mean annual budget, averaged over the whole farm area, is shown in Table P19.1. The totals are equivalent to an overall input of 7.7 t N year^{-1} with 2.0 t year^{-1} recovered in agricultural products. Although N fixation was the largest contributor of N to the farm, purchased concentrates also represented a major input. Only 25% of the total input was recovered in agricultural products. This represents a more efficient utilization of N than the 20% recovery estimated by Jarvis (1993) for a model of a typical conventional dairy farm in south-west England. However, the model assumes a more intensive management than at Tŷ Gwyn, with a stocking rate of 2.2 LU ha^{-1}. This results in a greater imbalance between N input and recovery in products, equivalent to 270 kg N ha^{-1}, compared with 91 kg ha^{-1} for Tŷ Gwyn.

Separate measurements indicated that leaching losses at Tŷ Gwyn were < 30 kg N ha^{-1}. If the possibility of significant changes in the quantity of N stored in soil organic matter is ignored, the values indicate that the equivalent of at least 60 kg N ha^{-1} year^{-1} (3.8 t year^{-1}) may have been lost from the farm as gaseous emissions. The data provide no information about what proportion of this loss was due to denitrification or to ammonia (NH$_3$) volatilization. In the case of the model farm described by Jarvis (1993), 55% of the total gaseous loss was attributed to denitrification and 45% to volatilization, although it was considered that the denitrification loss was probably an under-estimate.

Acknowledgements
This work was supported by the Ministry of Agriculture, Fisheries and Food with additional assistance from F. Pimpinella, University of Perugia, Italy.

References
IGER (1995) *Conversion to Organic Milk Production.* Confidential Report to Ministry of Agriculture, Fisheries and Food, London. Institute of Grassland and Environmental Research, Aberystwyth, 192 pp.

Jarvis, S.C. (1993) Nitrogen cycling and losses from dairy farms. *Soil Use and Management* 9, 99–105.

van der Werff, P.A., Baars, A. and Oomen, G.J.M. (1994) Nutrient balances and measurement of nitrogen loss on mixed ecological farms on sandy soils in the Netherlands. *Biological Agriculture and Horticulture* 11, 41–50.

Nitrogen and Dune Grasslands

M.N. Mohd-Said and E.R.B. Oxley
*School of Biological Sciences, University of Wales, Bangor, Gwynedd
LL57 2UW, UK*

The nitrogen (N) content of dune systems progressively increases as the dune 'succession' proceeds from newly blown sand, through various stages of fixed dunes and dune grassland, to mesotrophic grasslands (Figure P20.1). The accumulation of N occurs from a number of sources including atmospheric inputs and biological fixation.

This project is estimating the N flux and its impact on vegetation development and species composition in a vegetational continuum from dune grassland to mesotrophic agricultural grasslands. The Newborough site in Anglesey, North Wales also includes long-term mowing and grazing experiments, which provide an opportunity to investigate the effects of management on the accumulation of N. Total N content will be measured at a number of sample points and in experiments at varying distances from both

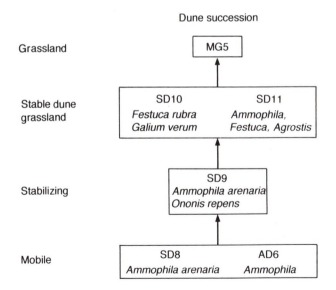

Fig. P20.1. A schematic representation of the dune succession showing plant communities as identified by the National Vegetation Classification.

the sea and at various distances from an intensive chicken farm which is believed to be an ammonia source.

Figures P20.2 and P20.3 show preliminary measurements of total N in above-ground plant material from a mowing experiment on dune grasslands (SD10 and MG5). These measurements are from above-ground plants and quadrat samples collected in February 1996. Vegetation samples were analysed for N by means of Kjeldahl procedures using the Kjeltec Auto 1030 Analyzer. Results were analysed for significance using ANOVA in MINITAB program in which it was found that cutting had a highly significant impact on N content (38.25 and 44.89 kg N ha^{-1}, respectively, in cut and uncut systems).

It is proposed to use diffusion tubes for the measurement of atmospheric inputs or availabilities, and a variety of analytical methods for investigating

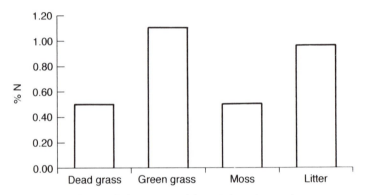

Fig. P20.2.   The total %N of plant material from February samples.

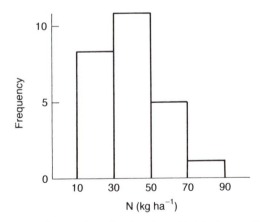

Fig. P20.3.   Frequency distribution of estimation of total N per unit area from dune grassland under different mowing regimes. Samples from 10-cm quadrats in February.

the N cycle. The response of experimental species mixtures will also be investigated using dune and mesotrophic grassland species with high, medium and low scores on the Ellenberg N indicator system.

Nitrate Leaching to the Groundwater in Two Grazed Pastures in Relation to Drought Susceptibility and Urine-affected Areas

M.J.D. Hack-ten Broeke[1], A.H.J. van der Putten[2], W.J. Corré[3], W.J.M. de Groot[1] and H.F.M. Aarts[2]
[1]DLO-Winand Staring Centre for Integrated Land, Soil and Water Research, PO Box 125, 6700 AC Wageningen, The Netherlands; [2]DLO-Research Institute for Agrobiology and Soil Fertility, PO Box 14, 6700 AA Wageningen, The Netherlands; [3]DLO-Research Institute for Agrobiology and Soil Fertility, PO Box 129, 9750 AC Haren, The Netherlands

The aim of the experimental farm De Marke is to develop a system for sustainable dairy farming. The emphasis of research at the farm has been on nutrient management. Monitoring of nitrate (NO_3^-) leaching, denitrification and nitrogen (N) uptake by the crop was carried out in two grazed pastures. Simulation models for unsaturated soil water behaviour and N dynamics in the soil were validated and used for the quantification of the effect of urine-affected areas in the pastures on NO_3^- leaching. Because urine-N is deposited in patches, it will result in locally high N inputs, which also has fertilizer added during the remaining part of the growing season. Therefore the possible impact of a site-specific technique, allowing omission of fertilizer or manure applications at urine-affected areas, on NO_3^- leaching was calculated.

In this study, monitoring and modelling took place in two rotationally grazed pastures on sandy soils. Groundwater levels varied between 0 and 2 m depth at site A and between 0.5 and 2.8 m depth at site B. The soil at site A had a silt + clay content of 17% throughout the soil profile and the organic matter content of the top soil was 3%. At site B the silt + clay content was 12% in the top soil and decreased to 3% in the subsoil. The organic matter content of the top soil was 6%. These characteristics resulted in a sufficient moisture supply at site A, due to higher groundwater levels and greater capillary rise than at site B. At the drought-susceptible site B, supplementary irrigation was required to ensure grass production. Nitrogen concentrations were measured either in soil water through suction cup sampling or in groundwater through piezometer sampling (Hack-ten Broeke et al., 1996). Subsequently, N leaching was calculated by multiplying the concentrations with calculated soil water fluxes. Denitrification was measured by incubation of soil cores (upper 20 cm of the soil) with acetylene. Dry matter production and N content in the grass sward was measured in strips at the start of each cut (mowed or grazed) to quantify N uptake.

The simulation models SWACROP (for unsaturated soil water behaviour) and ANIMO (for N dynamics in the soil) were used to calculate NO_3^- leaching and, after validation, to quantify the effect of urine deposition. Using a calculated spatial distribution of urine-affected areas in each field, a frequency distribution of NO_3^- concentrations in the soil water or ground-water of the field was calculated (Hack-ten Broeke *et al.*, 1996).

Finally, the possibility of applying spatially differentiated fertilization was studied. A combination of a technique to locate urine patches in pastures by surveying soil salinity and the use of machines for site-specific fertilization should allow the possibility of omitting fertilizer and manure applications at urine-affected areas. The possible reduction in NO_3^- leaching as a result of implementing this technique was calculated.

At site A N leaching was 1, 44 and 21 kg ha^{-1}, respectively, in the years 1992, 1993 and 1994 and at site B it was, respectively, 42, 110 and 96 kg ha^{-1}. Net (harvested) N uptake in these 3 years was 328, 371 and 327 kg ha^{-1} at site A and 335, 293 and 298 kg ha^{-1} at site B. On average, N uptake was 342 kg ha^{-1} at site A and 309 kg ha^{-1} at site B. Denitrification was only measured in 1994 and amounted to 31 kg ha^{-1} at site A and 14 kg ha^{-1} at site B. The higher moisture supply at site A thus resulted in less N leaching, more N uptake and a higher denitrification level compared to site B.

The comparison of simulation results and measured NO_3^- concentrations was considered satisfactory. The measured average NO_3^--N concentrations per hydrological year (1 April–1 April) for site A were 4.7, 7.5 and 8.4 mg l^{-1} in 1991/92, 1992/93 and 1993/94, respectively. The simulated values for these years were 5.5, 6.2 and 8.5 mg l^{-1} NO_3^--N. For site B the measured average values were 27.6, 33.3 and 24.5 mg l^{-1}, respectively, and the simulated concentrations 27.0, 28.2 and 24.4 mg l^{-1}. Thus the model was considered to be validated.

Next, the models were applied to urine-affected areas, resulting in a frequency distribution of NO_3^- concentrations. Measured peak values could be explained by modelling N transport under one or more overlapping urine patches, deposited in September or later. For site A the measured peak values were 28.4 mg l^{-1} NO_3^--N for 1992/93 and 22.3 mg l^{-1} for 1993/94 and the simulated maximum values were 19.0 and 16.3 mg l^{-1}, respectively. At site B measured peak values for these 2 years were 79.6 and 48.1 mg l^{-1}, respectively, corresponding with simulated maximum values of 82.9 and 54.9 mg l^{-1} (Hack-ten Broeke *et al.*, 1996). To avoid such high NO_3^- concentrations there should be no grazing after August.

The site-specific fertilization technique, which avoids fertilization of urine patches, is expected to result in a further reduction of NO_3^- concentrations. Figure P21.1 shows only the calculated reductions for situations with at least two subsequent fertilizations (on the x-axis). On average, 10–20% reduction of the NO_3^- concentrations was calculated and in some cases the reduction was as high as 40%.

The earlier in the growing season the urine patch is deposited, the higher the reduction in NO_3^- concentration can be. The highest calculated reduction

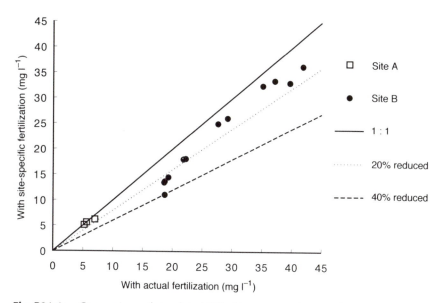

Fig. P21.1. Comparison of simulated NO₃-N concentrations under a urine patch as a result of site-specific fertilization and simulated NO₃-N concentrations using actual subsequent fertilizations for the years 1991–94.

was 41% at site B for a urine patch deposited on 13 May 1994, followed by five fertilizations in the remainder of the growing season. The lowest calculated reduction for site B within these simulations was 7% for a urine patch deposited on 11 July 1993, followed by two subsequent fertilizations. The reduction was also affected by weather conditions.

Reference

Hack-ten Broeke, M.J.D., De Groot, W.J.M. and Dijkstra, J.P. (1996) Impact of excreted nitrogen by grazing cattle on nitrate leaching. *Soil Use and Management* 12, 190–198.

Potential Nitrate Leaching Resulting from the Nitrogen Mineralization Dynamics in Grassland Soils

B. Sapek
Institute for Land Reclamation and Grassland Farming at Falenty 05-090, Raszyn, Poland

The pool of nitrate (NO₃) leached into the groundwater is closely related to the development and intensity of the mineralization process, which plays a particular role in grassland soils, which are enriched in organic matter (OM) (Jarvis *et al.*, 1995).

The results of the investigations concerning N mineralization determined by the *in situ* incubation method (Adams *et al.*, 1989) have been extended to some attempts to estimate the nitrogen (N) losses due to nitrate (NO_3^-) leaching (Sapek 1995a,b, 1996).The quantity of N released during the mineralization process was compared with the mineral N determined directly in the samples of soil under growing plants. The impact of N from fertilizers and precipitation was also considered. Potential NO_3^- leaching (PL) as a function of mineralization efficiency was estimated with the background of different soil pH and N fertilizer forms (ammonium nitrate (AN), calcium nitrate (CN)) on grassland experiments at Janki (J) and Laszczki (L) in 1995 (Table P22.1). The results indicated bigger PL values in the soil with smaller clay and OM content, where greater mineralization efficiency was observed (experiment J). The fertilization with CN enhanced the decrease of PL in acid soil, particularly in soil with a higher clay and OM content (experiment L) (Table P22.1). The increase in soil pH by CN fertilizer probably stimulated the nutrient uptake by plants.

The mean NO_3^--N concentration in water samples taken from the plastic tube wells installed on the experimental sites was higher in experiment J, where the PL values were also higher. Nitrate concentration in samples from a ditch on that site was higher during the period from August to December 1995 (Table P22.2).

The changes in NO_3^- content in the soil surface layer after (AV) and before (BV) the vegetative growth season were also investigated in meadows and

Table P22.1. Nitrogen mineralization and balances in grassland fertilizer experiments.

			N-mineralization (kg ha^{-1})					
Experiment	Form of fertilizer	Soil pH	Mineralization efficiency* a	N content in soil under plants b	$\Delta = a - b$	N in fertilizer[†] and precipitation[‡] c	N uptake by plants d	Potential nitrate leaching PI = $(\Delta + c) - d$
J	AN	3.8	148.4	101.1	47.3	255	232.4	69.9
		5.8	118.4	89.4	29.0	255	225.5	58.5
	CN	4.2	133.5	92.2	41.3	255	254.7	41.6
		6.3	130.9	89.6	41.3	255	221	75.3
L	AN	4.0	114.7	79.1	35.6	255	248.9	41.7
		6.5	119.0	68.8	50.2	255	257.5	47.7
	CN	4.4	130.3	71.5	58.2	255	304.2	9.6
		6.7	114.1	77.4	36.7	255	273.4	18.3

* *In situ* incubation.
[†]240 kg N ha^{-1}.
[‡]15 kg N ha^{-1}.

pastures situated in three small watersheds (Rupin, Szafranki and Lady) within the framework of the project 'Poland Agriculture and Water Quality Protection' sponsored by USEPA (Sapek, 1995b). The NO_3-N content in soil in spring (BV) was smaller than in autumn (AV).

The differences (Δ = AV − BV), which are the values of potential N losses due to leaching, were smaller on the pastures (Table P22.3). It can be supposed that on pastures the N losses were mainly a result of the denitrification process (Jarvis et al., 1995). Generally, the probability of NO_3 leaching below the root zone was larger on pastures. The mean values of NO_3-N concentration in the groundwater samples taken from plastic tube wells installed in three watersheds also show this phenomenon (Table P22.4).

Table P22.2. Nitrate-N concentrations in tube wells and adjacent ditches at experimental sites.

| | NO_3-N (mg l^{-1}) | | | |
| | Janki | | Laszczki | |
Sampling date (1995)	Plastic tube well	Ditch	Plastic tube well	Ditch
16.01–03.04	34.9	9.6	2.6	10.2
02.05–03.07	5.3	7.9	0.9	8.0
02.08–04.09	3.3	12.5	0.9	0.9
02.10–04.12	3.7	3.0	8.9	0.3

Table P22.3. Nitrate-N in surface soils from three catchments.

| | NO_3-N (kg ha^{-1}) | | | | | |
| | Rupin | | Szafranki | | Lady | |
Sampling time (1993–95)	Meadow ($n = 5$)	Pasture ($n = 5$)	Meadow ($n = 5$)	Pasture ($n = 4$)	Meadow ($n = 7$)	Pasture ($n = 3$)
AV (1993–94)	68.6	64.6	50.6	44.4	115.7	61.7
BV (1994–95)	33.7	45.5	48.1	48.3	83.0	44.4
Δ = AV − BV	34.9	19.1	2.5	−3.9	32.7	17.3

Table P22.4. Nitrate-N in tube wells in meadows and pastures in three catchments.

| | NO_3-N (mg l^{-1}) | | | | | |
| | Rupin | | Szafranki | | Lady | |
Sampling date	Meadow ($n = 5$)	Pasture ($n = 5$)	Meadow ($n = 5$)	Pasture ($n = 4$)	Meadow ($n = 7$)	Pasture ($n = 3$)
November (1993–95)	3.16	2.49	4.38	6.47	8.38	3.78
March (1994–95)	1.84	2.05	1.05	5.35	3.64	7.48

References

Adams, M.A., Polglase, P.J., Attivil, P.M. and Weston, C.J. (1989) *In situ* studies of nitrogen mineralization and uptake in forest soils; some comments on methodology. *Soil Biology and Biochemistry* 21, 423–429.

Jarvis, S.C., Scholefield, D. and Pain, B.F. (1995) Nitrogen cycling in grazing systems. In: Bacon P.E. (ed.) *Nitrogen Fertilization in the Environment.* Marcel Dekker, New York, pp. 381–419.

Sapek, B. (1995a) Nitrogen mineralization in the meadow soil depending on pH and nitrogen fertilization. *Zesz. Prob. Post. Nauk Rol.* 421a, 323–330 (in Polish).

Sapek, B. (1995b) The monitoring of water on a farm scale. In: *Proceeding of the Second International IAWQ Specialized Conference and Symposia on Diffuse Pollution.* Brno & Prague, Czech Republic, August 13–18 1995. Part I, Sec. 1, 14–19.

Sapek, B. (1996) Impact of soil pH on nitrogen mineralization in grassland soils. In: Van Cleemput, O., Hofman, G. and Vermoesen, A. (eds) *Progress in Nitrogen Cycling Studies.* Kluwer Academic Publishers, Dordrecht, pp. 271–276.

Progress and Challenges in Measuring and Modelling Gaseous Nitrogen Emissions from Grasslands: An Overview

36

O.T. Denmead

CSIRO Centre for Environmental Mechanics, Canberra, Australia

Summary. The main concern of this chapter is the gaseous nitrogen (N) emissions, ammonia (NH_3) and nitrous oxide (N_2O). Progress over the last 20 years in measuring and modelling these emissions is outlined briefly before examining present limitations and new developments.

Tremendous advances have been made in measuring NH_3 emissions from the landscape through the development of non-disturbing micrometeorological techniques which permit examination of NH_3 exchange in large fields or small plots. Small plot developments have been significant in quantifying NH_3 emissions from various practices in grassland management. Notable features have been the employment of mass balance techniques and the development of passive samplers for measuring NH_3 flux. Further progress, however, seems to be limited by the availability of suitable direct, on-line techniques for measuring atmospheric NH_3 concentrations. For the most part, wet-trapping techniques are still in vogue. The possibilities for new techniques, such as Fourier Transformed Infrared Spectroscopy (FTIR) and Tunable Diode Lasers (TDL), are examined briefly.

Measurement of N_2O emissions on an extensive field scale is still very difficult. Our knowledge of field processes remains heavily dependent on chamber studies. Instrumental requirements for conventional micrometeorological flux measurements are very demanding; resolutions of atmospheric N_2O concentrations to about 0.1 ppbv are necessary. New generation analysers (FTIR, TDL) will be capable of this, but at present there may be more benefit in exploring regional-scale fluxes through boundary-layer budgeting techniques. These are proving successful for carbon dioxide (CO_2) and methane (CH_4), and that success should carry over to N_2O.

Models of NH_3 exchange on a large scale are concerned with atmospheric chemistry and redeposition. Transport models incorporating these processes are proving useful in delineating the extent of NH_3 pollution. A challenge still is to develop short range transport models for assessing the importance of dry deposition close to the emitting source.

Models of N_2O production and emission are necessarily complex. They must be capable of modelling not only the microbiology and soil chemistry, but also the soil microclimate which has a large influence on N transformations and gas diffusion. Sensitivity analyses have indicated that correct prediction of soil temperature and moisture content is as important as the correct description of the production process. The challenge here seems to be to reduce the complexity of models to make them more suitable for everyday use.

Scope

Lee *et al.* (see Chapter 31) point out that grasslands are major contributors of nitric oxide (NO), nitrous oxide (N_2O) and ammonia (NH_3) to the atmosphere, and that all three have direct or indirect roles in the greenhouse effect. It is important, therefore, to quantify their emissions and to model their production, emission and dispersion. Emissions of NO are considered in this context in the contributions of Jarvis, Meixner *et al.* and Fowler *et al.* (see Chapter 1, 4 and 14). This chapter concentrates on measuring and modelling emissions of NH_3 and N_2O.

Historical Perspective

Ammonia

When my colleagues, John Freney and Jeff Simpson, and I started work on gaseous emissions of N from grasslands more than 20 years ago, information on the topic and measurement capabilities were fairly primitive. Ammonia exchange between the land and the atmosphere was thought of more in terms of deposition than emission. An influential paper was written by Eriksson (1966), which showed maps of atmospheric NH_3 concentration over Europe and characterized NH_3 exchange by a deposition velocity of about 1 cm s^{-1}. (Processes of plant uptake by stomatal diffusion were not yet contemplated; NH_3 compensation points were, in essence, set to zero.) The highest atmospheric concentrations then prevailing were reported to be about 20 ppvb and the data translated into an average annual dry deposition of NH_3-N over Western Europe of approximately 14 kg ha^{-1}. Measurement techniques were equally as simplistic, sometimes consisting of exposing dishes of acid near the ground and equating the rates at which the dishes absorbed NH_3 with fluxes from the atmosphere. The absorption rates were, of course, more measures of concentration than flux.

Other influential works at that time were those of Healy *et al.* (1970), which pointed to the overwhelming importance of animal sources in the NH_3 budget of the UK, and Hutchinson *et al.* (1972), which established the ability of plants to absorb NH_3 from the air through stomata.

Now, we have very detailed maps of atmospheric NH_3 concentrations and emission and deposition rates, particularly in Europe, and the emphasis is equally as much on emission as it is on deposition (Buijsman *et al.*, 1987; Asman and Drukker, 1988; Asman and van Jaarsveld, 1992; ECETOC, 1994; Sutton *et al.*, 1994). The importance of the animal sources has been established unequivocally. ECETOC (1994) estimates that they constitute about 3 of the 4 Tg NH_3-N now emitted annually to the atmosphere in Western Europe, and Lee *et al.* (see Chapter 31) estimate a global emission of 35 Tg NH_3-N year^{-1} constituting 59% of global emissions of N.

Measurement techniques have certainly improved from the early days of exposed acid dishes, but there are still some deficiencies, while mechanistic models embracing both emission and deposition processes as well as short-range transport of NH_3 are still required. These needs are addressed in later sections.

Nitrous oxide

There is much less confidence in our knowledge of N_2O emissions from grasslands. In the latest global inventory of N_2O sources, IPCC gives figures of 1 Tg N_2O-N year^{-1} for N_2O production from grasslands, and 0.4 Tg N_2O-N year^{-1} from animal wastes (Prather *et al.*, 1995). In total, these estimates account for 25% of the total anthropogenic N_2O source, but each has an uncertainty factor of 2. An equally authoritative inventory by Duxbury *et al.* (1993) gives a figure of 0.7 Tg N_2O-N year^{-1} for grasslands and nothing for animal wastes, while Lee *et al.* (see Chapter 31) estimate a global emission of 2.4 Tg N_2O-N year^{-1} for grasslands, including those fertilized with animal excreta.

Twenty years ago, N_2O emissions from fertilized soils were perceived as having a possible deleterious effect on stratospheric ozone due to the work of Crutzen (1976) and others, but the role of N_2O as a greenhouse gas was just beginning to be recognized. There was controversy over whether the soils of the earth constituted a net source or sink for atmospheric N_2O, and whether N_2O emissions could occur during nitrification, e.g. Freney *et al.* (1978, 1979). Chambers offered the only feasible means for measuring N_2O fluxes in the field, and new chamber systems were invented frequently, e.g. Ryden *et al.* (1978), Denmead (1979) and Matthias *et al.* (1980).

Twenty years on, knowledge of the microbiology and biochemistry of N_2O formation has increased considerably, and it is accepted that soils are net sources of N_2O for the atmosphere and that N_2O is produced over a wide range of soil moisture contents during both nitrification and denitrification. However, prediction of actual pathways and emission rates seems still to be elusive (see Chapters 1 and 2). In addition, our knowledge of field rates of N_2O emission continues to be based almost exclusively on chamber measurements, and the problem of soil variability that has bedevilled this approach in the past is still present (see Chapter 16).

This chapter returns to these problems in a later section. In what follows, our present capabilities for measuring and modelling emissions of NH_3 and N_2O are outlined, and future needs are touched upon.

Ammonia Exchanges

Measurement techniques

Unlike the situation for most other trace gases, micrometeorological methods have been by far the preferred approach for measuring emissions of NH_3 from agricultural systems. Probably the first attempt to measure NH_3 emissions from grasslands was the study of Denmead *et al.* (1974), who used a conventional micrometeorological gradient-diffusion approach to measure the NH_3 flux into the atmosphere from a pasture grazed by sheep. The same general approach has been used many times since. A particularly notable example was the monumental study by Lenhard and Gravenhorst (1980), who used aircraft to measure gradients of atmospheric NH_3 concentration up to 700 m above the ground in a rural area of Western Germany. Through some courageous assumptions, they calculated the corresponding upward fluxes of NH_3 on a number of occasions and concluded that they could be maintained by NH_3 volatilization from the excrement of domestic animals in the region.

A major development has been the adaptation of conventional micrometeorological approaches developed for hectare-sized plots to smaller treated areas with lateral dimensions of tens of metres. The development has been catalysed not only by economic needs, but also by the realization that because NH_3 volatilization proceeds rapidly, the time course of the emission is an important requirement of the experiment so that the time for treatment application should be short. Based on the conservation of mass, these small-plot methods equate the difference in the horizontal flux of gas across upwind and downwind boundaries of a test plot with the rate of emission of the gas from the surface of the plot along the line of the wind. (The horizontal flux is the product of wind speed and gas concentration integrated with respect to height.) Both the theory and the measurements are simple, the latter consisting of profiles of gas concentration on upwind and downwind boundaries, the wind speed profile and the wind direction which gives the fetch over the plot. A recent description of the theory has been given by

Denmead (1995) and practical applications to measuring emissions of NH_3 from grasslands are given, for instance, by Ryden and McNeill (1984), Jarvis et al. (1989) and a number of contributions to this volume. Sommer et al. (see Chapter 7) describe an application of the technique to measuring NH_3 emissions from a slurry storage tank.

A significant innovation has been the use of circular test plots with the downwind measurements made at the plot centre. Because the wind always blows to the centre of the plot regardless of compass direction, the experiment then becomes independent of wind direction and the test plot has a constant fetch, the plot radius. Example applications to grasslands are given by Denmead (1983), Wilson et al. (1983), Leuning et al. (1985), Pain et al. (1989) and Klempau et al. (1996, personal communication).

Yet another simplification to the experimental procedure has been made possible by the development of passive NH_3 samplers which permit direct measurement of the horizontal flux at any height with just one instrument (Leuning et al., 1985; Schjørring et al., 1992). The samplers are so designed that air flows through them in direct proportion to the wind speed, while their internal surfaces are coated with a chemical which absorbs all the NH_3 gas in the air flowing through. The mass of NH_3 trapped in the sampler over time is thus linearly related to the horizontal flux. The samplers dispense with the needs for power, anemometers, pumps, airlines and flow meters. Further, in many experiments they can be left in place for several days. Both these advantages make them very useful for long-term experiments in remote locations.

A final simplification, highly suitable for applications over short, uniform grass surfaces, is the concept of a stability independent height, labelled Z_{INST} by Wilson et al. (1982). For a given surface roughness and plot radius, the horizontal flux at Z_{INST} is in a nearly constant ratio to the surface flux density, regardless of atmospheric stability conditions. Thus the surface flux can be inferred from the horizontal flux measured at only one height. Further elaboration of this approach is given by Denmead (1983), Wilson et al. (1982, 1983) and Wilson and Shum (1992). Applications to grasslands are described in those papers and also in Pain et al. (1989) and Sherlock et al. (1989).

Sensors

Most field investigations to date have employed trapping techniques in which an air stream is passed for a known time and at a known flow rate through an acid trap (Denmead et al., 1976), or an acid-impregnated filter pack (Allen et al., 1988) or a denuder tube coated with an acid absorbent (Ferm, 1979). Ammonia concentrations are determined subsequently, after determination of the mass of NH_3 gas trapped. The trapping techniques commonly used have detection limits in the range 0.1–0.5 ppbv (Sutton et al., 1993;

Feshenfeld, 1995). This is a labour-intensive, tedious and slow business. There is a need in many investigations for more rapid, more interactive, on-line measurement such as is possible for, say, CO_2. New sensing techniques are under development, which will permit continuous on-line monitoring. One very successful instrument is the annular continuous-flow denuder described by Wyers *et al.* (1993), which is said to have detection limits of 5 pptv for a 30-min integration time or 10 pptv for a 1-min sampling time.

A difficulty with on-line analysis is the tendency of NH_3 to be absorbed easily onto surfaces that it contacts, which leads to high backgrounds and memory effects due to retention on sampling tubing and other plumbing (Feshenfeld, 1995). However, this is more of a problem for measuring *absolute* concentrations than it is for measuring *differences,* which are the essential ingredients of micrometeorological approaches to flux measurement. Then, offsets will often cancel out.

Kolb *et al.* (1995), Feshenfeld (1995) and Zahniser *et al.* (1995) discuss the possibilities for direct spectroscopic determination of gaseous NH_3 without the intervention of a collecting medium. Feshenfeld (1995) points out that such methods would be most desirable as they could provide both continuous measurements and unequivocal identification of the compound. They might also be made rapid enough to permit the use of the micrometeorological technique of eddy correlation. Methods currently under development include Fourier Transform Infrared Spectroscopy (FTIR) and Tuneable Diode Lasers (TDL). While likely to be useful in the future, these systems do not appear yet to have the sensitivity required for many applications. Feshenfeld (1995) quotes sensitivities of 1.5–5 ppbv for FTIR, but no such information seems to be available publicly yet for TDL systems.

Models of ammonia emission and deposition

Agricultural practices

As the contributions to this volume testify, many case studies of NH_3 emissions from grasslands have been made and the outcomes of many practices quantified. However, the resulting knowledge base is largely pragmatic, depending mostly on *ad hoc* predicting equations derived from statistical relationships developed from many field trials. Usually there is only one variable: the rate of N application. However, the chemistry and physics of NH_3 volatilization are well understood by now, and it is known that environmental factors such as wind speed, moisture state, radiation inputs, surface temperature and evaporation rate play important roles in determining rates of volatilization. However, only a handful of papers in this volume consider such variables: there is perhaps a message in this. Many of the single-variable regression equations have r^2 values > 0.9. The inference may be that in the long term, the rate of NH_3 volatilization is not important in determining the overall emission: loss processes can be fast or slow, but the same total amount of NH_3 is volatilized in the end.

It must be concluded, however, that while useful for inventory purposes, e.g. Buijsman *et al.* (1987), Jarvis and Pain (1990) and ECETOC (1994), and hence for identifying important emission sources, the simplified models are not generalizable to new situations or useful in formulating emission controls. The development of more comprehensive models will require more frequent time steps and more frequent observations of NH_3 fluxes in order to define the diurnal cycle, and more measurements of important controlling variables.

Long-range transport

Recent years have seen the development of models of long-range transport in order to predict the extent and consequences of the dispersion of NH_3 from agricultural, mainly grassland, sources over large distances, hundreds of kilometres. These incorporate meteorology, atmospheric chemistry and the mechanics of NH_3 exchange at plant, soil and water surfaces. Examples include Asman and van Jaarsveld (1992), ApSimon *et al.* (1994) and Sutton *et al.* (1994). While refinements are still necessary, these models should prove very useful in pollution studies and in understanding ecological consequences of NH_3 deposition to the landscape. The necessary refinements include linking to Geographical Information Systems in order to better define the nature of the deposition surfaces.

Current models appear to be well founded, but there is a pressing need for verification of their predictions. Concentration profiles appear to be predicted well (Asman and van Jaarsveld, 1992), but verifying dry and wet deposition of NH_3 and ammonium (NH_4^+) and emission inputs is difficult. As ECETOC (1994) puts it, progress in modelling work is impeded by the wide range of uncertainty in emission estimates, incomplete insight in NH_3 exchange processes between atmosphere and vegetation, and the difficulties inherent in diurnal variations in emissions and depositions.

Short-range transport

Because gaseous NH_3 can be readily absorbed by plants and soils, it is commonly believed that a significant proportion of NH_3 emitted from a source will be dry deposited in close proximity. The evidence for significant local recovery of volatilized NH_3 is conflicting. Denmead *et al.* (1993) report investigations of NH_3 recovery by sugarcane crops where there was a strong ground level source of NH_3 resulting from the breakdown of surface-applied urea fertilizer. The rate of NH_3 uptake by the crop F was calculated from the relationship:

$$F = \frac{N_c - \gamma}{r_s + r_b} \cdot L$$

Ammonia is assumed to diffuse from the turbulent air within the canopy across the leaf boundary layer (resistance r_b) and into the leaves through the stomata (resistance r_s). N_c is the NH_3 concentration at mid-canopy, γ is the NH_3 compensation point (Farquhar *et al.*, 1980), assumed to be 5 μg NH_3-N m^{-3}, and L is the leaf area index. For the most developed crop, L was 2.3 and a typical daytime value for the crop resistance $\{= (r_s + r_b)/L\}$ was 50 s m^{-1}. Despite very high NH_3 concentrations within the canopy (average daytime values exceeding 80 μg NH_3-N m^{-3} with a peak daytime average of 500 μg m^{-3}), the plants recaptured only 23% of the volatilized NH_3. Turbulent transfer out of the canopy space was too rapid in comparison with stomatal diffusion to permit a large recovery. The same qualitative results for deposition downwind of the emitting area might be expected. Because NH_3 concentrations decrease with distance from the source, vegetation downwind would be even less effective in capturing NH_3.

On the other hand, the transport model of Asman and van Jaarsveld (1992) predicts that about 10% of the NH_3 emitted from a source at 1 m above the ground could be dry-deposited within 100 m of the source and 30% within 10 km, while Hill *et al.* and Ross *et al.* (1996, personal communications) have results indicating substantial deposition of NH_3 within tens of metres of ground level sources. This is an important question to settle and more experimental and modelling work is needed.

N_2O Emissions

Measurement techniques

The lack of a cheap, sensitive rapid-response, on-line instrument for measuring atmospheric concentrations of N_2O has not allowed the ready adoption of micrometeorological techniques as has occurred for NH_3. Gradient-diffusion, eddy correlation or eddy accumulation techniques all require accurate measurement of small differences in a large background. Typically, one is seeking a resolution of order 0.1 ppbv in a background of about 300 ppbv. For NH_3, with a small background concentration, the requirements for flux measurement are less demanding. The necessary required precision is about five times less. However, new sensors for atmospheric N_2O are being developed. Their capabilities are examined further in a later section.

Chamber systems

So far, almost all our knowledge of N_2O emissions from grasslands comes from chamber measurements. Applications and difficulties of chambers have been expounded many times and need not be elaborated here. Recent reviews are those of Livingston and Hutchinson (1995) and Smith *et al.* (1995).

Apart from possible interferences to the microclimate and the flux itself due to chamber design and operation, the biggest problem is that of variability in soil emissions. In a study of N_2O emissions from fertilized grassland, Ambus and Christensen (1994) found coefficients of variation between chambers spaced only 0.11 m apart of as much as 55%, increasing to 139% for chambers 7.1 m apart. Velthof *et al.* (see Chapter 16) found comparable variability in chamber studies of N_2O emissions from grassland and found more than 10-fold variations in mean emission rates between sites in an area of only 80×96 m. They also found that variations in soil contents of nitrate (NO_3^-), NH_4^+ and moisture explained a large part of the variance in N_2O fluxes. Other researchers attribute some of the variability to the coexistence of aerobic and anaerobic microsites in the soil, variability in microbial populations and small-scale puddling resulting from grazing (see Chapter 1). Ambus and Christensen (1994) suggest that in their study, N_2O flux patterns at scales beyond 7 m were controlled by soil moisture variability, whereas a patchy distribution of denitrifying microsites governed N_2O emissions at scales below 1 m.

Characterizing variability and interpreting emission patterns in the light of that knowledge is obviously a very large challenge. Ambus and Christensen (1994) show the usefulness of geostatistics in this task. In order to reduce variability, Smith *et al.* (1994a) employed a very large chamber, 62 m^2 in area. From geostatistical analysis, they calculated that the chamber reduced small-scale variability by a factor of 3.

To conclude, it should be said that despite problems of spatial variability, chambers have an important part to play in studying N_2O emissions and will continue to be used for the foreseeable future. Perversely perhaps, one of their advantages is that by pinpointing small-scale variability, they can lead to fundamental investigations of the biophysics and biochemistry of N_2O formation. Other advantages include their portability, their relative cheapness, their abilities to permit study of controlling factors of N_2O production and to conduct experimentation on a scale not possible with large area methods and, importantly, their ability to detect small fluxes. It can be calculated that for comparable sensor precision, chambers can detect fluxes 100 times smaller than alternative micrometeorological approaches. For these reasons, chambers should be regarded as complementary to micrometeorological measurements, not alternatives.

Micrometeorological methods

New sensor developments are beginning to make micrometeorological techniques feasible. Both gradient diffusion and eddy correlation have been employed successfully to measure N_2O fluxes from grasslands (Galle *et al.*, 1994; Hargreaves *et al.*, 1994; Wienhold *et al.*, 1994). The first technique relies on measuring changes in atmospheric N_2O concentration with height above the surface, while the second relies on fast measurement of fluctuations in the instantaneous N_2O concentration at just one height above the emitting surface. A related technique is eddy accumulation. In this approach, air associated with updrafts and downdrafts is sampled into separate bins for N_2O analysis. The mean difference in concentration between the bins is combined with turbulence measurements to calculate the vertical flux at the sampling point. The precision required in concentration measurements is much the same for all three techniques, about 0.1% of background or tenths of a ppbv for a 10% resolution of the flux. Detailed accounts of those approaches are given in the references above and also in Fowler *et al.* (1995) and Lenschow (1995). Suffice it to say that the technology is expensive in comparison with chambers and the site requirements demanding – large, uniform, level areas, hundreds of metres in lateral extent. The attraction of micrometeorological approaches is that by integrating fluxes over large areas, they reduce the effects of point to point variability. This aspect has been examined in some detail by Smith *et al.* (1994b) and Fowler *et al.* (1995).

Sensors

The last 5 years have seen applications to N_2O flux measurement of TDL systems which have fast response (of order 10 Hz) and high precision (Hargreaves *et al.*, 1994; Wienhold *et al.*, 1994; Fowler *et al.*, 1995) and FTIR spectroscopy which, while not of fast response, permits simultaneous, high precision measurements of CH_4, N_2O, H_2O and CO_2 in one scan, and allows concentrations and gradients to be determined either at one location or over horizontal distances of 25 m or more (Galle *et al.*, 1994). The fast response of TDL systems makes them suitable for *in situ* eddy correlation (Wienhold *et al.*, 1994), but they can also be used remotely to measure gradients or concentration differences (Hargreaves *et al.*, 1994; Wienhold *et al.*, 1994). Zahniser *et al.* (1995) give a precision of 0.2 ppbv N_2O for TDL systems and examples given by Galle *et al.* (1994) suggest a similar precision for FTIR. The operating principles of TDL and FTIR systems are described in Zahniser *et al.* (1995) and Kolb *et al.* (1995).

Arah *et al.* (1994) describe an automated gas chromatographic system for micrometeorological measurement of N_2O fluxes. Fluctuations in ambient temperature apparently limit the precision attainable, but by using repeated analyses on bag samples, a resolution of 1 ppbv is possible. Arah *et al.* (1994) were able to measure 'coherent' gradients of N_2O over a fertilized grassland on more than half the sampling occasions.

Presently, TDL and FTIR systems are expensive and require specialist operators, but it can be anticipated that advancements in technology over the next few years will make them cheaper and more user friendly and will improve precision.

Boundary layer budgeting techniques

The need to extrapolate flux measurements from local to regional scales is just as pressing a problem as extrapolating from chamber to field. Budget techniques are being developed for both the daytime convective boundary layer (CBL) and the nocturnal boundary layer (NBL), e.g. Denmead *et al.* (1996). CBL budgeting techniques treat the well-mixed layer of air between heights of, say, 100 m and 1000 m as an integrator of surface fluxes along the path of a column of air moving over the landscape. They calculate the average surface flux from the scalar concentration in and above the mixed layer, and the CBL height. The flux estimates are averaged over regions of $10–10^4$ km^2, extending 10–100 km upwind.

NBL budgets are useful when low-level, radiative inversions inhibit vertical mixing. Surface gas fluxes can then be estimated from the rate of concentration change below the inversion. Both CBL and NBL budgets have good potential for estimating regional fluxes, but there is an urgent need for validation through direct measurements of fluxes and budget parameters. Denmead *et al.* (1996) give examples of CBL and NBL budgeting for estimating regional CO_2 fluxes. The sensor developments mentioned above will permit extension to regional fluxes of other trace gases, including N_2O. Choularton *et al.* (1995) describe use of NBL techniques to measure regional CH_4 fluxes. Choularton *et al.* (1995) also outline another boundary-layer budgeting technique which employs aircraft to collect air samples upwind and downwind of an emitting area. They used the technique to estimate regional CH_4 emissions from wetland areas of Scotland. Fowler *et al.* (see Chapter 14) give a similar example for N_2O.

Models

As pointed out at the beginning of this chapter, knowledge of the microbiology and biochemistry of N_2O formation is now quite extensive. Nonetheless, there seems still to be a large gap between laboratory and field. The difficulty in assessing whether N_2O is being produced through nitrification or denitrification, which has been touched on by several contributors, is a simple example. The problem of accounting for spatial variability in a mechanistic framework is another. These problems are discussed in detail in the chapters by Dendooven *et al.* and Langeveld *et al.* (see Chapters 2 and 29).

New models are appearing but these appear to be far more complex than previous attempts. The Denitrification-Decomposition (DNDC) model of Li *et al.* (1992, 1994) is a prime example. It contains four interacting submodels: soil climate, denitrification, decomposition and plant growth. The soil climate model simulates hourly vertical profiles of temperature and moisture in the soil and the flux of soil water. This information is fed to the other submodels.

The denitrification submodel calculates hourly denitrification rates and N_2O and N_2 production during periods when the soil has greater than 40% water-filled pore space. The decomposition submodel calculates daily carbon decomposition, nitrification, NH_3 volatilization and microbial CO_2 production. The plant growth submodel calculates daily root respiration, N uptake by roots and plant growth.

Validating models like DNDC in their entirety would seem to be a daunting if not impossible task. Wang *et al.* (1995) have applied a modified version of DNDC to observations of N_2O emission from a legume-pasture soil. The study revealed the model's sensitivity to soil water dynamics and temperature. Simulating them correctly was very important to the model's success. In turn, correct prediction of the surface energy balance of the soil and partitioning of the total water loss into soil evaporation and plant transpiration was required. A challenge for the future is to reduce the complexity of such models so as to make them useful workaday tools.

References

Allen, A.G., Harrison, R.M. and Wake, M.T. (1988) A meso-scale study of the behaviour of atmospheric ammonia and ammonium. *Atmospheric Environment* 22, 1347–1353.

Ambus, P. and Christensen, S. (1994) Measurement of N_2O emission from a fertilized grassland: an analysis of spatial variability. *Journal of Geophysical Research* 99, 16,549–16,555.

ApSimon, H.M., Barker, B.M. and Kayin, S. (1994) Modelling studies of the atmospheric release and transport of ammonia in anticyclonic episodes. *Atmospheric Environment* 28, 665–678.

Arah, J.R.M., Crichton, I.J., Smith, K.A., Clayton, H. and Skiba, Y. (1994) Automated gas chromatographic analysis system for micrometeorological measurements of trace gas fluxes. *Journal of Geophysical Research* 99, 16,593–16,598.

Asman, W.A.H. and Drukker, B. (1988) Modelled historical concentrations and depositions of ammonia and ammonium in Europe. *Atmospheric Environment* 22, 725–735.

Asman, W.A.H. and van Jaarsveld, H.A. (1992) A variable-resolution transport model applied for NH_x in Europe. *Atmospheric Environment* 26A, 445–464.

Buijsman, E., Maas, H.F. and Asman, W.A.H. (1987) Anthropogenic NH_3 emissions in Europe. *Atmospheric Environment* 21, 1009–1022.

Choularton, T.W., Gallagher, M.W., Bower, K.N., Fowler, D., Zahniser, M. and Kaye, A. (1995) Trace gas flux measurements at the landscape scale using boundary-layer budgets. *Philosophical Transactions of the Royal Society, Series A* 351, 357–369.

Crutzen, P.J. (1976) Upper limits on atmospheric ozone reductions following increased application of fixed nitrogen to the soil. *Geophysical Research Letters* 3, 169–172.

Denmead, O.T. (1979) Chamber systems for measuring nitrous oxide emission from soils in the field. *Soil Science Society of America Journal* 43, 89–95.

Denmead, O.T. (1983) Micrometeorological methods for measuring gaseous losses of nitrogen in the field. In: Freney, J.R. and Simpson, J.R. (eds) *Gaseous Loss of Nitrogen from Plant-Soil Systems*. Martinus Nijhoff, The Hague, pp. 133–157.

Denmead, O.T. (1995) Novel meteorological methods for measuring trace gas fluxes. *Philosophical Transactions of the Royal Society of London, Series A* 351, 383–396.

Denmead, O.T., Simpson, J.R. and Freney, J.R. (1974) Ammonia flux into the atmosphere from a grazed pasture. *Science* 185, 609–610.

Denmead, O.T., Freney, J.R. and Simpson, J.R. (1976) A closed ammonia cycle within a plant canopy. *Soil Biology and Biochemistry* 8, 161–164.

Denmead, O.T., Freney, J.R., Dunin, F.X., Jackson, A.V., Reyenga, W., Saffigna, P.G., Smith, J.B.W. and Wood, A.W. (1993) Effect of canopy development on ammonia uptake and loss from sugarcane fields fertilised with urea. *Proceedings of the Australian Society of Sugar Cane Technologists* 15, 285–292.

Denmead, O.T., Raupach, M.R., Dunin, F.X., Cleugh, H.A. and Leuning, R. (1996) Boundary layer budgets for regional estimates of scalar fluxes. *Global Change Biology* 2, 255–264.

Duxbury, J.M., Harper, L.A. and Mosier, A.R. (1993) Contributions of agroecosystems to global climate change. In: Harper, L.A., Mosier, A., Duxbury, J.M. and Rolston, D.E. (eds) *Agricultural Ecosystem Effects on Trace Gases and Global Climate Change*. ASA Special Publication Number 55, American Society of Agronomy, Madison, pp. 1–18.

ECETOC (1994) *Ammonia Emissions to Air in Western Europe*, European Centre for Ecotoxicology and Toxicology of Chemicals, Brussels, 196 pp.

Eriksson, E. (1966) Air and precipitation as sources of nutrients. In: Liser, H. and Schatter, K. (eds) *Handbuch der Pflanzenenahrang und Dungun*, Vol. 2, Part 1. Springer-Verlag, Berlin, pp. 774–792.

Farquhar, G.D., Firth, P.M., Wetselaar, R. and Weir, B. (1980) On the gaseous exchange of ammonia between leaves and the environment: determination of the ammonia compensation point. *Plant Physiology* 66, 710–714.

Ferm, M. (1979) Method for the determination of atmospheric ammonia. *Atmospheric Environment* 13, 1385–1393.

Feshenfeld, F. (1995) Measurement of chemically reactive trace gases at ambient concentrations. In: Matson, P.A. and Harris, R.C. (eds) *Biogenic Trace Gases: measuring emissions from soil and water*. Blackwell Science, Oxford, pp. 206–258.

Fowler, D., Hargreaves, K.J., Skiba, U., Milne, R., Zahniser, M.S., Moncrieff, J.B., Beverland, I.J. and Gallagher, M.W. (1995) Measurements of CH_4 and N_2O fluxes at the landscape scale using micrometeorological methods. *Philosophical Transactions of the Royal Society, Series A* 351, 339–356.

Freney, J.R., Denmead, O.T. and Simpson, J.R. (1978) Soil as a source or sink for atmospheric nitrous oxide. *Nature* 273, 530–532.

Freney, J.R., Denmead, O.T. and Simpson, J.R. (1979) Nitrous oxide from soils at low moisture contents. *Soil Biology and Biochemistry* 11, 167–173.

Galle, B., Klemedtsson, L. and Griffith, D.W.T. (1994) Application of an FTIR system for measurement of nitrous oxide fluxes using micrometeorological methods, an ultralarge chamber system, and conventional field chambers. *Journal of Geophysical Research* 99, 16,575–16,583.

Hargreaves, K.J., Skiba, U., Fowler, D., Arah, J., Wienhold, F.G., Klemedtsson, L. and Galle, B. (1994) Measurement of nitrous oxide emission from fertilized grassland using micrometeorological techniques. *Journal of Geophysical Research* 99, 16,569–16,574.

Healy, T.V., McKay, H.A.C., Pilbeam, A. and Scargill, D. (1970) Ammonia and ammonium sulfate in the troposphere over the United Kingdom. *Journal of Geophysical Research* 75, 2317–2327.

Hutchinson, G.L., Millington, R.J. and Peters, D.B. (1972) Atmospheric ammonia: absorption by plant leaves. *Science* 175, 771–772.

Jarvis, S.C. and Pain, B.F. (1990) Ammonia volatilization from agricultural land. *The Fertiliser Society, Proceedings No. 298*. The Fertiliser Society, Thorpe Wood, Peterborough, 35 pp.

Jarvis, S.C., Hatch, D.J. and Roberts, D.H. (1989) The effects of grassland management on nitrogen losses from grazed swards through ammonia volatilization: the relationship to excretal N returns from cattle. *Journal of Agricultural Science, Cambridge* 112, 205–216.

Kolb, C.E., Wormhoudt, J.C. and Zahniser, M.S. (1995) Recent advances in spectroscopic instrumentation for measuring stable gases in the natural environment. In: Matson, P.A. and Harris, R.C. (eds) *Biogenic Trace Gases: measuring emissions from soil and water*. Blackwell Science, Oxford, pp. 259–290.

Lenhard, U. and Gravenhorst, G. (1980) Evaluation of ammonia fluxes into the free atmosphere over Western Germany. *Tellus* 32, 48–55.

Lenschow, DL (1995) Micrometeorological techniques for measuring biosphere atmosphere trace gas exchange. In: Matson, P.A. and Harriss, R.C. (eds) *Biogenic Trace Gases: measuring emissions from soil and water*. Blackwell Science, Oxford, pp. 126–163.

Leuning, R., Freney, J.R., Denmead, O.T. and Simpson, J.R. (1985) A sampler for measuring atmospheric ammonia flux. *Atmospheric Environment* 19, 1117–1124.

Li, C., Frocking, S. and Frocking, T.A. (1992) A model of nitrous oxide evolution from soil driven by rainfall events. I. Model structure and sensitivity. *Journal of Geophysical Research* 97, 9759–9776.

Li, C., Frocking, S. and Harriss, R. (1994) Modelling carbon biogeochemistry in agricultural soils. *Global Biogeochemical Cycles* 8, 237–254.

Livingston, G.P. and Hutchinson, G.L. (1995) Enclosure-based measurement of trace gas exchange: application and sources of error. In: Matson, P.A. and Harriss, R.C. (eds) *Biogenic Trace Gases: measuring emissions from soil and water*. Blackwell Science, Oxford, pp. 14–51.

Matthias, A.D., Blackmer, A.M. and Bremner, J.M. (1980) A simple chamber for field measurement of emissions of nitrous oxide from soils. *Journal of Environmental Quality* 9, 251–256.

Pain, B.F., Phillips, V.R., Clarkson, C.R. and Klarenbeek, J.V. (1989) Loss of nitrogen through ammonia volatilisation during and following the application of pig or cattle slurry to grassland. *Journal of the Science of Food and Agriculture* 47, 1–12.

Prather, M., Derwent., Ehhalt, D., Fraser, P., Sanheuza, E. and Zhou, X. (1995) Other trace gases and atmospheric chemistry. In: Houghton, J.T., Meira Filho, L.G., Bruce, J., Hoesung, L. Callander, B.A., Haites, E., Harris, N. and

Maskell, K. (eds) *Climate Change 1994.* Cambridge University Press, Cambridge, pp. 72–126.

Ryden, J.C. and McNeill, J.E. (1984) Application of the micrometeorological mass balance method to the determination of ammonia loss from a grazed sward. *Journal of the Science of Food and Agriculture* 35, 1297–1310.

Ryden, J.C., Lund, L.J. and Focht, D.D. (1978) Direct in-field measurement of nitrous oxide flux loss from soils. *Soil Science Society of America Journal* 42, 731–737.

Schjørring, J.K., Sommer, S.G. and Ferm, M. (1992) A simple passive sampler for measuring ammonia emission in the field. *Water, Air and Soil Pollution* 62, 13–24.

Sherlock, R.B., Freney, J.R., Smith, N.P. and Cameron, K.C. (1989) Evaluation of a sampler for assessing ammonia losses from fertilized fields. *Fertilizer Research* 21, 61–66.

Smith, K.A., Clayton, H., Arah, J.R.M., Christensen, S., Ambus, P., Fowler, D., Hargreaves, K.J., Skita, U., Harris, G.W., Wienhold, F.G., Klemedtsson, L. and Galle, B. (1994a) Micrometeorological and chamber methods for measurement of nitrous oxide fluxes between soils and the atmosphere: overview and conclusions. *Journal of Geophysical Research* 99, 16,541–16,548.

Smith, K.A., Scott, A., Galle, B. and Klemedtsson, L. (1994b) Use of a long-path infrared gas monitor for measurement of nitrous oxide flux from soil. *Journal of Geophysical Research* 99, 16,585–16,592.

Smith, K.A., Clayton, H., McTaggart, I.P., Thomson, P.E., Arah, J.R.M. and Scott, A. (1995) The measurement of nitrous oxide emissions from soil by using chambers. *Philosophical Transactions of the Royal Society, Series A* 351, 327–338.

Sutton, M.A., Pitchairn, C.E.R. and Fowler, D. (1993) The exchange of ammonia between the atmosphere and plant communities. *Advances in Ecological Research* 24, 301–393.

Sutton, M.A., Asman, W.A.H. and Schjørring, J.K. (1994) Dry deposition of reduced nitrogen. *Tellus* 46B, 225–273.

Sutton, M.A., Schjørring, J.K. and Wyers, G.P. (1995) Plant–atmosphere exchange of ammonia. *Philosophical Transactions of the Royal Society, Series A* 351, 261–278.

Wang, Y.P., Meyer, C.P., Smith, C.J. and Galbally, I.E. (1995) Prediction of carbon dioxide and nitrous oxide fluxes from a legume-pasture soil at Wagga Wagga. In: Binning, P., Bridgman, H. and William, B. (eds) *Proceedings of the International Congress of Modelling and Simulation 1995, University of Newcastle, Australia,* Vol. 1. Modelling and Simulation Society of Australia, Canberra, pp. 83–88.

Wienhold, F.G., Frahm, H. and Harris, G.W. (1994) Measurements of N_2O fluxes from fertilized grassland using a fast response tunable diode laser spectrometer. *Journal of Geophysical Research* 99, 16,557–16,567.

Wilson, J.D. and Shum, W.K.N. (1992) A re-examination of the integrated horizontal flux method for estimating volatilisation from circular plots. *Agricultural and Forest Meteorology* 57, 281–295.

Wilson, J.D., Thurtell, G.W., Kidd, G.E. and Beauchamp, E.G. (1982) Estimation of the rate of gaseous mass transfer from a surface plot to the atmosphere. *Atmospheric Environment* 16, 1861–1867.

Wilson, J.D., Catchpoole, V.R., Denmead, O.T. and Thurtell, G.W. (1983) Verification of a simple micrometeorological method for estimating the rate of gaseous mass transfer from the ground to the atmosphere. *Agricultural Meteorology* 29, 183–189.

Wyers, G.P., Otjes, R.P. and Slanina, J. (1993) A continuous-flow denuder for the measurement of ambient concentrations and surface-exchange fluxes of ammonia. *Atmospheric Environment* 27A, 2085–2090.

Zahniser, M.S., Nelson, D.D., McManus, J.B. and Kebabian, P.L. (1995) Measurement of trace gas fluxes using tunable diode laser spectroscopy. *Philosophical Transactions of the Royal Society, Series A* 351, 371–382.

Index

Figures in **bold** indicate major references.
Figures in *italic* refer to diagrams, photographs and tables.